Recycling Land

Recycling Land

Understanding the Legal Landscape
of Brownfield Development

Elizabeth Glass Geltman

Ann Arbor

THE UNIVERSITY OF MICHIGAN PRESS

Copyright © by the University of Michigan 2000
All rights reserved
Published in the United States of America by
The University of Michigan Press
Manufactured in the United States of America
⊗ Printed on acid-free paper

2003   2002   2001   2000   4   3   2   1

A CIP catalog record for this book is available from the British Library.

Library of Congress Cataloging-in-Publication Data

Geltman, Elizabeth Glass.
    Recycling land : understanding the legal landscape of brownfield development /
Elizabeth Glass Geltman.
        p.   cm.
    Includes bibliographical references and index.
    ISBN 0-472-10919-7 (cloth : alk. paper)
    1. Brownfields—Law and legislation—United States.   2. Hazardous waste site remediation—Law and legislation—United States.   3. Industrial real estate—Environmental aspects—United States.   I. Title.
    KF3946 .G45 2000
    344.73'0462—dc21   99-049308

For my wonderful children
Andy, Jeff and Rachel

Contents

CHAPTER 1

Introduction: What Is a "Brownfield"?

Prior to 1970, transactions concerning the transfer of industrial property were governed by state legislation and common law and not by a web of federal and state environmental laws and regulations. Before the enactment of Comprehensive Environmental Response, Compensation, and Liability Act (CERCLA) in 1980, little thought was put into environmental matters when transferring property and the common law doctrine of caveat emptor or "buyer beware" prevailed.[1] Often property owners sold or transferred property without ever disclosing the presence of contaminants on-site. Absent certain common law causes of action (i.e., trespass, nuisance, ultrahazardous activities) prospective purchasers and developers of industrial and commercial property were largely without legal relief for site contamination discovered after the property transfer or sale occurred.

In 1980, CERCLA changed the law and gave the government, as well as current owners and operators of contaminated property, a cause of action against predecessors as well as against the generators and transporters who placed the hazardous substances on the property. Over the years since its enactment, the application of CERCLA's strict,[2] retroactive,[3] and joint and several liabil-

---

[1] *See generally* ELIZABETH GLASS GELTMAN, ENVIRONMENTAL LAW & BUSINESS: 1996 SUPPLEMENT (1996); ELIZABETH GLASS GELTMAN, ENVIRONMENTAL ISSUES IN BUSINESS TRANSACTIONS, VOL. 1 & 2 (1994 & 1995 Supp.); *Recycling Land: Encouraging the Redevelopment of Contaminated Property,* 10 NATURAL RESOURCES & ENV'T 3 (1996); Elizabeth Ann Glass, *The Modern Snake in the Grass: An Examination of Real Estate & Commercial Liability under Superfund & SARA and Suggested Guidelines for the Practitioner,* 14 B.C. ENVTL. AFF. L. REV. 381 (1987).

[2] *See* New York v. Shore Realty Corp., 759 F.2d 1032, 1042 (2d Cir. 1985) [hereinafter *Shore Realty*]; United States v. Conservation Chem. Co., 619 F. Supp. 162, 186-91 (W.D. Mo. 1985) [hereinafter *Conservation Chem. Co. III*]; Mardan Corp. v. C.G.C. Music, Ltd., 600 F. Supp. 1049, 1094 (D. Ariz. 1984), *aff'd,* 804 F.2d 1454 (9th Cir. 1986); Bulk Distrib. Ctrs., Inc. v. Monsanto Co., 589 F. Supp. 1437, 1442 (S.D. Fla. 1984) [hereinafter *Bulk Distrib. Ctrs.*]; Pinole Point Properties, Inc. v. Bethlehem Steel Corp., 596 F. Supp. 283, 289 (N.D. Cal. 1984); United States v. Conservation Chem. Co., 589 F. Supp. 59, 62 (W.D. Mo. 1984) [hereinafter *Conservation Chem. Co. I*]; United States v. NEPACCO, 579 F. Supp. 823, 827 (W.D. Mo. 1984), *aff'd in part and rev'd in part,* 810 F.2d 776 (8th Cir. 1986) [hereinafter *NEPACCO I*]; United States v. Price, 577 F. Supp. 1103, 1114 (D.N.J. 1983); Ohio *ex rel.* Brown v. Georgeoff, 562 F. Supp. 1300, 1305 (N.D. Ohio 1983). Congress has implicitly approved of the judicial interpretation that CERCLA imposes strict liability. *See* H.R. Rep. No. 253, 99th Cong., 1st Sess., pt. 1 (1985), *reprinted in* 1986 USCCAN 2835, 2856.

[3] *See, e.g.,* United States v. Waste Indus., Inc., 734 F.2d 159 (4th Cir. 1984); United States v. Shell Oil Co., 605 F. Supp. 1064 (D. Colo. 1985); Artesian Water Co. v. Government of New Castle County, 605 F. Supp. 1348 (D. Del. 1985); Jones v. Inmont, 584 F. Supp. 1425 (S.D. Ohio 1984); *NEPACCO I,*

ity[4] on those persons defined as "responsible parties" has chilled the transfer of industrial and commercial real estate throughout the country but especially in older cities in the Northeast and Midwest.[5] As one commentator explained:

> Environmental liability has become one of the leading legal problems of this decade. . . . [T]he effects of these developing legal rules are . . . severe: they threaten to disrupt and even to reorder established investments, longstanding methods of doing business and pre-existing expectations about legal rights and responsibilities.[6]

CERCLA targets potentially responsible parties (PRPs) that may be drawn into litigation over who bears the cost of cleaning up the contaminated site. PRPs include the site's current owners, past owners, and any party whose waste was disposed of at the site. CERCLA defines liability as:

> *retroactive*, which means that even those companies that obeyed the existing laws on hazardous-waste disposal before Superfund was passed may be held liable for cleaning up a site;

---

579 F. Supp. at 823; United States v. Stringfellow, 14 Envtl. L. Rep. (Envtl. L. Inst.) 20,388 (C.D. Cal. Apr. 9, 1984); United States v. Wade, 577 F. Supp. 1326 (E.D. Pa. 1983); Ohio *ex rel* Brown v. Georgeoff, 562 F. Supp. 1300 (N.D. Ohio 1983); Philadelphia v. Stepan Chem. Co., 544 F. Supp. 1135 (E.D. Pa. 1982); United States v. Price, 523 F. Supp. 1055 (D.N.J. 1981) *aff'd*, 688 F.2d 204 (3d Cir. 1982); *see generally* Amy Blaymore, *Retroactive Application of Superfund: Can Old Dogs Be Taught New Tricks?* 12 B.C. ENVTL. AFF. L. REV. 1 (1985). For a critical discussion of the retroactive application of joint and several liability, *see Superfund Improvement Act of 1985—Hearings on S. 51 Before the Senate Comm. on the Judiciary,* 99th Cong., 1st Sess. 165-72, 216-20 (1985) (statement of George Clemon Freeman, Jr.).

  [4]*See Shore Realty,* 759 F.2d at 1044; United States v. New Castle County, 642 F. Supp. 1258, 1266 (D. Del. 1986); United States v. Medley, 24 Env't. Rep. Cas. (BNA) 1856, 1857 (D.S.C. 1986); *Bulk Distrib. Ctrs.,* 589 F. Supp. at 1443; *Conservation Chem. I,* 589 F. Supp. at 63; *NEPACCO I,* 579 F. Supp. at 844; United States v. A & F Materials Co., 578 F. Supp. 1249, 1256 (S.D. Ill. 1984); United States v. Chem-Dyne Corp., 572 F. Supp. 802, 807-8 (S.D. Ohio 1983); United States v. Wade, 577 F. Supp. 1326, 1337 (E.D. Pa. 1983).

  For the position taken by the EPA, *see United States Environmental Protection Agency, Memorandum on Cost Recovery Actions under the Comprehensive Response, Compensation, and Liability Act* (Aug. 26, 1983), *reprinted in* [Fed. Laws] Env't. Rep. (BNA) No. 21, at 5531, 5535 (Oct. 30, 1983).

  [5]CHARLES BARTSCH & ELIZABETH COLLATON, INDUSTRIAL SITE REUSE CONTAMINATION & URBAN REDEVELOPMENT: COPING WITH THE CHALLENGES OF BROWNFIELDS (Northeast-Midwest Inst. 1994). *See generally Slants & Trends: It's Okay to Clean Up,* 20 HAZ. WASTE NEWS (May 15, 1995) ("Fear of Superfund Liability stops most brownfield cleanups in their tracks," reports independent consultant Win Porter). *See generally* KATHERINE N. PROBST ET AL., FOOTING THE BILL FOR SUPERFUND CLEANUPS: WHO PAYS AND HOW? (Brookings Inst. 1995); THOMAS W. CHURCH & ROBERT T. NAKAMURA, CLEANING UP THE MESS: IMPLEMENTATION STRATEGIES IN SUPERFUND (Brookings Inst. 1993); JOHN A. HIRD, SUPERFUND: THE POLITICAL ECONOMY OF ENVIRONMENTAL RISK (Johns Hopkins Univ. Press 1994); DANIEL MAZMAINIAN & DAVID MORELL, BEYOND SUPERFAILURE: AMERICA'S TOXIC POLICY IN THE 1990's (Westview Press 1992).

  [6]Kenneth S. Abraham, *Environmental Liability and the Limits of Insurance,* 88 COLUM. L. REV. 942 (1988).

strict, which means that a company may be liable for polluting a site, even if it was doing its best to avoid damage by using state-of-the-art waste management; and

joint and several, which means that any firm linked to a site may have to pay the full cost of cleaning it up, regardless of the amount of waste that it dumped, unless it can prove that other companies were polluting too.[7]

CERCLA allows a suit to be brought for hazardous substance cleanup costs against a person whose activities were legal and often permitted by the states when undertaken. No "fault" is required under the law. The disposal need not have caused any actual harm to any person or have constituted a health hazard. The action need not be considered negligent or extremely hazardous. The person ultimately held liable does not need to be the one who caused, contributed to, or benefited economically from the pollution at the site in need of cleanup if that person currently owns the property.

Similarly, under CERCLA, it is not considered a defense or mitigating factor that the defendant used state-of-the-art technology when disposal occurred. Moreover, CERCLA holds that even if a party contributed only a teaspoonful of hazardous substance to a site a mile wide, then that small contributor could be held legally responsible for the entire cost of cleanup of the site even though that party was not the factual cause of the majority of the site.

"Brownfields paralysis" normally occurs when the extent of contamination and cost of cleanup at urban and suburban industrial sites or gas stations is unknown. Uncertainty associated with the potential for CERCLA and other environmental liability especially impedes the sale or transfer of used industrial sites. As a matter of federal environmental law, CERCLA is but one of a greater patchwork of legal requirements creating uncertainty for property owners, prospective buyers, developers, and lenders regarding transactions involving brownfields. State legislation and common law are another major concern. As one commentator explained:

> America's obsolete and rusting hulks send a not-too-subtle message of dereliction and despair. Private owners often give up on their properties, allowing them to revert to the public domain. Local officials and civic leaders, usually unaware of the possibilities for reuse and the tools to achieve it, want to avoid the challenges. Better to forget these used complexes (sometimes referred to as "brownfields") and build new industrial parks far away from the scene of desolation (in what are called "greenfields").
>
> Particularly problematic are sites where no private or public steward ensures that pollution is contained and the property is secured. Congressional tours of abandoned and contaminated industrial sites in Chicago and Cleveland recently found that fences have been torn down, dumping has continued, and arsonists have burned buildings.

---

[7] THE ECONOMIST, December 3, 1994.

Lenders and prospective buyers also avoid reuse opportunities because of a natural reluctance to assume any added costs or liability for cleanup. Some lenders admit to a new version of redlining—referred to as "greenlining"—in which financial institutions deny credit to certain communities because of environmental concerns. Even sellers, including large corporations with the resources to complete a cleanup, often hesitate to market certain properties because of the potential risk of future liability. General Electric, for instance, "mothballed" a facility in Lynn, Massachusetts, rather than sell it to companies wanting to reuse the property.[8]

Brownfields typically are characterized as abandoned, inactive, or underutilized industrial or commercial sites located primarily in urban areas[9] and their surrounding suburbs. There is no one legal definition of what constitutes a "brownfield." Each state and EPA regions have slightly different legal definitions. "You cannot have a generic definition because every state is going to have a different need and they are going to blend differently their environmental program and economic development plan."[10] Nevertheless, there are certain common factors that are included in most state and federal definitions of brownfields. These include:

urban or urbanized site location;
industrial or commercial land use;
actual or perceived environmental contamination (though the amount
     may not be known or quantified);
not the subject of current or imminent environmental enforcement;
located in a distressed economic region.

In short, most brownfield sites are sites where there was a past or current commercial or industrial use that caused some environmental contamination in the

---

[8]Charles Bartsch & Richard Munson, *Restoring Contaminated Industrial Sites,* ISSUES IN SCIENCE AND TECHNOLOGY (March 22, 1994).

[9]*See* State Brownfields Policy & Practices: A Report (including key transcripts) of an IRM Conference for State Officials (June 27-28, 1994) ("The term *brownfields* refers generally to contaminated urban properties widely believed to have become abandoned, and not remarketed or reused in significant part because their owners or operators fear they may be contaminated—and if they are, the process of meeting regulatory obligations to clean them up will be prohibitively time consuming and expensive"). *See generally* DAVID ALLARDICE ET AL., INDUSTRY APPROACHES TO ENVIRONMENTAL POLICY IN THE GREAT LAKES REGION. *Chicago*: FEDERAL RESERVE BANK OF CHICAGO (July 1993); DEBBY COONEY ET AL., REVIVAL OF CONTAMINATED INDUSTRIAL SITES: CASE STUDIES. *Washington, D.C.*: NORTHEAST-MIDWEST INSTITUTE (1993); RICHARD L. HEMBRA, SUPERFUND: PROGRESS, PROBLEMS, AND REAUTHORIZATION ISSUES. *Washington, D.C.*: U.S. GENERAL ACCOUNTING OFFICE (April 21, 1993); VIRGINIA AVENI ET AL., CUYAHOGA COUNTY BROWNFIELDS REUSE STRATEGIES: WORKING GROUP REPORT. *Cleveland, Ohio*: CUYAHOGA COUNTY PLANNING COMMISSION (July 1993); CHARLES BARTSCH ET AL., NEW LIFE FOR OLD BUILDINGS: CONFRONTING ENVIRONMENTAL AND ECONOMIC ISSUES TO INDUSTRIAL REUSE. *Washington, D.C.*: NORTHEAST-MIDWEST INSTITUTE (1991).

[10]Charles W. Powers, STATE BROWNFIELDS POLICY & PRACTICE: A REPORT OF AN IRM CONFERENCE FOR STATE OFFICIALS at 14 (Held June 27-28, 1994 Baltimore, Maryland).

ordinary course of business. More significantly, the sites are locations where there is an expected or desired reuse of the property.

Some policy analysts believe that there are more than 500,000 brownfield sites nationwide that contain some level of environmental contamination.[11] The collective cleanup cost for these sites has been estimated as $650 billion.[12]

Not all brownfield sites are created equal. Typically there are four stages of brownfield concern. The stages range from slightly contaminated properties located in strong real estate markets to fully impaired properties that will be left to government cleanup if any cleanup is to occur.

The following brief table describes the four stages that form the brownfields continuum:

Stages of Brownfield Evolution

Stage 1:	Strong Real Estate Market—Private Sector Absorbs Costs
Stage 2:	Brownfield Traps—Business Owner Afraid to Sell/Expand
Stage 3:	Mothballed Property—Property Sits Idle and Dirty
Stage 4:	Tax-Deliquent Properties—Government Must Clean Up

The first-stage brownfields, those in a desirable location or an otherwise strong real estate market, are generally sold with little or insignificant delays and are cleaned up by the private sector. The cleanup is generally mandated by the lender financing the transaction—and is often held to a standard that far exceeds the legal cleanup level. Until very recently, these sites were almost always cleaned up without government knowledge or intervention. Today, with the increase in state voluntary cleanup programs, some of these sites are gaining more governmental oversight than ever occurred or was contemplated by any federal law.

The second stage I call the "brownfields trap." These generally consist of working businesses in which some level of environmental impairment is a likely fact. Many business owners are precluded from expanding, refinancing, or selling their businesses because of fear of discovering environmental contamination in the transaction process. These are working factories, gas stations, repair shops, and dry cleaning establishments that are hardly the focus of either state or federal environmental regulatory oversight. Yet, entering into a real estate transaction is likely to trigger sufficient inquiry by the private lending community that the owner is deterred from expanding the business to its optimum potential. Quite simply, many business owners are afraid that in finding money to expand they may be

---

[11]*See* CHARLES BARTSCH & ELIZABETH COLLATON, INDUSTRIAL SITE REUSE, CONTAMINATION AND URBAN REDEVELOPMENT: COPING WITH THE CHALLENGES OF BROWNFIELDS (Northeast-Midwest Inst. 1994) ("In framing the issues . . . it is important to distinguish between the approximately 1,200 Superfund high-priority sites—the worst of the bad with little prospect for economically viable use—and those brownfields sites characterized by low and medium level of environmental contamination").

[12]*Brownfields, Possible Liability Reforms Spur Growth in Environmental Insurance*, HAZARDOUS WASTE NEWS (Feb. 20, 1995).

turned down by new financiers for environmental reasons. Even worse, however, is the fear that the inquiry will cause their existing lender to decide that it needs the borrower to undertake additional environmental cleanup measures in order to keep the existing loan.

Some owners proceed from "brownfields paralysis" for fear of stepping into a "brownfields trap" to "mothballing" properties. At the third "mothball" stage some owners may decide that it is cheaper and less risky to let property sit completely idle than to attempt to sell it. These owners are usually large, very wealthy corporations. They usually continue to pay taxes on the property. They don't, however, undertake any formal cleanup measures or prepare the property for sale. Rather, these corporations hold onto the property to assure the property is not improperly used and the subject of future litigation.

General Electric (GE) is a classic example of one company that has developed this mothball approach to brownfields.[13] The company's policy is not to buy, sell, or lease any property that might be considered brownfields. As an old, well-established company it has gotten burned on past efforts to divest itself of properties it no longer needs.

Expansion or redevelopment of mothballed sites is inhibited by real or perceived contamination due to hazardous substances and/or petroleum. Such contamination generally occurs as a result of normal operating procedures, rather than bad faith acts demonstrating a reckless disregard for the environment (e.g., midnight dumping of toxic substances).

In Hoboken, New Jersey, for example, GE sold a property in the first part of this century. It was sold at least two to three times after GE first divested the property. A number of the subsequent owners contaminated the site with materials GE does not and has not ever used in its manufacturing. These subsequent polluters, however, were no longer in business. The property was eventually developed into residential property. When the residents discovered the contamination, they sued GE under federal CERCLA for the cost of cleanup. Because there were no other viable PRPs, GE settled the suit but left with an angry sense that it was paying for contamination it did not create. Nonetheless, under CERCLA's joint and several standard, GE could not prove that it did not contribute any hazardous substance to the site. Thus, even though the hazardous substance that was causing the health concerns clearly was not caused by GE, GE was left with the entire cost of cleanup. Moreover, as a direct result of the Hoboken experience (as well as others), GE has engaged in a campaign to change the laws to add environmental protections for sellers as part of brownfields remedies.[14]

These mothballed, idle properties possess valuable economic potential when cleaned up, reused, or redeveloped for industrial, commercial, or residential

---

[13] *See* Statement of Jane Gardner, Director Brownfields Program, General Electric, "Brownfields Redevelopment: A Blueprint for a Successful Program (Oct. 4, 1996, Boston, Massachusetts), excerpts reprinted in, Fourth Annual Fall Meeting of the American Bar Association Section of Natural Resources, Energy, and Environmental Law (Oct. 2-6, 1996, Boston, Massachusetts).

[14]*Id.* at 5.

purposes. Moreover, most present statistically low health risks. As one commentator explained:

> Not all brownfields are seriously contaminated with hazardous substances or petroleum, but the mere thought that they might be makes owners weak in the knees and scares off buyers and lenders. . . . Brownfields scare owners, buyers and lenders because the cost of cleaning them up to government standards may exceed market value. They scare seasoned industrialists looking for developable sites. Most would prefer the lower cost of locating in cities where utilities, interstate highways, deepwater ports, suppliers and labor pools already exist, but they often feel driven to build new, more expensive facilities on clean land in suburban or rural locations. That only worsens urban sprawl and strains tax revenues to pay for new infrastructure.[15]

The final stage in the brownfields continuum is abandonment. Some owners not only mothball the property, many also cease to pay property taxes. Others are forced into bankruptcy. Some simply walk away from the property and assume no further ownership responsibilities. If the former owner is a viable entity, then the state government or municipality may be able to sue that entity as a PRP under CERCLA. Where the former owner has no money, however, the government must either search for another viable PRP or bear the costs itself. This last stage is the problem CERCLA and the state equivalents were designed to address.

Brownfields can pose serious problems for cities and their residents because sites that are left undeveloped degrade the environment and represent lost opportunities to bring jobs and a broader tax base to the inner city.[16] Undeveloped brownfields also present environmental justice issues, as "no population of people should be forced to shoulder a disproportionate share of negative environmental impacts of pollution or environmental hazards. . . ."[17] If cleaned and rehabilitated, brownfields are capable of housing emerging technologies and manufacturing processes. Their adaption to new uses could restore not only buildings and their physical environment, but also jobs and the vitality of the communities surrounding them. Because many brownfields are located in urban areas, revitalization would particularly benefit low-income and minority residents who have suffered the economic and health consequences of living near blighted buildings and contaminated land. Reuse would also take advantage of existing infrastructure and reduce urban sprawl.

One of the most attractive features of brownfields redevelopment is that most brownfield sites have important infrastructure already in place. Many brownfields

---

[15]Johnine J. Brown, *Environmental Justice Conflict Could End with Justice If Brownfields Are Reclaimed,* ILLINOIS LEGAL TIMES, June 1995.

[16]*See* Charles Bartsch & Richard Munson, *Restoring Contaminated Industrial Sites,* ISSUES IN SCIENCE AND TECHNOLOGY (March 22, 1994).

[17]*See* Terms of Environment, EPA 175-B-94-015, *reprinted in* Elizabeth Glass Geltman, Environmental Law and Business Teacher's Manual, at 313 (Michie 1996).

have close, if not immediate, access to markets for labor and materials, as well as access to transportation facilities (water, rail, and highway). Existing infrastructure, among other considerations, plays a significant role in a company's decision-making process when determining whether to locate its operations at a "greenfield" site or at a "central city" location. Specific considerations that enter this decision-making process include:

access to markets for labor, materials, and final output;
access to transportation facilities;
site preparation costs;
infrastructure—including roads, water, sewer, and electric power;
"agglomeration of economies";
local land use and environmental regulation;
local taxation; and
prices of land and local labor.

In order to neutralize the costs of relocation, localities seeking to attract businesses away from the inner cities often use tax dollars to replicate the existing infrastructure in the inner city communities. This factor shifts the economic burden of developing new infrastructure from the company intending to relocate to the public. It causes greenfields (with their total lack of environmental liability) to become even more attractive and contributes to the increased pollution of property. Old factories merely abandon the old inner city property, without cleaning it up, and relocate to new pristine property—which becomes polluted.

To the extent that CERCLA is an important hindrance to redevelopment, four aspects of existing environmental liability law and the regulatory enforcement process may be identified as inhibiting property transfer decisions:

the effect of uncertain environmental liability costs on property transactions;
the effect of environmental liability on ability to obtain real estate financing;
the presence of uncertain cleanup standards once contamination is detected and its effect on the incentive to bring former industrial sites and their neighboring properties to the market; and
the difficulties in designing and enforcing contracts transferring future environmental liability both with the state and federal government and between private parties.

The stigma of brownfields site contamination and associated environmental liability has created a trend of withholding economically valuable industrial property from certain commercial real estate markets. Prospective purchasers and developers often forego redeveloping brownfields to avoid incurring unquantifiable environmental liability. Lenders are reticent to extend credit secured by brownfields for fear of:

devaluation of the collateral due to cleanup costs potentially exceeding the
value of the property;

directly assuming CERCLA liability as an owner or operator through workout activities if the loan goes bad or by foreclosure on such property;

inability to foreclose on contaminated property if the debtor creates or leaves
an environmental hazard and then defaults on the loan.

This book is designed to outline the legal basis for brownfield liability issues.
Chapter 2 presents an overview of common law and federal statutory causes of action and their potential application in the brownfield context. Chapter 3 provides
an introduction to state brownfield statutes that have been enacted to provide
brownfield stakeholders with limited protection from environmental liabilities.
Chapters 4 through 13 give a detailed, region by region analysis of each of the state
brownfields laws and programs. Chapter 14 provides a brief overview of current
federal brownfield initiatives. Finally, Chapter 15 presents various policy options
for addressing the brownfield redevelopment issues.

CHAPTER 2

The Unanticipated Effects of Environmental Law

Toxic Torts

Even if there were no CERCLA statute, past or present landowners may be liable to third parties under common law theories for the cleanup of hazardous substances as well as for personal injury and property damage resulting from the release of hazardous substances. Present owners have always been required by common law to maintain their property in a safe manner. Property owners have always been liable at common law for undertaking unreasonable activities on their property that cause harm to others. This obligation is called a "duty of care." It is the reason that current owners are often held liable to people who slip and fall on their property. When owners are held liable due to a slip-and-fall accident, the owner is deemed to have failed to maintain his property in a safe condition.

Past landowners present more difficult issues under common law doctrine because they often lack a "privity relationship." The past landowner is not expected to maintain the property in a safe condition once he no longer owns the property. That duty is now owed by the current owner. Thus, even if the past landowner actually created the hazard resulting in the fall, the plaintiff in a slip and fall case could not sue the past landowner for his injuries because there is no legal "privity." No duty is owed between the past owner and the public.

While allowing recovery of cleanup costs and natural resource damages, Superfund does not create a federal cause of action for recovery of personal injury damages. While Superfund sites attract a myriad of plaintiffs and lawyers, many are suing under common law toxic tort theories in addition to filing CERCLA causes of action. Thus, it is not unexpected that toxic tort litigation has increased dramatically since 1980 under state common law and continues to present a concern in brownfields redevelopment. As a result of this growth, there has been much study and talk of creating a new federal tort or administrative compensation system for personal injury and property damage due to toxins.

Litigation over exposure to toxic substances in groundwater is now frequent. Such suits are spurred by developments in the law of class actions and product liability, by efforts to develop new theories of liability (e.g., recovery for "cancer *risk*" and immune system injury or suppression), and by the willingness of some courts and juries to award large punitive damages. Because most brownfields are located in inner cities or highly populated suburbs, the possibility of large class action suits being brought (1) for real or perceived risks of health hazards due to being located near a contaminated site, or (2) for devaluation of property due to perceived environmental and health risks, continues to deter potential brownfields developers and, often more importantly, their respective lenders.

Nuisance

The common law theory that has been the most widely utilized in environmental law, including use regarding brownfields properties, is nuisance. To satisfy a claim of nuisance, the plaintiff must demonstrate that he has suffered a cognizable injury and is not complaining for mere "inconvenience, discomfort, or annoyance." The doctrine of nuisance is centered around the "reasonable use and enjoyment" of one's property:

> [L]iability [of a possessor of land] is not based upon responsibility for the creation of the harmful condition, but upon the fact that he has exclusive control over the land and the things done upon it and should have the responsibility of taking reasonable measures to remedy conditions on it that are a source of harm to others. Thus a vendee . . . of land upon which a harmful physical condition exists may be liable under the rule here stated for failing to abate it after he takes possession, even though it was created by his vendor, lessor or other person and even though he had no part in its creation.[1]

Activities that annoy or disturb the use of property, or substantially make its ordinary use uncomfortable, may be considered a nuisance. The determination of whether a nuisance exists is fact specific and will depend upon the extent of the injury suffered. In analyzing nuisance claims, courts traditionally utilize a balancing approach: the harm suffered by the plaintiff must be greater than the social usefulness of the activity engaged in by the defendant.

Nuisance can be divided into interferences that invade either public or private interests. This division creates two distinct tort causes of action. The distinction between the two causes of action depends not on the nature of the interference with property rights, but on the number of individuals affected by the offensive conduct. A public nuisance involves a "right common to the public" rather than a specific individual. A private nuisance affects only limited individuals and does so in a way not shared by the public at large. In *New York v. Shore Realty Corp.*,[2] the Second Circuit held a land developer liable for cleanup costs when he purchased land knowing that hazardous waste was stored on the site and that cleanup would be expensive. The court held the purchaser liable for both public and private nuisance even though he had not participated in the generation or transportation of the hazardous waste.

---

[1]RESTATEMENT OF TORTS § 839, comment d (1979) (describing the liability of a landowner for a nuisance on the landowner's property).

[2]759 F.2d 1032, 1051 (2d Cir. 1985), quoting Restatement (Second) of Torts § 839(d) (1979).

Therefore, the current owner of a site containing hazardous wastes may be liable to third parties for cleanup costs under either a public or private nuisance theory *even if he was not involved in the creation of the hazardous waste contamination.*[3] Nuisance law thus imposes a duty on current owners to abate hazardous conditions created by a party that previously owned the site.

Recent decisions have confirmed the effective use of the common law nuisance cause of action in the brownfields context. In *Wilshire Westwood Associates v. Atlantic Richfield Co.,*[4] for example, the court held that appellants were "entitled to pursue their nuisance claim against respondents as former lessees of the contaminated property."[5] The court considered nuisance principles in the context where the plaintiffs purchased a gasoline service station parcel in 1982, intending to build an office building. Their soils consultant negligently advised them that they would have no difficulty excavating the site. When construction began in 1985, however, the plaintiffs discovered gasoline contamination; the state required its removal at a cost of $3 million. Plaintiffs sued the former service station operators and the soils consultant for reimbursement, alleging various theories, including nuisance. The court held that the action, filed within three years of abatement, was timely under the statute of limitations for a continuing nuisance.

Recently, several states have enacted legislation that allows the state to enter the property and abate a nuisance or other imminent hazard and attach a lien for the cost of cleanup on the property. Under most of these statutes, the lien arises at the time the state government begins spending money, but it is not perfected until it is filed in the appropriate records office.

Trespass

Trespass is also implicated in brownfields redevelopment. A trespass is defined as "any intentional invasion of the plaintiff's interest in the exclusive possession of property." In the environmental context, plaintiffs sued a chemical plant in trespass for damages from chemicals falling on their farm.[6] Historically, a trespass cause of action was found where:

coal dirt was deposited in stream;[7]
distillery slops were discharged into a stream;[8]
deposits of ash from coke works were made into stream;[9] and
deposits were made of culm and dirt.[10]

---

[3]Consequently, the present owner seeks recovery from the former owner, who actually created the nuisance.

[4]1993 Cal. App. LEXIS 1188 (Cal. App. 4th Nov. 23, 1993).

[5]*Id. See also* Mangini, 230 Cal. App. 3d 1125.

[6]Maddy v. Vulcan Materials Co., 737 F. Supp. 1528, 1540 (D. Kan. 1990).

[7]Keppel v. Lehigh Coal & Navigation Co., 50 A. 302 (Pa. 1901).

[8]Hileman v. Hileman Distilling Co., 33 A. 575, 577 (Pa. 1896).

[9]Lentz v. Carnegie Bros. & Co., 23 A. 219, 220 (Pa. 1892).

[10]Gallagher v. Kemmerer, 22 A. 970 (Pa. 1891).

When such physical invasion violates private rights, it is no defense that the defendant went to great expense to clean emissions or that activity was not a nuisance per se.[11]

The doctrine of trespass is closely related to nuisance. The basic distinction is that nuisance causes a "substantial and unreasonable interference," whereas trespass causes an actual "encroachment by 'something' upon the plaintiff's exclusive rights of possession." Distinguishing between what conduct constitutes a trespass rather than a nuisance can be difficult. Therefore, both torts are often alleged in tandem. In the environmental context, the general rule for trespass requires an actual invasion of a pollutant onto the plaintiff's property before a cause of action in trespass exists. Thus, if airborne particulates fall onto another's land a trespass suit may be brought.

Many brownfields properties are created by physical encroachment of a pollutant from a neighbor's land. At common law, the only remedy for a neighbor to complain about the migration of pollution from one property to another was to show that it constituted a nuisance or to claim trespass. The causation requirements at common law are, however, a continuing impediment to the viability of a trespass case. Often it is hard to prove where the source of contamination emanated. There could be more than one source. Thus, the trespass is not clear because it is difficult to fingerprint the trespasser.

Negligence

Long before CERCLA was enacted, it was well settled that the owners of land (and those occupying the land) owe certain duties with regard to the safety of individuals who come onto the land (the extent of the duty being dependent on the status of the visitor, i.e., as an invitee or licensee). Similarly, the landowner owes a duty to his neighbors to use care when conducting activities on the land to avoid causing harm to his neighbors. In order to state a cause of action in negligence, the plaintiff must show the following:

(1) the defendant was under a duty to protect the plaintiff from injury;
(2) the defendant breached that duty;
(3) the plaintiff suffered actual injury or loss; and
(4) the loss or injury proximately resulted from the defendant's breach of the duty.

This "duty of care" requires an actor to conform to a certain standard of conduct to protect the other from unreasonable risk of harm.

A brownfields case considering the question of negligence was brought in Maryland. In *Rosenblatt v. Exxon Co.*,[12] the court considered whether, under

---

[11] Wente v. Commonwealth Fuel Co., 83 N.E. 1049, 1051 (1908).
[12] 642 A.2d 180 (Md. 1993).

Maryland law, the current tenant of commercial property has a cause of action in strict liability, negligence, trespass, or nuisance for economic losses sustained, against a prior tenant whose activities during its occupancy allegedly caused the property to become contaminated. The plaintiff asserted that the former tenant owed a duty because it was foreseeable that contamination would harm subsequent tenants. Moreover, because the prior tenant was in the business of producing, handling, storing, and marketing petroleum products, the plaintiff asserted that the prior tenants (Exxon) should be held to a high degree of care in conducting its business.

Exxon asserted that the doctrine of negligence should not be extended to subsequent occupants of property. Exxon argued that, unlike a contemporary occupier of land, a subsequent occupant can avoid harm simply by investigating prior to occupying the land. Thus, Exxon argued that it owed no duty to plaintiff because there was no relationship between Exxon and plaintiff. The court agreed with Exxon and concluded that "there exists no relationship between the parties which would have made it foreseeable that an act or failure to act by Exxon would result in harm to Rosenblatt."[13] The court said it was "unwilling to impose upon a lessee of commercial property a duty to remote successor lessees for losses resulting from a condition on the property that could have been discovered with reasonable diligence prior to occupancy and thus could have been avoided."[14] The court explained:

> The imposition of a duty upon one to another serves to balance the burdens between the parties in avoiding the harm. . . . [W]e have recognized that care must be taken to avoid unduly burdening an occupant of land in the use of the land. But we have imposed a duty upon the occupant of land where it would be unreasonable to expect the other party to be able to protect itself from the harm, as in the case of occupiers of adjacent land or guests who come onto the land.[15]

The failure of negligence suits for a lack of privity is emphasized by courts' explanation of current owners' and lessors' need to investigate at the time of the transaction. The court emphasized in *Exxon*:

> A lessee of commercial property, however, is expected to make basic inquiry and inspection of the property prior to entering into a lease. When Rosenblatt entered into the lease, he was aware that the property had been used for a gas station. Thus, he knew or should have known that gasoline contamination was possible. He could have required that the property be tested for contamination; he could have negotiated express warranties in the lease. He was in a position to avoid completely the harm he now alleges.[16]

---

[13]*Id.*
[14]*Id.*
[15]*Id.*
[16]*Id.*

The absence of privity among the parties as evidence in the *Exxon* case is typical in the brownfields context. Moreover, many federal courts have dismissed common law causes of action as being preempted by federal legislation such as RCRA, CERCLA, and the Clean Water Act.

Although a party purchasing brownfields property is likely to defeat a negligence action in a court of law, there is no way for the party purchasing a brownfields site to eliminate the risk of litigation based on such a claim. These feared transaction costs are just as great a deterrent to many lenders and developers as the actual finding of liability.

Strict Liability for Abnormally Dangerous Activities

The abnormally dangerous activity doctrine is premised on the principle that "one who carries on an abnormally dangerous activity is subject to liability for harm to the person, land or chattels of another resulting from the activity, although he has exercised the utmost care to prevent the harm."[17]

The abnormally dangerous activity doctrine emphasizes the dangerousness and inappropriateness of the activity. Despite the social utility of the activity, the doctrine imposes liability on those who, for their own benefit, introduce an extraordinary risk of harm into the community. The rule reflects a policy determination that such "enterprise[s] should bear the costs of accidents attributable to highly dangerous [or unusual activities]." Because some conditions and activities can be so hazardous and of "such relative infrequent occurrence," the risk of loss is justifiably allocated as a cost of business to the enterpriser who engages in such conduct.[18] Thus, the doctrine recognizes the additional policy consideration that enterprises are generally in a better position to administer the unusual risk.

The policies underlying the tort of strict liability for abnormally dangerous activities do not rest on property rights. Rather, the policy arguments require businesses to "internalize" the external costs of business and shift the loss potential onto the party deemed best able to shoulder it. Liability for the harm caused by abnormally dangerous activities does not necessarily cease with the transfer of property because the former owner of the property whose activities caused the hazard might be in the best position to bear or spread the loss. In assessing whether a particular activity is ultrahazardous, courts will view the magnitude of the risk presented, the probability of harm, and the resulting severity. In order to maintain a cause of action of strict liability for abnormally dangerous activities the plaintiff must show "several" of six independent criteria, including:

> the existence of a high degree of risk of some harm to the person, land or chattels of others;

---

[17]RESTATEMENT (SECOND) OF TORTS, § 520 (1977).

[18]*Id.* at 556; *see Berg,* 181 A.2d at 487; Kenney v. Scientific, Inc. 497 A.2d 1310 (N.J. Super Ct. Law Div. 1985).

the likelihood that the resulting harm will be great;
the inability to eliminate the risk by the exercise of reasonable care;
the extent to which the activity is not a matter of common usage;
the inappropriateness of the activity to the place where it is carried on; and
the extent to which its value to the community is outweighed by its dangerous attributes.[19]

The theory behind such strict liability is that the actor is in a "better position than the victim 'to insure against the risk, allocate the cost, and reduce or warn against the dangers'."

An obvious example of what may constitute an ultrahazardous activity is the operation of a hazardous waste disposal site.[20] The risk of contaminated ground-water, increased risk of certain diseases to any nearby residents, and severity of the harm may deem such activities "ultrahazardous." Those engaged in activities (including any industrial activities) that have the potential for severe harm to the public are required to take precautions to mitigate the risk of harm. Otherwise, such persons and their successors face the possibility of strict liability and the resulting damages that must be paid to injured parties. Of course, what constitutes an abnormally dangerous activity will change over time as technologies, neighborhoods, and expectations change. As activities become more common, they are less likely to be subject to strict liability.

Accordingly, most brownfields sites that were used as gas stations or dry cleaners are not likely to incur common law strict liability. As urban sprawl spreads and housing moves closer to factories, even industrial uses may cease to be considered ultrahazardous for purposes of the common law strict liability doctrine.

Aider and Abettor Liability

A person may incur liability "[f]or harm resulting to a third person from the tortious conduct of another . . . if he . . . knows that the other's conduct constitutes a breach of duty and gives substantial assistance or encouragement to the other so to conduct himself. . . ."[21] This so-called aider and abettor liability has never been directly applied in a brownfields context. Nevertheless, many lenders are concerned about aider and abettor liability when loaning money to redevelop brownfields sites.

Much of this concern is due to the case of *O'Neil v. Q.L.C.R.I., Inc.*[22] In that case, the state attorney general alleged that the mortgage holder aided and abetted violations of the federal Clean Water Act. The state alleged that the lender had "influence and control" over the principal polluters because the lender knew of the

---

[19]RESTATEMENT (SECOND) OF TORTS, § 519.

[20]*See also* State v. Ventron, 468 A.2d 150 (1983); T & E Indus., Inc. v. Safety Light Corp., 587 A.2d 1249 (N.J. 1991).

[21]RESTATEMENT (SECOND) OF TORTS § 876.

[22]*Id.*

sewage problem and could have conditioned the loans on the fixing of the sewage problem. The court determined that the common law concept of aiding and abetting could be used to determine liability under the Clean Water Act. The court explained that, "Nothing in the FWPCA [Federal Water Pollution Control Act of 1972] rules out the use of the aiding and abetting doctrine." Since the state alleged sufficient involvement and possible control, the court held that "all the aiding and abetting claims can withstand a motion to dismiss and are not futile."[23]

In order for one person or entity to be held accountable as an aider and abettor for the environmental problems caused by another, the former must have some sort of actual or contractual control over the latter. In addition, there must be a knowing failure to exercise that control to ensure environmental compliance. Such factors place lenders in a dilemma. To protect the collateral, most lenders want to monitor the environmental as well as the financial affairs of the debtor. If a lender does not monitor the environmental affairs of its debtor, the lender could incur substantial environmental liabilities if the debtor does not adhere to environmental or other laws causing the loan to go bad and the lender to foreclose. Alternatively, if the lender does monitor the environmental affairs of the debtor and fails to call the note or otherwise force the debtor to come into environmental compliance, the lender could be held liable as an aider and abettor. Rather than resolve this dilemma contractually, conservative lenders often decline to make loans on such properties due to the mere risk of litigation. Thus, aider and abettor law can directly impact on brownfields redevelopment.

Vicarious Liability

At common law, a landowner can be held responsible for the conduct of an independent contractor, including a remediation contractor, if that activity is inherently dangerous.[24] Whether an activity is inherently dangerous is a question of law that is to be decided by applying a three-part test:

(1)   the activity must be an activity relating to land or to other immovables;
(2)   the activity itself must cause the injury and the defendant must have been engaged directly in the injury-producing activity; and
(3)   the activity must not require the substandard conduct of a third party to cause injury.[25]

Courts have interpreted the third part of the test to require that the activity can cause injury "even when conducted with the greatest prudence and care."[26]

---

[23]750 F. Supp. 551 (D.R.I. 1990). *See generally* Thunder, Liability of Quasi-Sources, Harv. Envt. L. Rev. (1986).

[24]Olsen v. Shell Oil Co., 365 So.2d 1285 (La. 1978); Updike v. Browning-Ferris, Inc., 808 F. Supp. 538, 541 (W.D. La. 1992).

[25]Perkins v. F.I.E. Corp., 762 F.2d 1250 (5th Cir. 1985).

[26]*Updike,* 808 F. Supp. at 541 (quoting Ainsworth v. Shell Offshore, Inc., 829 F.2d 548 (5th Cir. 1987), *cert. denied,* 485 U.S. 1034, 108 S. Ct. 1593, 99 L. Ed. 2d 908 (1988)).

In *Duckworth v. Barrios*,[27] the court considered whether a landowner could be held liable for contracting to have lead paint removed from the house by pressure washing. The court found "that high pressure washing a personal residence to remove lead paint may not be deemed inherently dangerous for purposes of establishing vicarious liability of the home owners."[28] Therefore, the homeowners who hired the contractor to remove paint from their home could not be held vicariously liable for the acts of their independent painting contractor in pressure washing their home.

The question of what remediation activities are inherently dangerous for purposes of incurring vicarious liability in the brownfields cleanup context has yet to be addressed.

Caveat Emptor

The law of caveat emptor (or "let the buyer beware") has only gradually assimilated other principles of consumer protection law. For years, "courts continued to cling to the notion that a seller had no duty whatsoever to disclose anything to the buyer."[29] The traditional "principle of caveat emptor dictated that in the absence of express agreement, a seller was not liable to the buyer or others for the condition of the land existing at the time of transfer."[30]

Exceptions to the broad immunity of caveat emptor inevitably developed in the sale of land. The doctrine has eroded over the years to require the seller to disclose to the buyer certain facts in regard to the sale of the property. As one court explained:

> Caveat emptor developed when the buyer and seller were in an equal bargaining position and they could readily be expected to protect themselves in the deed. Buyers of mass produced development homes are not on an equal footing with the builder vendors and are no more able to protect themselves in the deed than are automobile purchasers in a position to protect themselves in a bill of sale.[31]

Many jurisdictions have limited the doctrine of caveat emptor to commercial properties (including most brownfields sites) and have only imposed duties on sellers and brokers of residential property (including any residential properties built on former industrial sites) through consumer-protection laws. Thus, although the

---

[27]1995 U.S. Dist. LEXIS 5599 (E.D. La. 1995).

[28]*Id.*

[29]John H. Scheid, Jr., Note, *Mandatory Disclosure Law: A Statute for Illinois,* 27 J. MARSHALL L. REV. 155, 160 (1993) ("the law offers more protection to a person buying a dog leash than it does to the purchaser of a house") (citing Paul G. Haskell, *The Case for an Implied Warranty of Quality in Sales of Real Property,* 53 GEO. L.J. 633 (1965)).

[30]T & E Indus., Inc. v. Safety Light Corp., 123 N.J. 371, 387, 587 A.2d 1249 (1991).

[31]Wawak v. Stewart, 449 S.W.2d 922, 924 (Ark. 1970).

doctrine of caveat emptor is eroding, it still retains certain vitality in the brownfields context where the purchaser is a developer or other person deemed sophisticated in business.

Federal Environmental Law and the Brownfields Redevelopment Dilemma

Numerous federal environmental statutes may affect brownfields redevelopment. The following table illustrates other major federal statutes that may be invoked in the brownfields context:

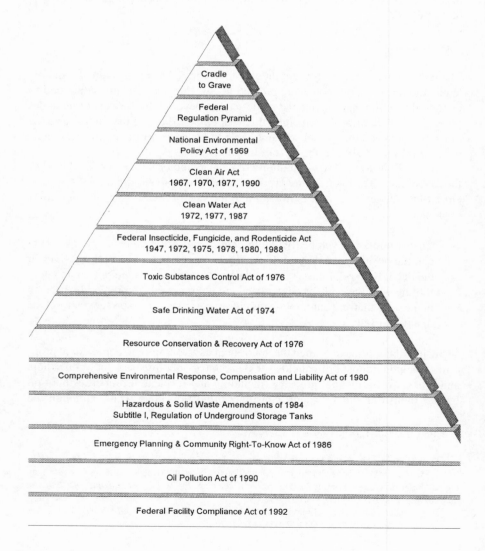

Cradle to Grave

Federal Regulation Pyramid

National Environmental Policy Act of 1969

Clean Air Act
1967, 1970, 1977, 1990

Clean Water Act
1972, 1977, 1987

Federal Insecticide, Fungicide, and Rodenticide Act
1947, 1972, 1975, 1978, 1980, 1988

Toxic Substances Control Act of 1976

Safe Drinking Water Act of 1974

Resource Conservation & Recovery Act of 1976

Comprehensive Environmental Response, Compensation and Liability Act of 1980

Hazardous & Solid Waste Amendments of 1984
Subtitle I, Regulation of Underground Storage Tanks

Emergency Planning & Community Right-To-Know Act of 1986

Oil Pollution Act of 1990

Federal Facility Compliance Act of 1992

The Clean Air Act (CAA)

National Ambient Air Quality Standards

The Clean Air Act (CAA)[32] regulates air pollution above threshold amounts generated by both "stationary sources" and "mobile sources." A "stationary source" of air pollution is a "building, structure, facility, or installation" that "emits or may emit any air pollutant."[33] Thus, a manufacturing plant might be a "stationary source"; a motor vehicle, a "mobile source."

EPA has implemented CAA programs that (1) set goals for certain levels of air quality in specified areas, (2) attempt to prevent substantial deterioration in the air quality in areas that have reached the goals, (3) prohibit air pollutant emissions exceeding certain standards, (4) reduce emission levels of certain hazardous air pollutants, and (5) protect the stratospheric ozone layer.

The National Ambient Air Quality Standards (NAAQS) program requires certain levels of air quality in specified areas.[34] EPA has set NAAQS, which establish maximum allowable concentration levels, for certain "criteria" pollutants, such as nitrogen oxide.[35] Areas that meet these criteria are called "attainment areas." Areas failing to meet these criteria are called "nonattainment" areas. Brownfields tend to be located in nonattainment areas.

State implementation plans (SIPs) are the principal means to enforce the CAA NAAQS program. SIPs require the creation of air permitting systems for stationary sources that force nonattainment areas to make reasonable further progress toward attainment. Similarly, the SIP must contain a permitting system that will prevent significant deterioration (PSD) of areas in attainment—so that they are not degraded into nonattainment areas.[36] The state SIP must establish requirements for "the installation, maintenance, and replacement of equipment, and the implementation of other necessary steps, by owners or operators of stationary sources to monitor emissions from such sources."[37]

EPA has established maximum permissible increments and concentrations of sulfur oxides, particulate matter, carbon monoxide, hydrocarbons, and other photochemical oxidants, and nitrogen oxides for designated areas in the country.[38] Importantly, development increasing emissions of such regulated pollutants may occur only if the increment is not violated. Most brownfields sites are located in inner cities and, therefore, in nonattainment areas. Thus, air pollution permitting requirements operate as another disincentive to brownfields redevelopment.

---

[32]42 U.S.C. §§ 7401 to 7671q.
[33]42 U.S.C. § 7411(a)(3).
[34]42 U.S.C. § 7409.
[35]42 U.S.C. § 7409(c).
[36]42 U.S.C. § 7470(1).
[37]42 U.S.C. § 7410(a)(2)(A), (C), (D)(I)(I), (F)(I).
[38]42 U.S.C. §§ 7473(a), 7476(a).

In contrast to nonattainment brownfields, it is often cheaper and easier to acquire the necessary air permits in greenfields, which tend to be located in attainment areas.

In addition to the contribution of stationary sources of pollution, a great deal of air pollution in nonattainment areas, such as cities, comes from automobiles. Thus, EPA has established emission standards for motor vehicles and other engines.[39] These requirements force many inner city and suburban residents to get the exhaust of their car inspected on an annual basis. Exhaust from mobile sources is greatest in the densely populated inner cities—where there are more cars and buses. The proliferation of mobile sources contributes to the nonattainment and makes permitting of new sources—or factories to be developed on brownfields sites—more difficult to obtain.

National Emission Standards for Hazardous Air Pollutants (NESHAP)

In addition to regulation of the "criteria" pollutants emitted by stationary and mobile sources, EPA has established emission standards to reduce the emission levels of certain hazardous air pollutants, such as asbestos, chlorine, and mercury.[40] On EPA approval, states may implement a program to help achieve and to enforce these emission standards.[41]

Asbestos removal is a common problem at brownfields sites. A facility that must be torn down or renovated is likely to contain asbestos. If the building is demolished, the asbestos must be removed in accordance with the Clean Air Act's emissions standards for hazardous air pollutants (as well as meeting OSHA requirements). Even if the building is merely renovated, the Americans with Disabilities Act may require sufficient asbestos removal to trigger EPA and OSHA requirements.

Regulations to Prevent Depletion of the Stratospheric Ozone Layer

The Clean Air Act Amendments of 1990 (CAAA) and the resulting regulations implement the Montreal Protocol on Substances that Deplete the Ozone Layer, a protocol to the Vienna Convention for the Protection of the Ozone Layer. The CAAA requires the gradual phase-out of ozone-depleting chemicals, such as chlorofluorocarbons.[42] Most buildings will have old HVAC systems that use ozone-depleting substances. In renovating the building, a new owner or operator would be required to remove the old HVAC system and install a new system in accordance with the CAAA. This cost of removal, like the problem with asbestos removal, dra-

---

[39] 42 U.S.C. §§ 7521, 7541.
[40] 42 U.S.C. § 7412(b), (d).
[41] 42 U.S.C. § 7412(l)(1).
[42] 42 U.S.C. §§ 7671c, 7671d.

matically increases the cost of redeveloping a brownfield compared to the cost of building on a greenfield.

The Clean Water Act (CWA)

The Federal Water Pollution Control Act of 1972,[43] commonly known as the Clean Water Act (CWA), regulates any discharge of pollutants into the surface waters of the United States. *Pollutants*, broadly defined under the CWA, includes solid waste, sewage, chemical wastes, rock, and sand.[44]

Under the CWA, the national pollutant discharge elimination system (NPDES) program specifically regulates direct discharges into surface waters. Under the NPDES, EPA may issue a permit allowing the discharge of pollutants if certain requirements and standards are met.[45] The following are requirements and standards: (1) effluent restrictions, such as prohibiting the discharge of a radiological agent into navigable waters; (2) water quality-related effluent restrictions; (3) performance standards for controlling effluent discharges through the use of, for example, technology or operating methods; (4) toxic and pretreatment effluent standards; (5) records, reports, and inspections; and (6) ocean discharge standards.[46] If a state wants to regulate discharges into navigable waters within the borders of the state, the state may petition EPA to approve its administration of a state plan.[47]

Similar to the problem associated with CAA permits, NPDES permits are often more difficult to obtain in brownfields sites. These sites are disproportionately located in areas with poor water quality. Thus, it is often more time consuming, expensive, and difficult to obtain NPDES permits to operate a facility in brownfields locations versus greenfields locations.

Wetlands Protection

The CWA requires a permit "for the discharge of dredged or fill material into navigable waters at" designated disposal sites.[48] The Army Corps of Engineers (Corps) issues these permits, commonly known as "404 permits."[49] The CWA does not prohibit the discharge of nonpolluting dredged or fill material, such as resulting from normal farming and ranching activities and building or maintaining an irrigation ditch.[50] The practical effect of the program is, however, to preclude building in wetlands without getting a permit from the Corps.

---

[43]33 U.S.C. §§ 1251 to 1387.
[44]33 U.S.C. § 1362(6).
[45]33 U.S.C. § 1342(a).
[46]33 U.S.C. §§ 1311, 1312, 1316-18, 1343.
[47]33 U.S.C. § 1342(b).
[48]33 U.S.C. § 1344(a).
[49]33 U.S.C. § 1344(a), (d).
[50]33 U.S.C. § 1344(f).

Many brownfields sites in the thirteen original colonies are located in wetlands. This is because when our country was founded the law encouraged filling what were then called swamps—due to concern about mosquito breeding and the spread of malaria. As a result, swampland was sold cheap and often filled with contaminated soils.

Driving up the Interstate 95 corridor on the east coast there are ample examples of brownfields sites located in swamplands. These properties would require all the permit requirements of the 404 program in addition to meeting cleanup requirements due to hazardous waste.

In addition, states like New Jersey that have a great deal of contaminated wetlands also tend to have state statutes regulating wetlands fill that are even stricter than the federal law. These well-meaning laws can slow cleanup and redevelopment if a brownfield is located in marshy or partially marshy land.

The Oil Pollution Act of 1990

Aside from the Resource Conservation and Recovery Act (RCRA) and CERCLA, the federal act that may deter brownfields redevelopment most significantly is the Oil Pollution Act of 1990 (OPA).[51] Enacted after the Exxon *Valdez* oil spill, OPA was designed to provide a comprehensive regulatory and liability scheme governing all forms of petroleum pollution affecting the navigable waters of the United States[52] to the extent they are not covered by CERCLA. In general, OPA dictates that "each responsible party for a . . . facility from which oil is discharged, or which poses the substantial threat of a discharge of oil, into or upon the navigable waters or adjoining shorelines . . . is liable for the removal costs and damages specified in subsection (b) that result from such incident."[53] The damages referred to include "damages for injury to, or economic losses resulting from destruction of real or personal property, which shall be recoverable by a claimant who owns or leases that property."[54] Thus, like CERCLA, OPA threatens strict, joint and several liability based on mere ownership of a site (as a prior or current owner or operator, generator or transporter).

Like CERCLA, OPA is considered a remedial statute broadly designed to protect against pollution. OPA is applicable to "waters of the United States," including but not limited to creeks that are not themselves navigable waters. A "facility" for purposes of OPA includes "any structure, group of structures, equipment, or device (other than a vessel) which is used for one or more of the following purposes: exploring for, drilling for, producing, storing, handling, transferring, processing, or transporting oil." The "responsible party" under OPA includes "any person owning or operating the facility." Unlike CERCLA with its petroleum ex-

---

[51]33 U.S.C. § 2701, *et. seq.* (OPA). *See generally* Environmental Law Institute, LAW OF ENVIRONMENTAL PROTECTION § 12.10 (1994).

[52]For a general discussion of water resources law, *see* A. DAN TARLOCK, LAW OF WATER RIGHTS AND RESOURCES (1994).

[53]33 U.S.C. § 2702(a).

[54]*Id.* § 2702(b).

clusion, however, the pollution OPA is designed to address is that resulting from oil exploration activities. As such, judicial constructions of OPA that would limit its application are generally disfavored.

OPA specifically creates a private right of action for direct recovery from the polluter, provided the claimants have, with certain permissive exceptions, first presented their claims for cleanup cost recovery or damages to the responsible party.[55] OPA holds responsible parties liable for "any removal costs incurred by any person for acts taken by the person which are consistent with the National Contingency Plan." Since the National Contingency Plan (NCP) is primarily an administrative scheme, it is not entirely clear what actions a private litigant should take to be "consistent" with the NCP. Because many brownfields sites are contaminated with petroleum products, OPA may be a source of liability.

The Endangered Species Act of 1973 (ESA)

Wetlands are, disproportionately, the home of endangered species. Thus, where there is a wetlands problem, an Endangered Species Act issue will often follow. Enacted in 1973, the ESA[56] "provide[s] a means whereby the ecosystems upon which endangered species and threatened species depend may be conserved."[57] While the focus is on the ecosystem, most action is at the species level. Applied to any federal action, domestic or foreign, the ESA provides for protection of species subject to extinction and their critical habitats. The Department of the Interior's Fish and Wildlife Service and the Department of Commerce's National Marine Fisheries Service jointly administer the ESA (collectively the "Secretary").

An "endangered species" is "any species . . . in danger of extinction throughout all or a significant portion of its range."[58] A "threatened species" is "any species . . . likely to become . . . an endangered species within the foreseeable future throughout all or a significant portion of its range."[59] The Secretary determines whether or not a species is endangered or threatened.[60] That determination is based on "the best scientific and commercial data available to" the Secretary.[61]

"Critical habitats" are those specific areas within or outside the geographical areas occupied by a threatened or endangered species that are necessary for the conservation of the species.[62] After considering economic effects, the Secretary must designate critical habitats of endangered or threatened species based on "the best scientific data available."[63] Most wetlands are considered critical habitats.

---

[55]*Id.* § 2713(a).
[56]16 U.S.C. §§ 1531 to 1544.
[57]16 U.S.C. § 1531(b).
[58]16 U.S.C. § 1532(6).
[59]16 U.S.C. § 1532(20).
[60]16 U.S.C. § 1533.
[61]16 U.S.C. § 1533(b)(1)(A).
[62]16 U.S.C. § 1532(5)(A).
[63]16 U.S.C. § 1533(b)(2).

No action approved, financed, or implemented by a federal agency may "jeopardize the continued existence of any endangered or threatened species," or may destroy or may harm the critical habitat of any such species unless the seven-member cabinet-level Endangered Species Committee (also known as the "God Committee" or "God Squad") exempts the action.[64] A federal court of appeals may review their decision.[65]

Persons developing property that may contain an endangered species must obtain an incidental take permit from the Department of Interior before beginning construction. The ESA imposes civil and criminal penalties for violations.[66] After notice and an opportunity for a hearing, the Secretary may impose a civil penalty of up to $25,000 for each violation.[67] A person convicted of knowingly violating the ESA may be fined up to $50,000 upon conviction, may be imprisoned for a year, or both.[68]

Since many brownfields in the east coast are located in wetlands and adjacent to estuaries, many of these properties are considered or might be considered critical habitats. Thus, it is possible to run into the Department of the Interior (DOI) permitting, as well as section 404 permitting from the Corps, when redeveloping brownfields in wetlands or estuarine environments.

Out west, the problem of endangered species act permitting has been seen regularly in Texas in the context of redevelopment of property on or near petroleum refinery sites.

The Toxic Substances Control Act (TSCA)

Enacted in 1976, TSCA[69] regulates chemicals, including PCBs, asbestos, and indoor radon. All three substances are likely to be found in old buildings. PCBs are found in old transformers. Asbestos is found in insulation, floor, and ceiling tiles. Radon, which is a naturally occurring substance emanating from the earth, may also be found in older buildings that were unlikely to have been insulated against radon.

Removal of PCBs and asbestos can be expensive. Since they are often found in old buildings, most brownfields projects that have buildings as infrastructure are likely to need both PCB and asbestos removal or abatement.

The Resource Conservation and Recovery Act (RCRA)

In addition to common law and CERCLA concerns, "the current RCRA law contains provisions that present barriers to remediating sites under the RCRA correc-

---

[64] 16 U.S.C. § 1536(a)(2), (e), (h).
[65] 16 U.S.C. § 1536(n).
[66] 16 U.S.C. § 1540.
[67] 16 U.S.C. § 1540(a)(1).
[68] 16 U.S.C. § 1540(b)(1).
[69] 15 U.S.C. §§ 2601 to 2671.

tive action program; State cleanup programs; and voluntary cleanup efforts."[70] As one commentator and former EPA official explained:

> The fundamental flaw in RCRA that hinders cleanup is that the law was primarily designed to regulate process wastes, not cleanup wastes. As a result, the law requires stringent treatment standards, usually based on combustion, for most waste streams; establishes lengthy permit requirements; and otherwise presumes that process wastes are continuously generated and disposed of at an ongoing manufacturing facility. Consequently, the law emphasizes proper waste management and prevention of future contamination. However, most cleanups are one-time events, and generally address vast quantities of contaminated soil and debris—not chemical process wastes. As a result, RCRA's requirements are awkward, expensive, and in some cases hinder and prevent cleanup.[71]

These RCRA provisions could be corrected with a series of targeted legislative changes that would enhance protection of human health and the environment, while potentially saving hundreds of millions of dollars. Proper reform requires that the remedy selection schemes under RCRA and CERCLA be consistent and similar. "Without reform of RCRA as a companion to Superfund reauthorization, our nation's hazardous waste remediation problem will continue to waste resources, frustrate environmental cleanup progress, and reward inaction."[72]

One of the most powerful means for redressing past dumping of hazardous waste is the "imminent hazard" provision of RCRA.[73] That provision authorizes EPA to bring suit for injunctive relief against any person who has contributed, or is contributing, to the past or present handling, storage, treatment, transportation,

---

[70]Capitol Hill Hearing Testimony of Don Clay, President, Don Clay Associates, Inc., and former Assistant Administrator, EPA Office of Solid Waste and Emergency Response (1989-93) before the House Commerce, Trade, and Hazardous Material Waste Site Cleanups under Superfund (July 20, 1995).

[71]*Id.* As explained to Congress:

> In recent years, a consensus has emerged among many RCRA stakeholders that statutory changes are needed to streamline the program and remove obstacles to cleanup. Fundamentally, society cannot afford to continue to be needlessly "burning dirt."

Id.

[72]*Id.*

[73]As the primary regulator of *active* hazardous waste sites, RCRA can be implicated in brownfields redevelopment. Both past and present owners of waste disposal sites must contend with regulatory and liability risks under RCRA. When the statute was first enacted in 1976, it focused primarily on the problems surrounding the increasing volume and haphazard disposal of *nonhazardous solid waste*. The focus, however, shifted dramatically to *hazardous waste* management following the Love Canal tragedy and the surfacing of hazardous waste problems at other sites around the country. As a result, RCRA was amended in 1980 and again in 1984 to significantly expand and strengthen the requirements relating to the management of hazardous wastes.

or disposal of solid or hazardous waste that may present an "imminent and sub-stantial endangerment to health or the environment."[74] Such persons may be ei-ther restrained from the activities causing the imminent and substantial endanger-ment or ordered to take other action that may be necessary to deal with the problems, including, among other things, reimbursement of funds expended by EPA in connection with the site, site cleanup, and the restoration of groundwater and other contaminated resources. Since the cost of cleaning up a hazardous waste site is considerable, this provision can result in millions of dollars of liability. How-ever, liability of subsequent landowners is limited under RCRA to situations where they "could not reasonably be expected to have actual knowledge of the presence of hazardous waste at such facility or site and of its potential for release."[75]

The 1984 amendments expanded EPA's authority to require cleanup of envi-ronmental problems caused by hazardous waste facilities. Corrective action is re-quired for "releases" of hazardous waste or constituents, including those that have migrated "beyond the facility boundary,"[76] from any "solid waste management unit" at a "treatment, storage or disposal" (TSD) "facility" subject to RCRA.[77] All RCRA permits issued after enactment of the Hazardous and Solid Waste Amend-ments (HSWA) require an operator to take corrective action for all releases of haz-ardous constituents at the TSD facility regardless of when the waste was placed in a unit or whether the unit is currently active. Thus, these corrective action re-quirements can be invoked at brownfields properties.

EPA may issue an administrative order requiring compliance whenever "any person" has violated or is violating any RCRA requirement and may assess a *civil penalty up to $25,000 per violation per day* for any past or current violation. The administrator also may revoke a permit for failing to comply with the permit require-ments under RCRA § 3005 or the performance standards applicable to owners and operators under RCRA §§ 3004 and 3005(d). A permit also may be terminated for failing to comply with permit conditions. In addition to administrative proceedings, EPA may seek injunctive relief and civil penalties up to $25,000 per day in a federal district court for any violation of the hazardous waste management requirements.

A person is also subject to criminal liability for certain knowing violations of the law.[78] Certain criminal offenses are subject to *a fine of up to $50,000 per day of violation and/or up to five years in prison.* A second conviction doubles the po-tential fine and prison sentence. If the violator knows that the violation places an-

---

[74]RCRA § 7003(a), 42 U.S.C. § 6973(a).

[75]RCRA § 3013(b), 42 U.S.C. § 6934(b).

[76]RCRA § 3004(v), 42 U.S.C. § 6924(v).

[77]RCRA § 3004(u), 42 U.S.C. § 6924(u); 40 C.F.R. § 264.101, 50 Fed. Reg. 28747 (July 15, 1985); 52 Fed. Reg. 45788 (Dec. 1, 1987). The scope of the 3004(u) authority is defined, in large part, by the terms *releases, solid waste management unit,* and *facility.* EPA construes these terms broadly.

[78]These include:

    (1) treats, stores, or disposes of hazardous waste without a permit or transports a hazardous waste to a facility that does not have a permit;

    (2) generates, handles, stores, treats, transports, or disposes of hazardous waste and knowingly de-stroys, alters, or conceals required records;

other person in imminent danger of death or serious bodily injury, he is subject to a fine up to $250,000 and/or imprisonment up to fifteen years. "Organizations" that knowingly endanger persons as a result of a violation are subject to fines up to $1,000,000.

In addition to enforcement by EPA, almost every state has been delegated authority to operate a state administered RCRA program.[79] RCRA also authorizes citizens to bring suits against persons who violate any RCRA permit, regulation, or provision, or whose past or present handling, storage, treatment, transportation, or disposal of hazardous waste may present an "imminent and substantial endangerment to health or the environment."

Problems with RCRA and Remediation

RCRA poses several distinct problems with respect to remediation wastes. While RCRA hazardous waste standards are geared toward manufacturing process wastes, current laws subject cleanup remediation wastes (that are defined as hazardous) to the same requirements that apply to process wastes. For example, contaminated soil that is excavated cannot be placed back on the ground without triggering the land ban treatment standards, the need for a RCRA permit, and facilitywide corrective action. As a result, RCRA regulations have:

> made cleanups very costly;
> imposed (in some cases) unnecessary requirements; and
> delayed remediation of many sites.

"In some cases, RCRA has had the anomalous effect of actually preventing cleanup from taking place."[80]

Specific examples abound. Citgo Petroleum owns a refinery in Lake Charles, Louisiana, that is subject to the RCRA corrective action requirements. On the facility property are several old surface impoundments containing an enormous amount—625,000 tons—of old refinery sludges. Under the land disposal restriction, if Citgo were to remove that sludge, it would have to be treated to stringent land ban standards prior to being placed on the ground. Treating such a vast quan-

---

(3) makes a false statement, misrepresentation, or material omission in any permit application, permit, manifest, record, report, label, or manifest;

(4) transports hazardous waste without a manifest; or

(5) exports hazardous waste without the consent of the receiving country or in violation of an international agreement.

[79]Almost every state has a state-delegated program that has received final EPA approval. Authorized states are required to go through the approval process again before they can administer any of the 1984 amendments. Until approval is obtained by a state, EPA is authorized to administer the 1984 amendments and issue permits containing new provisions required by the amendments in that state.

[80]Prepared Testimony of Don R. Clay, President, Don Clay Associates, Inc. and former Assistant Administrator, EPA Office of Solid Waste and Emergency Response (1989-1993) before the House Commerce Committee, Subcommittee on Commerce, Trade, and Hazardous Materials (July 20, 1995).

tity of petroleum-contaminated sludge to land ban levels could cost almost three-quarters of a trillion dollars. Obviously, Citgo is better off leaving the sludge right there. In an effort to avoid such a costly and environmentally unsound result, Citgo requested and EPA proposed a "treatability variance." Under the proposed variance, Citgo would have treated the sludge (though to less stringent levels than land ban required), and then disposed of the waste in a Subtitle C double-lined landfill. However, at the present, that variance has not been finalized, and its future seems uncertain. In the meantime, the sludge just sits there. It is precisely this type of situation that our reforms would address.

A second example is an old fuel transfer station in New York state. As is common at such sites, considerable soil contamination was caused by past oil spills. In an effort to reclaim this brownfields site, the owner proposed to clean it up and turn it into a harbor redevelopment project (similar to Harborplace in Baltimore). The soil was deemed hazardous under the toxicity characteristic test and thus, subject to RCRA's hazardous waste regulations. Because the RCRA permitting provisions posed a real obstacle, the site was neither cleaned up nor redeveloped.

Another example in New York state that is mirrored all across the country is the problem posed by groundwater contaminated with floating petroleum product. In New York, a major refinery discovered large quantities of petroleum product, which had leaked from decades of operation, floating on the groundwater beneath the facility. The owner wanted to recover this product, both to refine it and also to clean up the groundwater. However, the owner could not pump out the groundwater, recover the product, treat the remaining water, and reinject it into the groundwater without triggering the RCRA land ban, permit, and facilitywide corrective action requirements. Thus, the floating product remained in place, even though the recovery process would have removed the valuable product and replaced it with higher quality treated groundwater.

In California, there were two sites with a single owner separated by a right of way. One side was a surface impoundment, on the other side was a lined landfill. The owner wanted to remove the sludges from the unlined surface impoundment to the landfill; this was supported by the local community, which opposed an on-site incinerator. However, as best we know, the cleanup has never occurred because the land ban requirements would have mandated such treatment for the removed sludges.

A last example involves a federal facility in the Midwest that was highly contaminated from years of munitions production and explosion. In the midst of this contaminated wasteland were two RCRA-regulated units that were subject to closure requirements. Under RCRA, the federal facility operator would have been required to "clean close" the units (remove all of the waste and certify that it was gone) even though they were surrounded by contamination.

As these examples demonstrate, RCRA could be reformed to better enable safe brownfields redevelopment. The fundamental flaw that hinders RCRA cleanups is that the law was primarily designed to regulate newly generated process wastes and not remediation wastes. As a result, RCRA's requirements are "awkward, expensive, and, in some cases, hinder and prevent cleanups." The Association of State and Territorial Solid Waste Management Officials cites the same primary RCRA provisions as discussed in the outline at the beginning of this chapter as "counterproductive" in the remediation setting. Testimony before Congress confirmed these concerns:

> Indeed, RCRA requirements today have a negative impact on brownfields redevelopment and voluntary cleanups. Several witnesses note that changes will aid in restoring brownfields and should be applied for both Federal and State cleanups. . . . These are just a few examples of situations where RCRA prevents—not encourages—sensible, environmentally-protective action. There are many, many more.[81]

"Often Congress has created one-size-fits-all requirements that don't make sense in Alaska, or Texas, or don't meet the reality at a specific site. The result is federal programs that are inefficient and counterproductive." Statutory preference for permanence and treatment often are inconsistent with speedy and reasonable site-specific remedies.[82]

Another needed reform is greater integration of RCRA with wastewater treatment programs under the Safe Drinking Water Act and the Clean Water Act. RCRA's land ban program currently requires any hazardous waste that is placed on the land be treated using best demonstrated available technology (BDAT)— even if that land disposal unit is part of a treatment system that is controlled by another statute. Thus, wastewater treatment ponds used to comply with a Clean Water Act permit, or deepwell injection units regulated under the Safe Drinking Water Act, have been deemed by the courts as still subject to the additional treatment requirements of RCRA. The result is potentially disruptive, costly, and duplicative regulation with very little environmental benefit.[83]

---

[81]Testimony of Michael Oxley—Chairman, House Commerce Committee, Subcommittee on Commerce, Trade and Hazardous Material Waste Site Cleanups under Superfund (July 20, 1995).

[82]As part of the RCRA Reform Initiative in 1992, it was estimated that contaminated soil reforms alone would save $3.2 billion annually. Those numbers are undoubtedly higher today, but even without including the value of 1995 dollars and the addition of other reforms, changes encouraging brownfields redevelopment will save the economy hundreds of millions of dollars per year. *See* Prepared Testimony of Don R. Clay, President, Don Clay Associates, Inc., and former Assistant Administrator, EPA Office of Solid Waste and Emergency Response (1989-1993) before the House Commerce Committee, Subcommittee on Commerce, Trade, and Hazardous Materials (July 20, 1995).

[83]EPA believes that such regulation will provide minimal human health and environmental benefit. As stated in the preamble to the Phase III Land Disposal Restrictions rule, "That being said, the risks addressed by the rule, particularly deep wells, are very small relative to the risks presented by other environmental conditions or situations. In a time of limited resources, common sense dictates that we deal with higher risk activities first, a principle on which EPA, members of the regulated community, and the public can all agree. Nevertheless, the Agency is required to set treatment standards for these relatively low risk wastes and disposal practices during the next two years, although there are other actions and projects with which the Agency could provide greater protection of human health and the environment." 60 Fed. Reg. 11,704/2 (March 2, 1995).

RCRA should be amended to minimize unnecessary overlap where other environmental statutes safely control the same risks or releases as are controlled under the hazardous waste provisions of RCRA. In addition to the Clean Water Act and Safe Drinking Water Act, laws such as the Clean Air Act and Toxic Substances Control Act overlap with RCRA, causing duplicative regulation and confusion by the states and regulated community. RCRA should be amended to specify that if other environmental statutes provide EPA the authority to control releases and EPA has a mandate to control such releases within five years from date of enactment, that regulation under RCRA may not be used.

In addition, many RCRA experts propose an amendment to specify a reasonable "point of generation" at which wastes are deemed hazardous and therefore subject to the land ban. RCRA is currently unclear on this issue, creating confusion in the regulated community and requiring many wastes (including those wastes managed in wastewater treatment systems) to be treated to a level beyond which they are no longer considered hazardous. Adding a clarifying provision would provide greater certainty to EPA, states, and the regulated community as to when and where the land ban requirements actually apply. At the same time, this provision would prevent unnecessary controls, such as treatment of nonhazardous wastes, as well as eliminate redundant regulation.

Another proposed amendment concerns the existing definition of solid waste. This definition imposes often rigid controls and the onerous label of "hazardous" for many materials that are recycled. In order to simplify these rules, reduce unneeded burden on both regulators and regulated industry, and encourage legitimate recycling, most hazardous recyclable materials should be removed from the hazardous waste provisions of RCRA and, instead, subjected to a tailored set of technical standards that removes stigma concerns and enhances resource conservation.

Finally, the current hazardous waste regulations control many wastes and treatment residues that present low risks. In addition, EPA has questionable legal authority to modify or differentiate standards based on risks. In order to make RCRA more risk based, exit levels below which a waste is no longer subject to RCRA's hazardous waste provisions should be clearly defined in the law. In addition, RCRA should be amended to require EPA to modify the hazardous waste management requirements for wastes that pose a slightly higher risk than those below the exit level. This would allow EPA to tailor the standards to the risks involved.

Regulation of Underground Storage Tanks under RCRA

Underground storage tanks (USTs) are often present on brownfields sites because many sites are actually former gasoline filling stations or other industries that had their own private pumps. Before EPA regulations were issued, most tanks were constructed of bare steel and were not equipped with release prevention or detection features. Thus, most tanks leak. In addition, spills often occur when tanks are overfilled. Current EPA regulations require UST owners and operators to ensure that their tanks are protected against corrosion and equipped with devices that prevent spills and overfills no later than December 22, 1998.[84]

With certain limited exceptions,[85] UST owners and operators must report the discovery of any release or suspected release of regulated UST substances to the implementing agency within twenty-four hours, or another reasonable time period specified by the implementing agency. This includes owners or operators of brownfields sites that discover a regulated UST on their premises.

In addition, owners or operators who would like to take tanks out of operation, including those acquiring a brownfields site with an abandoned tank on it, must either temporarily or permanently close them.[86]

The RCRA UST provisions contain a secured creditor exemption similar to that in CERCLA. RCRA excludes from the corrective action requirements applicable to owners any person "who, without participating in the management of an UST, and otherwise not engaged in petroleum production, refining, and marketing, holds indicia of ownership primarily to protect the owner's security interest in the tank." This statutory provision is intended to exempt from cleanup responsibility a person whose only connection with a tank is as the holder of a security interest. In other words, a bank or other secured creditor who has extended credit to a borrower (commonly the tank's owner) and who has in return secured the loan or other obligation by taking a security interest in the tank is excluded from the definition of owner for purposes of RCRA subtitle I. EPA has proposed rules interpreting the RCRA secured creditor exemption.

---

[84]40 C.F.R. § 280.21. *See generally* MICHAEL L. ITALIANO ET AL., LIABILITY FOR STORAGE TANKS (1992). Subtitle I of RCRA allows state UST programs approved by EPA to operate in lieu of the federal program. EPA's state program approval regulations set standards for state programs to meet.

[85]42 U.S.C. § 6991(1).

[86]When UST systems are temporarily closed, owners and operators must continue operation and maintenance of corrosion protection and, unless all USTs have been emptied, release detection. If temporarily closed for three months or more, the UST system's vent lines must be left open and functioning, and all other lines, pumps, manways, and ancillary equipment must be capped and secured. After twelve months, tanks that do not meet either the performance standards for new UST systems or the upgrading requirements (excluding spill and overfill device requirements) must be permanently closed, unless a site assessment is performed by the owner or operator and an extension is obtained from the implementing agency. To close a tank permanently, an owner or operator generally must:

(1) notify the regulatory authority thirty days before closing (or another reasonable time period determined by the implementing agency);

(2) determine if the tank has leaked and, if so, take appropriate notification and corrective action;

(3) empty and clean the UST; and

(4) either remove the UST from the ground or leave it in the ground filled with an inert, solid material.

The Comprehensive Environmental Response, Compensation, and Liability Act of 1980

Congress enacted CERCLA[87] in the wake of "Love Canal" and the growing public concern about the dangers posed by hazardous waste sites. In Love Canal, a residential neighborhood was built over an abandoned hazardous substance site. Existing law then precluded the homeowners from suing Hooker Chemical Company (the company factually responsible for the site contamination) due to lack of privity relationship between the homeowners and the polluter. Similarly, RCRA was unavailable as a remedy for the homeowners because the site was an abandoned, inactive, hazardous waste site and RCRA regulates only active hazardous waste sites. Congress enacted CERCLA to remedy this gap in the law.

CERCLA serves two "essential" yet independent purposes. First, the law gives the federal government the tools necessary for prompt and effective response to the problems created by abandoned hazardous wastes. Second, CERCLA holds those "responsible parties" accountable for the costs and responsibility of cleanup. The stated purpose of CERCLA was "to force polluters to pay for costs associated with remedying their pollution."[88] The law was drafted so broadly, however, that many people who never participated in pollution (such as current owners of properties polluted by past owners) can be held liable for cleanup costs.

The following table illustrates the key provisions of CERCLA:

Section	Description
§ 101	Definitions, including (1) owner and operator, and (2) scope of innocent purchaser defense
§ 103	Release notification requirements for hazardous substances
§ 104	Authorizes EPA to conduct removal or remedial action consistent with NCP
§ 105	NCP requires establishing NPL through the hazard ranking system (HRS) and requires periodic revision of the NCP
§ 106	Authorizes issuance of unilateral administrative orders (UAO) requiring the abatement of releases or threat of release creating an imminent and substantial endangerment to health, welfare, or the environment
§ 107 (a)	Establishes liability for four classes of PRPs
§ 107 (b)	Establishes limited statutory defenses, including the third party defense, innocent purchaser defense, and the inheritance defense
§ 111	Creates the Superfund
§ 113	Bars pre-enforcement review
§ 116	Establishes timetable for reviewing the NPL sites, commencement of RI/FSs, and remedial action
§ 121	Establishes guidelines for setting cleanup standards
§ 122	Establishes standards for settlements with PRPs, including de minimis contributors

---

[87]42 U.S.C. §§ 9601-75. *See generally* DONALD W. STEVER, LAW OF CHEMICAL AND HAZARDOUS WASTE (1994); MICHAEL DORE, LAW OF TOXIC TORTS: LITIGATION/DEFENSE/INSURANCE (1994).

[88]United States v. Alcan Aluminum Corp., 964 F.2d 252, 259-60 (3d Cir. 1992).

Elements of a Cause of Action

To trigger liability, CERCLA section 107[89] requires a plaintiff to allege the following four elements:[90]

(1) the waste disposal site is a "facility;"[91]

(2) a "release" or "threatened release" of any "hazardous substance" from the facility has occurred;

(3) such "release" or "threatened release" has caused the plaintiff to incur response costs that are "consistent with the national contingency plan"; and

(4) the defendant must qualify as one of four categories of "covered persons" subject to CERCLA liability.[92]

CERCLA defines facility as "any building, structure, installation, equipment, pipe, . . . well, pit, pond, lagoon, impoundment, ditch, landfill, storage container . . . or any site where a hazardous substance has been deposited, stored, disposed of, or placed, or otherwise come to be located."[93] With only limited exceptions, courts have interpreted this definition very broadly to include "every place where hazardous substances come to be located."[94] In order to show that an area is a 'facility' a plaintiff need only show that a hazardous substance has "otherwise come to be located" there.[95]

A release is defined as "any spilling, leaking, pumping, pouring, emitting, emptying, discharging, injecting, escaping, leaching, dumping, or disposing into the environment."[96] Courts have repeatedly rejected any attempt to limit the broad coverage of this definition.[97] The definition was meant to "encompass the entire universe of ways in which hazardous substances may come to exist in the envi-

---

[89]42 U.S.C. § 9607.

[90]*See* Town of Munster v. Sherwin-Williams Co., 27 F.3d 1268 (7th Cir. 1994); Kerr-McGee Chemical v. Lefton Iron & Metal, 14 F.3d 321 (7th Cir. 1994).

[91]42 U.S.C. § 9601(9).

[92]42 U.S.C. § 9607(a); *Ascon Properties,* 866 F.2d at 1152.

[93]42 U.S.C. § 9601(9).

[94]Elf Atochem North America, Inc. v. United States, 1994 WL 559219 (E.D. Pa. 1994); United States v. Conservation Chem. Co., 619 F. Supp. 162, 185 (W.D. Mo. 1985); Brookfield-North Riverside Water Comm'n v. Martin Oil Mktg., 1992 U.S. Dist. LEXIS 2920 at *16. *But see* Kane v. United States 15 F.3d 87 (8th Cir. 1994) (house containing asbestos is a "consumer product in consumer use" exempt under CERCLA).

[95]Brookfield-North Riverside Water Commission v. Martin Oil Marketing, 1992 U.S. Dist. LEXIS 2920 at *16; United States v. Metate Asbestos Corp., 584 F. Supp. 1143, 1148 (D. Ariz. 1984).

[96]42 U.S.C. § 9601(22).

[97]Lincoln Properties, Ltd. v. Higgins, 823 F. Supp. 1528, 1536 (E.D. Cal. 1992); Amoco Oil Co. v. Borden, Inc., 889 F.2d 664, 669 (5th Cir. 1989); Amland Properties Corp. v. Aluminum Co. of America, 711 F. Supp. 784, 793 (D.N.J. 1989); United States v. Hardage, 761 F. Supp. 1501, 1510 (W.D. Okl. 1990). *See also* New York v. Shore Realty Corp., 759 F.2d 1032, 1045 (2nd Cir. 1985).

ronment."[98] Most courts have interpreted the term as including both active and passive releases of hazardous substances.[99] A release of hazardous substances is not limited to just one instance, but can occur each time a hazardous substance is disposed.[100]

"Response costs" are defined as those costs incurred in the removal of hazardous substances or any remedial action to clean up hazardous substances as well as any related enforcement activities.[101] "Removal" is defined as "the cleanup or removal of released hazardous substances from the environment, such actions may be necessary taken in the event of the threat of release of hazardous substances into the environment."[102] "Remedial action" is defined as "those actions consistent with permanent remedy taken instead of or in addition to removal actions in the event of a release or threatened release of a hazardous substance into the environment, to prevent or minimize the release of hazardous substances so that they do not migrate to cause substantial danger to present or future public health or welfare or the environment."[103] In other words, removal actions are the short-term activities necessary to stop contained contamination or adverse health effects from the site. Remedial activities are the long-term cleanup strategies. Whether "response costs" have been incurred or not is a case by case determination. Response costs have been held to include:

> medical monitoring;[104]
> security fencing, or other measures to secure the site;[105]
> investigating, monitoring, testing, and evaluation costs;[106]
> alternative water supplies;[107]
> prejudgment interest;[108]
> enforcement costs;[109]
> remedial actions, including storage, confinement, clay cover, neutralization,
> recycling or reuse, destruction, dredging or excavation, collection of leachate or runoff, and on-site treatment or incineration;[110] and
> sometimes EPA oversight costs.

---

[98]Lincoln Properties, Ltd. v. Higgins, 823 F. Supp. 1528, 1536 (E.D. Cal. 1992).

[99]*See, e.g.,* United States v. Petersen Sand and Gravel, Inc., 806 F. Supp. 1346, 1351 (N.D. Ill. 1992).

[100]823 F. Supp. at 1537 (citing Tanglewood East Homeowners v. Charles-Thomas, Inc., 849 F.2d 1568 (5th Cir. 1988)). *But see* Brookfield-North Riverside Water Comm'n. v. Martin Oil Mktg., Ltd., 1993 U.S. Dist. LEXIS 2920.

[101]42 U.S.C. § 9601(25).

[102]42 U.S.C. § 9601(23).

[103]42 U.S.C. § 9601(24).

[104]*See* Brewer v. Ravan, 680 F. Supp. 1176, 1179 (M.D. Tenn. 1988). *But see* Hopkins v. Elano Corp., 30 E.R.C. 1782, 1786 (S.D. Ohio 1989).

[105]*See* Amoco Oil Co. v. Borden, Inc., 889 F.2d at 672.

[106]*See* United States v. Wade, 577 F. Supp. 1326, 1333 n.4 (E.D. Pa. 1983).

[107]*See* Artesian Water Co. v. New Castle County, 659 F. Supp. 1269, 1287-88 (D. Del. 1987).

[108]United States v. Mexico Feed & Seed Co., 729 F. Supp. 1250, 1253-54 (E.D. Mo. 1990).

[109]42 U.S.C. § 9601(25); *see also* United States v. NEPACCO, 579 F.2d at 85.

[110]42 U.S.C. § 9601(24).

One of the critical issues in CERCLA liability at brownfields sites is whether or not the cleanup must be "consistent with the NCP" in order to be actionable.[111] Some courts have held that cleanup costs must be consistent with the NCP in a private cause of action.[112] Other courts, however, have held that consistency with NCP is only relevant as to apportionment of liability and not to the question of liability itself.[113] Absence of public notice and comment will not necessarily preclude cost recovery when the PRP in charge of the cleanup has notified the relevant regulatory agencies.[114] There is a rebuttable presumption that costs incurred pursuant to a consent decree will be deemed consistent with the NCP.[115]

In addition to cleanup costs, potentially responsible parties (PRPs) are also liable for "natural resource damages."[116] While the price tag for these damages is not clear, this form of Superfund liability at National Priority List (NPL) listed sites could ultimately prove to be even more expensive than cleanup costs. For example, at the Rocky Mountain Arsenal site the initial claim for natural resource damages was $1.8 billion.[117] Natural resource damages have not yet been considered an issue in private CERCLA sites (such as brownfields properties), but the statute does not differentiate between private and NPL listed sites when addressing natural resource damages. Moreover, CERCLA settlements are subject to a "reopener," a provision that allows the government to sue a PRP if unknown natural resource damage is later discovered.

Unlike RCRA and other environmental statutes, CERCLA liability does not require a threshold quantity or concentration of hazardous substance. A PRP may be liable regardless of how low the percentage of hazardous substance may be.[118]

---

[111]§§ 9607(a)(4) & (a)(4)(B); *see also* Ascon Properties Inc. v. Mobil Oil Co., 866 F.2d 1149, 1152 (9th Cir. 1989); New York v. Shore Realty Corp., 759 F.2d 1032, 1043 (2d Cir. 1985).

[112]*See, e.g.,* Pierson Sand & Gravel, Inc. v. Pierson Township, 851 F. Supp. 850 (W.D. Mich. 1994).

[113]*See, e.g.,* Charter Township of Oshtemo v. American Cyanamid Co., 1994 U.S. Dist. LEXIS 8544 (W.D. 1994). To evaluate consistency with the NCP, a court must evaluate the following factors:

the nature of the action taken;
the imminence or the release or threatened release;
whether a state or federal agency found a threat to public health or safety;
whether an agency has recommended action to eliminate the threat; and
the costs, complexity, and duration of the activity.

Hatco Corp. v. Amtreco, Inc., 846 F. Supp. 1578 (M.D. Ga. 1994).

[114]*Id.*

[115]*See, e.g.,* United States v. Colorado & Eastern Railroad Co., 1994 WL 647329 (10th Cir. 1994).

[116]*See generally* KEVIN M. WARD ET AL., NATURAL RESOURCES: LAW AND ECONOMICS (1992); Anderson, *Natural Resource Damages, Superfund, and the Courts,* 16 B.C. ENVTL. AFF. L. REV. 405 (1989).

[117]*See* United States v. Shell Oil Company, 605 F. Supp. 1064, 1084-85 (D. Colo. 1985).

[118]*Murtha,* 958 F.2d at 1200; New York v. Exxon, 744 F. Supp. 474, 485-86 n.16 (S.D.N.Y. 1990).

Liable Parties

With its liability provisions based on strict liability, causation is not an element of a CERCLA cause of action.[119] Thus, the plaintiff is not required to link the defendant's conduct or the defendant's waste firmly to the release or threat of release. Nor must the defendant's conduct be considered unlawful in order to incur liability. "The release or threat of release only need have emanated from a facility which [defendant] owned, or to which [defendant] transported."[120] Causation may appear as a peripheral issue under CERCLA at the apportionment phase of a multiparty CERCLA case. Causation, however, is almost never an issue in brownfields sites, where litigation is usually directed only at the current owner and his lender.

Under CERCLA, four classes of persons may be held liable:

(1) current owner or operators;
(2) past owners and operators at the time of disposal;
(3) generators; and
(4) transporters of hazardous waste that selected the disposal site.[121]

Owner and Operator Liability

Ownership liability may be based on mere title ownership of the site.[122] Thus, a PRP could be held liable as an owner if a release or threat of release occurs during the PRP's ownership of the property even if that person did not cause or contribute to the release.[123] In contrast, operator liability is determined by virtue of

---

[119]*See* Northwestern Mutual Life Insurance Co. v. Atlantic Research Corp., 847 F. Supp. 389 (E.D. Va. 1994); United States v. Maryland Sand, Gravel, and Stone Co., 1994 WL 541069 (D. Md. 1994).

[120]Dedham Water Co. v. Cumberland Farms Dairy, Inc. ("*Dedham Water II*"), 889 F.2d 1146, 1153 (1st Cir. 1989).

[121]*Id.* at 1044. *See generally* JENNIFER L. MACHLIN & R. YOUNG, MANAGING ENVIRONMENTAL RISK: REAL ESTATE AND BUSINESS TRANSACTIONS § 4 (1994); BAXTER'S ENVIRONMENTAL COMPLIANCE MANUAL § 15 (1994). Specifically, the statute reads:

(1) the owner and operator of a vessel (otherwise subject to the jurisdiction of the United States) or a facility,
(2) any person who at the time of disposal of any hazardous substance owned or operated any facility at which such hazardous substances were disposed of,
(3) any person who by contract, agreement, or otherwise arranged for disposal or treatment, or arranged with a transporter for transport for disposal or treatment, of hazardous substances owned or possessed by such person, by any other party or entity, at any facility owned or operated by another party or entity and containing such hazardous substances, and
(4) any person who accepts or accepted any hazardous substance for transport to disposal or treatment facilities or sites selected by such person.

42 U.S.C. § 9607(a).

[122]*See, e.g.*, Kerr-McGee Chemical Corp. v. Leftron Iron & Metal Co., 14 F.3d 321 (7th Cir. 1994).
[123]United States v. Taylor, 1994 WL 512758 (W.D. 1994). *But see* Long Beach Unified School District v. Godwin California Living Trust, 32 F.3d 1364 (9th Cir. 1994) (holders of an easement to oper-

control over the facility.[124] Courts have developed various theories concerning shareholder and corporate officer liability for environmental problems. They are as follows:

(1) *the personal participation theory* allows a court to hold a shareholder liable for CERCLA response costs if that individual personally and actively participated in disposal of the hazardous substance concerned;[125]

(2) under *the authority-to-control test*, operator liability is imposed as long as one corporation had the capability to control, even if it was never utilized;[126]

(3) under the actual control standard, operator liability will be incurred for the environmental violations when there is evidence of substantial control that is actually exercised over the activities;[127]

(4) under *the prevention test*, the court reviews evidence of an individual's authority to control waste-handling practices, including distribution of power in the corporation, percentage of shares owned, responsibility for waste disposal practices and neglect in that regard;[128] and

---

ate a pipeline over land found to contain hazardous substances were not, by virtue of interest alone, owners under CERCLA); Grand Trunk Western Railroad Co. v. Acme Belt Recrating, Inc., 859 F. Supp. 1125 (W.D. Mich. 1994) (defendant that held an easement was not an owner under CERCLA); Atlantic Richfield Co. v. Blosenski, 847 F. Supp. 1261 (E.D. Pa. 1994) (in order to be liable as a coowner under CERCLA, a defendant must manifest some intent to own the property).

For a discussion of "intervening owner liability," *see* Joslyn Manufacturing Co. v. Koopers Co., Inc., 40 F.3d 750 (5th Cir. 1994) (court dismissed motion to hold PRP liable as a past owner of a site simply because the PRP was a sophisticated purchaser who knew the site was contaminated; the court said owner liability would only apply if disposal occurred at the site while the PRP owned the property).

[124]*See* New York v. Shore Realty Corp., 759 F.2d at 1043-44; *see also* Edwards Hines Lumber Co. v. Vulcan Materials Co., 861 F.2d 155 (7th Cir. 1988).

[125]*See* United States v. Northeastern Pharmaceutical & Chem. Company, 579 F. Supp. 823 (W.D. Mo. 1984) ("*NEPACCO I*"), *aff'd in part and rev'd in part,* 810 F.2d 726 (8th Cir. 1986) ("*NEPACCO II*"), *cert. denied,* 484 U.S. 848 (1987). *See also* United States v. Carolina Transformer Co., 739 F. Supp. 1030, 1038 (E.D.N.C. 1989), *aff'd,* 978 F.2d 832 (4th Cir. 1992) (holding two directors who were consecutive presidents of a corporation personally liable for cleanup of PCB contamination because they personally supervised the activities causing it); United States v. Bliss, 20 Envtl. L. Rep. (Envtl. L. Inst.) 20,879, 20,883 (E.D. Mo. 1988) (holding corporate president personally liable for cleanup of hazardous waste after he personally authorized its improper disposal and physically assisted in unloading some of it); United States v. Conservation Chem. Co., 628 F. Supp. 391 (W.D. Mo. 1986) (holding corporate president, a chemical engineer who owned over 90 percent of the stock and was personally responsible for environmental controls, personally liable for cleanup of corporation's industrial chemical waste disposal facility); United States v. Ward, 618 F. Supp. 884 (E.D.N.C. 1985) (holding corporate president personally liable for improper disposal of hazardous substances after he personally negotiated, and received personal gain from, contract with transporter).

[126]*See* Nurad, Inc. v. William E. Hooper & Sons Co., 966 F.2d 837, 842 (4th Cir. 1992); Idaho v. Bunker Hill Co., 635 F. Supp. 665, 670-71 (D. Idaho 1986); United States v. Nicolet, Inc., 712 F. Supp. 1193 (E.D. Pa. 1989).

[127]United States v. Kayser-Roth Corp., 910 F.2d 24, 27 (1st Cir. 1990).

[128]Quadion Corp. v. Mache, 738 F. Supp. 270 (N.D. Ill. 1990); Kelley v. ARCO Industries Corp., 723 F. Supp. 1214, 1219-20 (W.D. Mich. 1989).

(5) under *the piercing the corporate veil theory*, there can be no direct operator liability of shareholders, officers, or directors under CERCLA unless the elements necessary to pierce the corporate veil are met.[129]

Most courts subscribe to one of the first four theories and allow direct operator liability under CERCLA without the necessity of piercing the corporate veil.

Transporter and Generator Liability

Only transporters who accept hazardous substances for shipment and participate in disposal site selection are held liable under CERCLA. The liability attaches when the transporters advice is a substantial contributing factor in the decision to dispose of hazardous waste at a particular facility.

By contrast, generators of hazardous substances need not have participated (or even known about) the site of disposal in order to be held liable.[130] Generator liability attaches even where the generator intended the wastes to be disposed elsewhere.[131] All that is needed to prove generator liability is that the generator possessed the hazardous substance and it ultimately was disposed of at the facility.[132]

Defenses to CERCLA Liability

CERCLA allows only a few, limited affirmative defenses.[133] These defenses include (1) an act of God; (2) an act of war;[134] and (3) an act of an unrelated third party. This third-party affirmative defense includes what has been termed the

---

[129]Joslyn Corp. v. T.L. James & Co., 696 F. Supp. 222, 224-45 (W.D. La. 1988), *aff'd,* 893 F.2d 80 (5th Cir. 1990), *cert. denied,* 498 U.S. 1108 (1991).

[130]*See, e.g.,* Tippins, Inc. v. USZ Corp., 37 F.3d 87 (3d Cir. 1994) (transporter must actively and substantially participate in the decision-making process that ultimately identifies a facility for disposal in order to incur liability); Atlantic Richfield Co. v. Blosenski, 847 F. Supp. 1261 (E.D. Pa. 1994); United States v. Hardage, 32 Env't. Rep. Cas. (BNA) 1073, 1081 (W.D. Okla. 1990).

[131]*See, e.g.,* Missouri v. Independent Petrochemical Corp., 610 F. Supp. 4,5 (E.D. Mo. 1985); O'Neill v. Picollo, 441 F. Supp. 706 (D.R.I. 1988).

[132]*See* United States v. Consolidated Rail Corp., 729 F. Supp. 1461 (D. Del. 1990).

[133]Section 107(a) further provides that liability is imposed "subject only to the defenses set forth in subsection (b) of this section." Section 107(b) lists these defenses as follows:

(b) There shall be no liability under subsection (a) of this section for a person otherwise liable who can establish by a preponderance of the evidence that the release or threat of release of a hazardous substance and the damages resulting therefrom were caused solely by—(1) an act of God; (2) an act of war; (3) an act or omission of a third party other than an employee or agent of the defendant . . . ; (any combination of the foregoing paragraph).

42 U.S.C. § 9607(b).

[134]For a discussion of the act of war defense, *see* United States v. Colorado & Eastern Railroad Co., 1994 WL 647329 (9th Cir. 1994).

"innocent purchaser" defense. Courts have interpreted these three statutory defenses very narrowly. For example, courts have found that heavy rains[135] and high floodwaters[136] were not acts of God for purposes of escaping CERCLA liability.

Third-Party Defense

The "third-party" defense[137] provides that an owner may escape liability if the release of the hazardous substances was caused by a third party and the owner can show that it took due care with regard to the risk of environmental contamination. Both past and current "owners and operators" may assert this defense. To successfully shield the defendant from liability, the defendant must show that:

(1) the hazardous contamination at the site was caused *solely* by other parties;

(2) no *direct or indirect contractual relationship existed* between it and the parties responsible for contaminating the site (for purposes of this element, "contractual relationship" is defined by statute to include land contracts, deeds, or other instruments transferring title or possession);[138]

(3) it *did not know or have reason to know* of the contamination at the site at the time of purchase; and

(4) it *exercised due care* with respect to the hazardous substance at the site.

Failure to prove any of these elements by a "preponderance of the evidence" precludes use of the defense.

The requirement that there be no contractual relationship has led to a great deal of litigation. For example, in *United States v. Monsanto Co.*,[139] the Fourth Circuit held that two site owners had not established the absence of a direct or indirect contractual relationship with the sublessee (SCRDI), the party causing the release, in part because they had accepted rent payments from SCRDI. In *O'Neil v. Picillo*,[140] generators of hazardous wastes ultimately disposed of at the contaminated site argued that the release had to have been caused solely by third

---

[135]United States v. Stringfellow, 661 F. Supp. 1053, 1061 (C.D. Cal. 1987) (the rains "were foreseeable based on normal climatic conditions and any harm caused by the rain could have been prevented through design of proper drainage channels").

[136]Lumbermens Mutual Cas. Co. v. Bellville Indust., Inc., 938 F.2d 1423 (1st Cir. 1991) (continuous polluting activity punctuated by a tropical rainstorm, a fire, and sporadic spills is not covered under insurance policy).

[137]42 U.S.C. § 9607(b)(3).

[138]42 U.S.C. § 9601(35).

[139]858 F.2d 160, 167 (4th Cir. 1988).

[140]682 F. Supp. 706, 728 (D.R.I. 1988).

parties for whom they had no responsibility, claiming that the licensed waste transporters to whom they had consigned their wastes had no contact with the contaminated site. The court rejected the defendants' argument, however, finding that:

> The simple fact is that during the time the defendants consigned their waste to licensed disposers, some of that waste, in identifiable containers, came to rest at the Picillo site. The defendants must demonstrate by a preponderance of the evidence that "a totally unrelated third party is the sole cause of the release." Absent any evidence along these lines, I must conclude that it is equally likely that either the licensed disposers or a subcontractor of the disposers deposited the waste at the site.

In *Washington v. Time Oil Co.*,[141] the court denied a motion for summary judgment by the defendant landowner, Time Oil, based on the innocent landowner defense. The court found that because Time Oil's subsidiary was also responsible for the contamination of the property, Time Oil had failed to establish that the release was caused solely by a third party for whom it was not responsible.

Innocent Purchaser Defense

An "innocent purchaser" may nonetheless escape liability if he purchased after the disposal of the hazardous substance and he did not know, or have reason to know, that the hazardous substance was disposed of on, in, or at the facility. "To establish that the defendant had no reason to know [of the hazardous substance] the defendant must have undertaken, at the time of acquisition, all appropriate inquiry into the previous ownership and uses of the property consistent with good commercial or customary practice in an effort to minimize liability." In interpreting that standard, the court "shall take into account any specialized knowledge or experience on the part of the defendant, the relationship of the purchase price to the value of the property if uncontaminated, commonly known or reasonably ascertainable information about the property, the obviousness of the presence or likely presence of contamination at the property, and the ability to detect such contamination by appropriate inspection."[142]

To qualify as an "innocent purchaser," one must have undertaken "all appropriate inquiry" into the previous ownership and uses of the property, consistent with "good commercial or customary practice" at the time of transfer. "Good commercial practice" is not defined in the statute. The innocent landowner defense has been repeatedly criticized for being vague regarding the extent of investigation

---

[141]687 F. Supp. 529, 532 (W.D. Wash. 1988).
[142]*Id.*

necessary to show "due diligence."[143] Numerous congressional efforts to define this term and the term "all appropriate inquiry" have failed.

The legislative history of this provision of the statute is also vague on the definition of "good commercial practice," indicating only that it requires that "a reasonable inquiry must have been made in all circumstances, in light of best business and land transfer principles." In deciding whether a defendant has complied with this standard, courts consider any specialized knowledge or expertise the defendant has, whether the purchase price indicated awareness of the presence of a risk of contamination, commonly known or reasonable in formation about the property, the obviousness of the presence of contamination at the property, and the ability to detect such contamination by appropriate inspection.

The statutory definition of "contractual relationship" includes land contracts, deeds, and other instruments transferring title, and thereby prohibits protection under this defense, unless the real property was acquired "after the disposal or placement of the hazardous substance" on it, and defendant establishes by a preponderance of the evidence:[144]

(1) At the time the defendant acquired the facility, the defendant did not know and had no reason to know that any hazardous substance which is the subject of the release or threatened release was disposed of on, in, or at the facility;

(2) the defendant is a government entity that acquired the facility by escheat, or through any other involuntary transfer or acquisition, or through the exercise of eminent domain authority by purchase or condemnation (the "municipal defense"); and

(3) the defendant acquired the facility by inheritance or bequest (the "inheritance defense").

The legislative history shows that Congress meant to establish "a three-tier system" in these innocent landowner provisions: "Commercial transactions are held to the strictest standard; private transactions are given a little more leniency; and inheritances and bequests are treated the most leniently" of all.[145] Although a large industry has developed providing services to ensure compliance with the dictates of the "innocent purchaser" defense, the courts have construed both the third-party and the innocent purchaser defense extremely narrowly.

---

[143]*See, e.g.,* United States v. Pacific Hide & Fur Depot, Inc., 716 F. Supp. 1341 (D. Idaho 1989).

[144]42 U.S.C. § 9601(35)(A)(I)-(iii).

[145]United States v. Pacific Hide & Fur Depot, Inc., 716 F. Supp. 1341, 1348 (D. Idaho 1989) (the three defendant inheritees all obtained their initial interest by familial gift and their ultimate interest by a corporate event beyond their control).

CERCLA Statutory Exclusions

In addition to the express statutory exemptions in CERCLA, there are certain statutory exemptions from CERCLA liability. These include:

the secured creditor exemption;[146]
the petroleum exclusion;[147]
the consumer product exemption;[148] and
the pesticide exclusion.[149]

The secured creditor exemption protects lenders who, without participating in the management of their debtor, hold indicia of ownership primarily to protect a security interest. This exemption was carved out to protect banks that hold legal title to land in title theory states. Since lenders in such states are actual owners of the property, Congress wanted to clarify that this type of ownership alone would not give rise to CERCLA liability. Recent cases upholding the secured creditor exemption include *Northeast Doran, Inc. v. Key Bank of Maine*,[150] which held that mortgagee that promptly divests itself of title property is deemed to have held it only as security interest for a mortgage was protected by the exemption. Similarly, in *Kemp Industries v. Safety Light Corp.*,[151] the prior titleholder who financed a construction project through a sale leaseback was held to be protected by the exemption. In *North Carolina v. W.R. Peele, Sr. Trust*,[152] the court held that the secured creditor exemption does not pertain to property held in trust and that a CERCLA action could be maintained against a trust. This exemption will be discussed in greater detail under the subpart discussing lender liability under CERCLA.

CERCLA defines "hazardous substance"—those materials regulated under CERCLA—by reference to substances listed under various other federal statutes.[153] However, CERCLA expressly excludes from its "hazardous substance" definition "petroleum, including crude oil or any fraction thereof which is not otherwise specifically listed or designated as a hazardous substance. . . ." Both the EPA and the courts interpret the petroleum exclusion to apply to petroleum products, even if a specifically listed hazardous substance, such as chrysene, is indigenous to such

---

[146]For the most recent discussions of the CERCLA secured creditor exemption, *see* Kelly v. EPA, 15 F.3d 1100 (D.C. Cir. 1994) (court vacated EPA's lender liability rule and determined that the courts, not EPA, is the adjudicator of liability issues under CERCLA).

[147]For a discussion of the petroleum exclusion, *see, e.g.,* Wilshire Westwood Assoc. v. Atlantic Richfield Corp., 881 F.2d 801 (9th Cir. 1989).

[148]*See* Kane v. United States 15 F.3d 87 (8th Cir. 1994) (house containing asbestos is a "consumer product in consumer use" exempt under CERCLA).

[149]*See* 42 U.S.C. § 9607(I).

[150]15 F.3d 1 (1st Cir. 1994).

[151]857 F. Supp. 373 (D.N.J. 1994).

[152]1994 U.S. Dist. LEXIS 16335 (E.D. N.C. 1994).

[153]*See* 40 C.F.R. § 302.4 (comprehensive listing of CERCLA hazardous substances).

products.[154] Furthermore, the petroleum exclusion in CERCLA "does apply to unrefined and refined gasoline even though certain of its indigenous components and certain additives during the refining process have themselves been designated as hazardous substances within the meaning of CERCLA."[155] EPA interprets the petroleum exclusion to apply to materials such as crude oil, petroleum feedstocks, and refined petroleum products, even if a specifically listed or designated hazardous substance is present in such products. EPA does not, however, consider materials such as waste oil to which listed CERCLA substances have been added to be within the petroleum exclusion.[156]

The statutory definition of facility under CERCLA excludes a "consumer product in consumer use."[157] This exclusion has been termed the "consumer products exclusion." For example, courts have held that CERCLA does not apply to asbestos that is incorporated into walls and ceilings of buildings based on this exclusion.[158] Certain courts, however, have found this exclusion limited. "The exception is for facilities that are consumer products in consumer use, not for consumer products contained in facilities."[159]

Finally, the pesticide exemption provides a very limited exception to CERCLA liability under Section 107 by indicating "no person . . . may recover . . . for any response costs or damages resulting from the application of a pesticide product registered under the Federal Insecticide, Fungicide, and Rodenticide Act [FIFRA]." The exclusion covers only liability for "field application" of a pesticide. This exclusion is limited to pesticide application of pesticides registered under FIFRA.[160] This is intended to mean the use of a pesticide in accordance with its purpose. Thus, pesticide releases, including spilling of pesticides, are covered under CERCLA.[161] Claimants who are injured by spilling, dumping, disposal, or

---

[154]*See* EPA Memorandum, July 31, 1987; Wilshire Westwood Assoc., 881 F.2d 801 (leaded gasoline falls within the petroleum exclusion despite hazardous indigenous components and additives).

[155]881 F.2d at 810.

[156]50 Fed. Reg. 13,460 (April 4, 1985). *See also* Mid Valley Bank v. North Valley Bank, 764 F. Supp. 1377, 1384 (E.D. Cal. 1991) (waste oil containing CERCLA hazardous substances does not fall under CERCLA's petroleum exclusion); United States v. Western Processing Co., 761 F. Supp. 713, 722 (W.D. Wash. 1991) (tank bottom sludge does not fall within the petroleum exclusion in part because the sludge at issue contained contaminants that were not indigenous to the crude oil itself).

[157]42 U.S.C. § 9601(9).

[158]*Id. See* Amcast Industrial Corp. v. Detrex Corp., 2 F.3d 746, 750 (7th Cir. 1993).

[159]Dayton Indep. Sch. Dist. v. U.S. Mineral Prods. Co., 906 F.2d 1059 (5th Cir. 1990). "The legislative history reinforces . . . that Congress intended to provide recovery only for releases or threatened releases from inactive and abandoned waste sites, not releases from useful consumer products in the structure of buildings. The sale of asbestos-containing products for useful consumption is not the 'arranging for disposal' of a hazardous substance at a 'facility,' Section 107(a) of CERCLA, that the statute is designed to combat."

[160]*See* Jordan v. Southern Wood Peidmont Co., 805 F. Supp. at 1581-82 (1992) (rejecting in dicta wood treatment pesticide seller's arguments that pesticides are generally excluded from CERCLA); United States v. Hardage, 733 F. Supp. 1424.

[161]*See* In re Sundance Corp., 149 B.R. at 663 (1993) (pesticide exemption unavailable for spillage of pesticides).

leaking of pesticides, whether intentional or accidental, have recourse under CER-CLA and are not protected by the pesticide exemption.

The Standard of Liability under CERCLA

CERCLA holds any or all of the Potentially Responsible Parties (PRPs) liable:

> *strictly* (i.e., without regard to fault or negligence);[162] in the normal case of "indivisible" injury;
> *jointly and severally* (i.e., any one contributing PRP can be held liable for the *entire* cleanup by EPA, the state, or the private plaintiff, and is left to his own devices to subsequently collect a "fair share" from other contributing PRPs);[163] and
> for *preenactment conduct* and cleanup costs as well as postenactment conduct and cleanup costs.[164]

Thus, an owner or operator may be liable whether or not his conduct was legal under state and federal law at the time undertaken, unless the release meets the narrow definition of a "federally permitted release." Furthermore, a PRP may be liable even if a PRP contributed only a small amount of hazardous substances to a facility, whether or not the hazardous substances contributed to the site by the PRP are the ones the release of which required cleanup, and whether or not the PRP *caused* the release of the hazardous substances actually requiring the cleanup. In short, EPA (or the state or private party cleanup plaintiff) can extract cleanup costs from any PRP or group of PRPs that they choose to sue, and those PRPs are left to use the original action or separate legal actions among themselves (or among themselves and other PRPs that they find and join or sue) to sort out, under evolving principles of "contribution" law, who ultimately bears how much of the responsibility for the cleanup costs.

Mechanisms of Suit under CERCLA

There are numerous methods by which CERCLA allows cleanup of hazardous sites. First, CERCLA grants the president of the United States broad authority to

---

[162]*See, e.g.,* United States v. Monsanto, 858 F.2d 160, 167-68 (4th Cir. 1988), *cert. denied,* 490 U.S. 1106 (1989); United States v. Northeastern Pharmaceutical & Chemical Co., 810 F.2d 726 (8th Cir. 1986), *cert. denied,* 484 U.S. 848 (1987).

[163]*See, e.g.,* United States v. Monsanto, 858 F.2d at 171-73; New York v. Shore Realty Corp., 759 F.2d 1032 (2d Cir. 1985).

[164]*See, e.g.,* United States v. Monsanto, 858 F.2d at 173-75; United States v. Northeastern Pharmaceutical & Chemical Co., 810 F.2d at 733-34. *But see* Freeman, *Inappropriate and Unconstitutional Retroactive Application of Superfund Liability,* 42 Bus. Law. 215 (Nov. 1986).

provide for the cleanup of sites contaminated by hazardous substances. Most of this authority has been delegated to EPA[165] via the regional EPA administrators.

Second, CERCLA also authorizes the United States to use "Superfund" monies to clean up a site. The government may then recover those response costs from the parties defined under the statute as responsible for the pollution.[166]

Third, private parties may also maintain a cause of action for recovery of cleanup "response" costs and interest from other potentially responsible parties. Section 113 of CERCLA makes clear that one held liable under Section 107 may seek contribution from others also responsible for contaminating the site.[167] Generally, § 113(f)(1) provides a right of contribution for any party against any other party who is or may be liable for the release or threatened release of hazardous substances.[168] This private cause of action was designed to encourage the cleanup of environmental hazards by private individuals, who then may recover the costs of the cleanup from the parties responsible for the hazard.[169]

It is the contribution provisions that cause the greatest uncertainty in the brownfields context. Even if a state has an effective voluntary remediation program in place and a party has received a covenant not to sue or no further action letter, the party redeveloping the brownfields site cannot be shielded from a private cause of action under CERCLA. This is true even if the regional EPA administrator joins in such an agreement with the state through a memorandum of understanding (MOU) (which binds the federal government and prevents the United States from filing suit).

CERCLA Contribution Actions

Contribution is the method by which a joint tortfeasor can compel a sharing of the liability burden with other parties responsible for the hurt.[170] In the 1986 amendments, Congress codified the right of contribution for response costs among PRPs.[171] CERCLA § 113(f)(1) allows a party who has incurred response costs to seek contribution from any person who is liable or potentially liable under Section 107. A party may bring a contribution claim during or after a CERCLA § 106 proceeding or CERCLA § 107 cost recovery action, or at any time after they have incurred response costs.

The legislative history of SARA shows that Congress did try to soften the harshness of joint and several liability by expressly allowing a cause of action for contribution.[172] Contribution is not always an issue at brownfields sites, since

---

[165]*See Alcan,* 964 F.2d at 258.

[166]*See* 42 U.S.C. § 9607, 9611-12.

[167]42 U.S.C. § 9613.

[168]*See, e.g.,* Alloy Briquetting Corp. v. Niagara Vest, Inc., 802 F. Supp. 943, 944 (W.D. N.Y. 1992).

[169]Nurad, Inc. v. William E. Hooper & Sons Co., 966 F.2d 837, 839 (4th Cir. 1992).

[170]United States v. Cannons Engineering Corp., 720 F. Supp. 1027, 1051 (D. Mass. 1989), *aff'd* 899 F.2d 79 (1st Cir. 1990).

[171]42 U.S.C. § 9613.

[172]REPORT OF THE COMMITTEE ON ENERGY, H.R. REP. NO. 253(I), 99th Cong., 2d Sess. 59 (1985), *reprinted* in 1986 USCCAN 2835, 2861.

most are single party sites and often the other PRPs that are eligible for suit are either unknown, no longer in existence or insolvent. Contribution may become an issue for the current or past owner of a brownfields site where that party wants to sue others for their past relationship to the site.

The majority rule is that where the harm is indivisible (as is usually the case with cleanup of CERCLA sites) liability under CERCLA is joint and several. The burden of demonstrating divisibility is placed on the defendant, who is the potentially responsible party (PRP). Congress allows EPA to enter into de minimis[173] and de micromis settlements, but this is within EPA's discretion and is beyond the defendant's control. It is important to note that, even if EPA takes such measures, private plaintiffs are unaffected and can still sue.

Certain equitable factors, especially the relative culpability of the PRPs, can be considered by a court when allocating response costs among liable parties. These factors are not legal defenses to liability, but are equitable criteria that the court may consider in apportioning CERCLA response costs.[174] In allocating liability, many courts utilize the factors set forth in *United States v. A&F Materials Co.* These are:

(1) the ability of the parties to demonstrate that their contribution to a discharge, release or disposal of a hazardous waste can be distinguished;
(2) the amount of the hazardous waste involved;
(3) the degree of toxicity of the hazardous waste involved;
(4) the degree of involvement by the parties in the generation, transportation, treatment, storage, or disposal of the hazardous waste;
(5) the degree of care exercised by the parties with respect to the hazardous waste concerned, taking into account the characteristics of such waste; and
(6) the degree of cooperation by the parties with federal, state or local officials to prevent any harm to the public health or the environment.[175]

Once a court determines a party is liable under CERCLA § 107(a), that party must contribute its equitable share of response costs under CERCLA § 113(f)(1). In allocating CERCLA liability, a liable party's share may range from 0 to 100 percent of the CERCLA liability for the site. The first, fourth, and fifth "Gore factors" require a comparison of fault and causation among the liable parties. Thus, although CERCLA is a strict liability statute, culpability is considered in equitable apportionment of multiparty Superfund sites under CERCLA § 113.

CERCLA allocation cases suggest that the remaining Gore factors are primarily utilized when there is not a glaring distinction between the fault of numerous liable parties. Certain Gore factors appear primarily relevant to allocation of

---

[173]CERCLA § 122(g) authorizes EPA to seek an early settlement with two types of PRPs: (1) those whose contribution to the problem was minimal; and (2) those whose allocable share of the response costs are minimal. 42 U.S.C. § 9722(g).

[174]*See generally* CHARLES ALAN WRIGHT AND ARTHUR R. MILLER, FEDERAL PRACTICE AND PROCEDURE: CIVIL 2D §§ 1270, 1271.

[175]*Id.* at 1256.

liability among generators rather than between generators and owners/operators. These factors generally do not mitigate liability at brownfields sites because such sites generally are not NPL listed and are rarely multiparty sites. They may be used, however, where the owner of a brownfields site sues another party for contribution to the site.

CERCLA expressly reserves the rights of private parties to contractually indemnify or release one another from liability.[176] This right is very important in the brownfields context because it is the only way parties can attempt to shift environmental liabilities when a sale of property occurs. The right to shift liability applies to the parties to the contract. A private indemnity agreement will not affect the right of the government to sue and hold any party liable under CERCLA. The private indemnity agreement will, however, allow a party sued to join the indemnitor in any action brought by the federal government for cleanup costs.

Whether a party agreed to release the other private party from CERCLA liability is, therefore, a matter of contract interpretation.[177] Factors that have been considered include:

(1) whether the contract language addressed CERCLA-type liabilities,
(2) whether the scope of the contractual language is so broad that it permits an inference regarding assumption of future-arising liabilities,
(3) whether the agreement predated or postdated CERCLA's enactment,
(4) whether the parties knew of the presence of hazardous wastes on the site,
(5) whether the cleanup issues were addressed in the parties' negotiations, and
(6) whether separate consideration was paid for the release of liability.

Courts have not been consistent in interpreting private indemnity agreements shifting environmental liabilities. Thus, even where private parties contractually allocate liability at brownfields sites, they cannot be certain that the liability-shifting provisions will be enforced by courts between the parties (or in a contribution action brought by a private third-party plaintiff). Moreover, since such contractual liability shifting will have no bearing on CERCLA liability to any federal governmental action, there is no way to contractually shift the risk of litigation (even if the indemnitor ultimately pays all costs required by the contract).

Analysis of Stakeholder Liability

Landowners: Current Owners and Operators

CERCLA Section 107(a)(1) provides that the current "owner and operator" of a facility may be liable. Unless the landowner can successfully invoke the Section 107(b)(3) third-party defense (which includes the "innocent" landowner

---

[176]*See* 42 U.S.C. § 9607(e)(1).
[177]Southland Corp. v. Ashland Oil, Inc., 696 F. Supp. 994, 1001 (D. N.J. 1988).

defense, the inheritance defense, and the municipal foreclosure defense), the present owner of a site may be liable under CERCLA *even if the release of the hazardous substance involved was caused entirely by a prior owner, lessee, or former lessee.*

Past Owners and Operators

Persons or businesses that owned land years ago are liable under Superfund for the cleanup of that land if: (1) hazardous substances were disposed of on the land while they owned it, or (2) they learned of contamination during ownership and failed to disclose this information when the property was transferred (even if disposal did not occur there during their ownership of the property).

Prior to enactment of the Superfund Amendments and Reauthorization Act (SARA) in 1986, past owners of a site were liable under CERCLA only if waste disposal occurred while they owned the site. CERCLA Section 101(35)(C), as amended by SARA Section 101(f), however, provides that a prior owner also may be liable under Section 107(a)(1) where he acquired actual knowledge of a release or threatened release while he owned the property and subsequently "transferred ownership" to another person "without disclosure."[178] Thus, a past owner not otherwise liable under Section 107(a) who discovers contamination during ownership must disclose the contamination to a purchaser or he will be held liable as a current owner or operator under Section 107(a)(1). Thus, a current "innocent landowner" who discovers contamination and wishes to sell finds himself faced with two undesirable alternatives: if he fails to disclose, he may be liable for cleanup costs; if he discloses, however, a prospective purchaser might insist on a lower price or may decide not to buy the property at all.

The Scope of "Owner" or "Operator"

Past and present "owners" or "operators" of a contaminated site may be liable for cleanup costs and natural resource damages under CERCLA Section 107(a). The scope of the terms *owner* and/or *operator* is very broad and continues to broaden under evolving Superfund case law. Persons not obviously connected to the risk, such as passive lessors and lessees, real estate brokers, lenders, parties to project finance transactions, insurance companies, parent corporations, successor corporations, officers, shareholders, employees, or other responsible corporate personnel of any of the above, may be liable as current or past "owners" and/or "operators" under CERCLA. It is this ambiguity that causes such reticence in the brownfields redevelopment context.

Certain types of "owners," however, are exempt from Superfund liability, including secured creditors, innocent landowners who have no knowledge of present contamination, persons inheriting property, and municipalities foreclosing on

---

[178]42 U.S.C. § 9601(35)(C).

property. A great deal of litigation has arisen concerning the scope of these exclusions from liability. Even the best structured transaction taking advantage of a state's brownfields legislation and other stakeholder protection laws cannot shield a developer or his lender from the risk of extensive CERCLA litigation, even if the owner is ultimately found not liable.

Successor Corporations

Superfund expressly includes corporations in the definition of *persons* subject to Superfund liability. A corporation that merges with or purchases the contaminated assets of a *present* owner or operator becomes a present owner or operator and will be strictly liable under CERCLA § 107(a)(1). Where a corporation merges with, or purchases the assets of, a *past* owner or operator, the successor corporation is not automatically liable under CERCLA. Liability must be analyzed under traditional common law rules of corporate liability under which liability of a successor corporation depends on the structure of the corporate acquisition.

The traditional American corporate rule regarding successor liability is that the successor corporation is liable for all of the obligations and liabilities of its predecessor if the new corporation acquires ownership by merger, consolidation, or purchase of all the outstanding stock. Thus, a corporation that acquires another through merger or consolidation generally will acquire all of the predecessor's CERCLA liabilities, including those that arise from any assets ever owned or activities ever conducted by the predecessor (including Superfund liability for disposal of hazardous substances away from the premises [off-site]) even if the assets were sold or the activities ended long before the acquisition. If the acquisition is through the sale or transfer of assets, however, the successor corporation generally does not acquire the predecessor's liabilities unless (1) the acquiring corporation expressly or impliedly agrees to assume such obligations, (2) the transaction amounts to a "de facto" consolidation or merger, (3) the purchasing corporation is merely a continuation of the selling corporation, or (4) the transaction was fraudulently entered into to escape liability. A fifth exception to the general rule is sometimes made where the transfer was without adequate consideration and provisions were not made for creditors of the transferor.

Courts rely heavily on the de facto merger exception to impose liability on successor corporations and are inclined to find that a particular transaction amounts to a de facto merger if (1) there is a continuation of the enterprise of the seller corporation, so that there is continuity of management, personnel, physical location, assets, and general business operations; (2) there is a continuity of shareholders that results from the purchasing corporation paying for the acquired assets with shares of its own stock, this stock ultimately coming to be held by the shareholders of the seller corporation so that they become a constituent part of the purchasing corporation; (3) the seller corporation ceases its ordinary business operations, liquidates, and dissolves as soon as practically possible; and (4) the purchasing corporation assumes those obligations (e.g., contracts) of the seller or-dinarily necessary for the uninterrupted continuation of normal business opera-tions of the seller corporation.

Although courts generally have taken a cautious approach to judicial evolu-tion of successor liability under the environmental statutes, there appears to be a trend toward expanding the liability of successor corporations. One federal appel-late court stated that "national uniformity" must be considered in resolving suc-cessor liability issues to avoid circumvention of CERCLA goals by some state laws that unduly restrict successor liability and that "when choosing between the tax-payers or a successor corporation, the successor should bear the cost [of cleanup]."[179] Several other federal courts have simply assumed the existence of successor liability in CERCLA cases without establishing successor liability stan-dards.[180]

In adopting a federal common law rule for piercing the corporate veil and holding a parent liable under Superfund for the activities of its subsidiary, the court in *United States v. Nicolet, Inc.*,[181] stated that:

Where a subsidiary is or was at the relevant time a member of one of the classes of persons potentially liable under [Superfund]; and the parent had a substantial financial or ownership interest in the subsidiary; and the parent corporation controls or at the relevant time controlled the management and operations of the subsidiary, the parent's separate corporate existence may be disregarded.

The government has interpreted this language to mean that parent corporations may be held directly liable under Superfund whenever a parent actively partici-pates in the management of a subsidiary that owns or operates a hazardous waste facility. Private industry, however, has taken the position that such a broad inter-pretation of the court's language is implausible—every parent would be liable if the test merely required a showing of control of a subsidiary's management and operations. Industry argues that the court's language must be read with traditional state law regarding corporate liability.

Although the majority of courts have held a parent corporation directly liable as an "owner or operator" under CERCLA § 107(a) for cleanup costs at a site op-erated by the corporation's subsidiary, in *Joslyn Corp. v. T.L. James & Co.*,[182] the Fifth Circuit expressly rejected the notion that parent corporations and corporate

---

[179]*See* Smith Land Improvement Corp. v. Celotex Corp., 851 F.2d 86, 92 (3d Cir. 1988), *cert. de-nied*, 488 U.S. 1029 (1989).

[180]*See, e.g.*, United States v. Bliss, 667 F. Supp. 1298 (E.D. Mo. 1985); United States v. Conserva-tion Chemical Co., 619 F. Supp. 162, 253 (W.D. Mo. 1985); Missouri v. Independent Petrochemical Corp., 610 F. Supp. 4 (E.D. Mo. 1985).

[181]712 F. Supp. 1193 (E.D. Pa. 1989).

[182]696 F. Supp. 222 (W.D. La. 1988), *aff'd*, 893 F.2d 80 (5th Cir. 1990).

officers may be held liable as owners or operators under CERCLA without first piercing the corporate veil under traditional corporate law.

Lessors And Lessees

Although CERCLA does not expressly address landlord-tenant liability, courts have found that lessors and lessees may be held jointly and severally liable as current and past "owners" and/or "operators" for cleanup costs under the act. Lessors who lease their property to a lessee who creates an environmentally hazardous condition may be liable under CERCLA even if the owner did not create or contribute to the contamination. In addition, lessees, as tenants who are entitled to the exclusive use and enjoyment of the leased premises, may be liable as an "owner" or "operator" under CERCLA. Courts have held that a lessee is an "owner" for purposes of liability under CERCLA. Since these rules attach to current as well as past lessors and lessees, there is as great a reticence to lease brownfields property as there is to purchase it.

Corporate Officers And Shareholders

A general principle of corporate law is that, absent special circumstances, a shareholder's liability is limited to the value of the shareholder's investment. Even with today's increased pressure to hold shareholders liable for environmental problems, courts continue to hold that shielding shareholders from environmental liability remains a legitimate reason for incorporation. In imposing personal liability for CERCLA response costs, the federal courts have been neither consistent nor predictable. As long as a shareholder retains the "capacity to control" the hazardous waste disposal practices or the corporation itself, then the shareholder will remain at risk for litigation asserting the shareholder's liability under CERCLA or other environmental theories.

This presents another great impediment to brownfields redevelopment. Not only does a corporate developer need to be concerned about the extent of liability the business may suffer, but he must also be concerned with personal liability for cleanup of brownfields sites as well.

Developers

Developers have been held potentially liable for CERCLA cleanup costs in the redevelopment of brownfields sites. For example, in *Tanglewood East Homeowners v. Charles-Thomas, Inc.*,[183] the Fifth Circuit upheld the district court's refusal to dismiss a CERCLA claim against a lender, residential developers, construction companies, and real estate agents and agencies, all of whom participated in the development of a subdivision built on contaminated property. It is not at all clear

---

[183]849 F.2d 1568 (5th Cir. 1988).

from the opinion on what basis the court concluded that specific parties (especially the lender and real estate brokers) could be held liable under CERCLA. The court did not specifically address the developers in its discussion of "present owners," "past owners," or "past arrangers and transporters"; indeed, it is difficult to see how brokers or lenders can fall into any of these categories, even given the court's broad interpretation. Nevertheless, the fact that the Fifth Circuit upheld the district court's refusal to dismiss the claims against the defendants, which included the developers, real estate brokers and others, has created great discomfort for developers of brownfields sites. This fear of liability led one commentator to conclude that, "Unless Congress authorizes the EPA to grant developers releases from liability, new inner city cleanup programs may be of limited value."[184]

Another case illustrating the concerns of developers is *City of North Miami v. Berger*.[185] In that case, a CERCLA action began out of the ill-fated efforts of the city of North Miami, Florida, and various entities and individuals to develop a municipal recreational complex on city-owned property. Those efforts included the 1974 to 1980 operation of a state permitted landfill at the site, a landfill that was later alleged to be the source of hazardous substance releases. From its inception, the Munisport development project faced intense regulatory scrutiny as well as pressure from various private interest groups. The city eventually sought recovery from the developer of the property, his attorney, the key shareholders in the development corporation, the demolition company working on the site, and the construction company hired to undertake the development. Although not all parties were ultimately held liable, the case demonstrated the great litigation risk inherent in development of brownfields sites.

Parties to Project Finance Transactions

Many large project financing involving real estate and industrial facilities are structured as sale/leasebacks, combining sale, lease, and loan transactions, each of which must be analyzed from the standpoint of environmental risks. In each type of transaction, the legal risk to the various participants normally turns on their status as "owners" or "operators" of an industrial facility or a piece of real estate.

In the traditional sale/leaseback, the owner of an industrial facility (who is usually also its operator) sells the project by transferring legal title to an entity, usually a bank or trust company serving as trustee under a grantor trust formed by an investor, which will provide the equity in the transaction and will become the beneficial owner of the project through the trust. The owner-trustee, as lessor, leases the facility back to the original owner, as lessee, who will continue to operate it under a long-term lease.

---

[184]Stephen C. Jones, *Unless Congress Authorizes the EPA to Grant Developers Releases from Liability, New Inner-City Cleanup Programs May Be of Limited Value*, THE NATIONAL LAW JOURNAL (May 15, 1995).

[185]1993 U.S. Dist. LEXIS 11015 (E.D. Va. Aug. 4, 1993).

In a "leveraged" lease deal, the owner-trustee obtains a loan from one or more financial institutions (sometimes the debt is raised through a public offering) secured by an assignment of the lessor's rights under the lease and by a mortgage on the lessor's interest in the project, in order to finance that portion of the purchase price and transaction expenses not covered by the investor's equity contribution. The lender's recourse is restricted to its collateral, and neither the owner-trustee nor the equity investor will have any personal liability for repayment of the loan. The lease assignment and mortgage typically introduces another bank or trust company into the transaction, which will hold the collateral for the lenders' benefit as indenture trustee under the terms of an indenture.

It is fundamental in sale/leasebacks that the owner-trustee/lessor be treated as the owner of the project for federal income tax purposes so that the equity investor, through the grantor trust, will be entitled to depreciation, interest deductions, and other significant tax benefits. From the standpoint of environmental law, however, this creates an immediate tension with the objective that the owner-trustee and the equity investor (not to mention the lenders) avoid, to the extent possible, the environmental responsibilities imposed on the "owner" or "operator" of the facility (or "persons" engaged in certain conduct).

When project financing is structured to include a sale, a lease, a mortgage, or all three, the impact of "owner" and "operator" (and "person") liability on each party to the transaction in his capacity as seller, buyer, lessor, lessee, mortgagor, and mortgagee must be considered. Normally, borrowers in project financing will fall into several of these categories and will have direct regulatory compliance and civil liability risks. Owner-trustees and equity investors also run certain risks of falling into one or more of these categories. Lenders, while not normally falling into any of them directly, will bear derivative risks related to the possibility of (1) borrower insolvency and (2) diminished collateral value due to the impact of environmental laws on the borrower. Further, under CERCLA, RCRA, the state equivalents, and other environmental laws, lenders themselves have been considered "owners," "operators," and perhaps "persons" upon foreclosure[186] or in difficult "workout" situations.[187] Thus, these lenders can become directly liable for

---

[186]*See* United States v. Maryland Bank and Trust Co., 632 F. Supp. 573 (D. Md. 1986). *Cf.* Guidice v. BFG Electroplating and Mfg. Co., 732 F. Supp. 556 (W.D. Pa. 1989).

[187]*See* United States v. Mirabile, 15 Envtl. L. Rep. (Envtl. L. Inst.) 20994 (E.D. Pa. June 6, 1985). *Cf.* Tanglewood East Homeowners v. Charles-Thomas, Inc., 849 F.2d 1568 (5th Cir. 1988) (court upheld district court's refusal to dismiss a complaint brought by residential landowners against a lending institution that supplied funding for a subdivision built on contaminated land since the lender may be a past owner at the time of disposal). *But cf.* United States v. Nicolet, 712 F. Supp. 1193 (E.D. Pa. 1989) (court held that a mortgagee may be held responsible for cleanup costs under CERCLA only if the mortgagee participated in the operational or managerial aspects of the facility); U.S. v. Fleet Factors Corp., 724 F. Supp. 955 (S.D. Ga. 1988), *aff'd and remanded,* 901 F.2d 1550 (11th Cir. 1990), *reh'g denied,* 911 F.2d 742 (11th Cir. 1990), *cert. denied,* 111 S.Ct. 752 (1991) (secured creditors may provide financial assistance and general, and even isolated instances of specific, management advice to its debtor without risking CERCLA liability if the creditor does not participate in the day-to-day management of the business).

regulatory compliance costs or civil liabilities of unprecedented magnitude, well beyond the value of the loan or the collateral involved.

Public and Private Lenders

Although CERCLA contains a "security interest" exemption that excludes from the definition of "owner or operator" any "person, who, without participating in the management of a vessel or facility, holds indicia of ownership primarily to protect his security interest in the vessel or facility,"[188] a lender may nevertheless become liable as a present or past "owner" or "operator" under Superfund by:

> assuming too much control over his debtor,[189]
> foreclosing on contaminated land,[190] or
> causing a release of hazardous substances.[191]

As an owner or operator, the lender could be held liable for cleanup costs or natural resource damages well beyond the amounts it originally had at risk in the lending transaction.

Although the security interest exemption has in fact protected most lenders from Superfund liability, the lending community became greatly concerned about potential Superfund liability when one court held a foreclosing lender liable as a present owner and operator under CERCLA[192] and another court suggested in dicta that a "capacity to control" the operations at a facility might be sufficient to void the statutory protection.[193] In addition, in 1989 the Federal Home Loan Bank Board highlighted the following concerns that environmentally contaminated properties could pose for both the public and private lender:

> reduced value of collateral;
> inability of borrowers to repay loans if they must also cover site cleanup costs;
> preemption of a mortgage loan security by state environmental cleanup liens
>   (so-called superlien laws) enforced in certain states;

---

[188]*See* CERCLA § 101(20)(A), 42 U.S.C. § 9601(20)(A). *See generally* RICHARD H. MAYS, ENVIRONMENTAL LAWS: IMPACT ON BUSINESS TRANSACTIONS 152-64 (1992); OWEN T. SMITH, ENVIRONMENTAL LENDER LIABILITY (1991); JOEL S. MOSKOWITZ, ENVIRONMENTAL LIABILITY AND REAL PROPERTY TRANSACTIONS: LAW AND PRACTICE 77-93 (1989).

[189]*See* United States v. Mirabile, 15 Envtl. L. Rep. (Envtl. L. Inst.) 20994 (E.D. Pa. June 6, 1985).

[190]*See* United States v. Maryland Bank and Trust Co., 632 F. Supp. 573 (D. Md. 1986). *Cf.* Guidice v. BFG Electroplating and Mfg. Co., 732 F. Supp. 556 (W.D. Pa. 1989).

[191]*See* United States v. Fleet Factors Corp., 724 F. Supp. 955 (S.D. Ga. 1988), *aff'd and remanded,* 901 F.2d 1550 (11th Cir. 1990), *reh'g denied,* 911 F.2d 742 (11th Cir. 1990), *cert. denied,* 111 S.Ct. 752 (1991).

[192]*See* United States v. Maryland Bank and Trust Co., 632 F. Supp. 573 (D. Md. 1986).

[193]United States v. Fleet Factors Corp., 724 F. Supp. 955 (S.D. Ga. 1988), *aff'd and remanded,* 901 F.2d 1550 (11th Cir. 1990), *reh'g denied,* 911 F.2d 742 (11th Cir. 1990), *cert. denied,* 111 S.Ct. 752 (1991).

potential for the lender to become directly liable for the cost of cleanup of the site if it engages in workout activities or forecloses on the property;

the concern that the lender may be forced to choose between foregoing collateral interest by not foreclosing on property and incurring significant cleanup costs under CERCLA; and

the possibility that the borrower would not maintain the facility in an environmentally safe manner and corresponding fears of liability if the lender either monitors or fails to monitor the environmental affairs of the debtor.

These concerns and the decline of available capital for certain high-risk industries led to increased lender surveillance of environmentally contaminated properties as well as difficulties for certain businesses to secure loans on properties at risk for environmental problems. As one commentator observed: "Private developers, even if determined to acquire an old property, often are stymied by lenders concerned about their inheritance of liability, devaluation of collateral, and the effect of cleanup costs on the project's vitality."

Efforts to address lenders' concerns remain uncertain. EPA issued a "Lender Liability Rule" to calm lenders' fears and provide them with guidance about what actions would and would not constitute "participation in management," a condition causing the lender to lose its statutory protection. Although the rule addressed only Superfund liabilities (and did not address any other state or federal causes of action), it was "welcomed by the banking community as a good solution to their problem." EPA's rule was, however, struck down by the D.C. Court of Appeals on the grounds that EPA lacked statutory authority to issue the rule. Recent efforts to revise CERCLA to grant authority to promulgate a new lender liability rule have failed. The administration continues to support such efforts.

New federal regulations, however, should help get polluted urban properties off of lenders' untouchable list.[194] On May 4, 1995, regulations were amended to provide an incentive for bankers and developers to help rescue cities with polluted industrial properties. "For the first time, lenders subject to the federal Community Reinvestment Act (CRA)—aimed at directing capital into poor neighborhoods—can claim CRA credit for loans made to help clean up and redevelop urban, industrial property."[195] The new rule, orchestrated by Environmental Protection Agency administrator Carol Browner, was designed to complement the EPA's new Brownfields Action Agenda (BAA).[196] The BAA strives to redevelop abandoned and contaminated property. As part of the BAA, EPA removed 27,000 sites from its Superfund list, releasing 12,000 from potential liability.

---

[194]Reed D. Rubinstein, *Shortening the 10-Foot Pole,* THE CONNECTICUT LAW TRIBUNE (May 15, 1995).

[195]*See* Community Reinvestment Act Regulations, 60 Fed. Reg. 22156 (May 4, 1995).

[196]*Id.*

Remediation Contractors

In 1986, Congress enacted certain limited protections under SARA for remediation contractors hired by the federal government.[197] CERCLA allows EPA to relieve contractors from liability for their negligent acts under certain very restricted circumstances.[198] The indemnity provided by EPA must be written[199] and will only be for the contractor's negligence. There is no protection for gross negligence or intentional misconduct.[200] In order to be eligible for this "indemnity and hold harmless agreement," the contractor must show that insurance is not available to the contractor at a fair and reasonable price despite the contractor's diligent effort to obtain insurance coverage.

In addition, § 119 (a)(1) states:

> A person who is a response action contractor with respect to any release or threatened release of a hazardous substance or pollutant or contaminant from a vessel or facility shall not be liable under this title or under any other Federal law to any person for injuries, costs, damages, expenses, or other liability (including but not limited to claims for indemnification or contribution and claims by third parties for death, personal injury, illness or loss of or damage to property or economic loss) which results from such release or threatened release.

Indemnification applies only to response action contractor liability that results from a release of any hazardous substance or pollutant or contaminant if such release arises out of EPA contracted response action activities. Furthermore, where EPA exercises its discretionary authority to grant indemnity and hold harmless agreements with Remedial Action Contractors (RACs), EPA generally requires the indemnification agreement to include deductibles and to place limits on the amount of indemnification to be made available.[201]

Although SARA exclusions for EPA-hired RACs provided certain limited protections where EPA agrees to give an "indemnity and hold harmless agreement," these exclusions apply in the CERCLA context almost solely to NPL listed sites. In addition, remediation contractors hired by private PRPs enjoy no such protection. For example, the Ninth Circuit squarely addressed the question of independent contractor liability in *Kaiser Aluminum & Chemical Corp. v. Catellus De-*

---

[197]42 U.S.C. § 9619(c). *See generally* RANDALL L. ERICKSON, ENVIRONMENTAL REMEDIATION CONTRACTING 245-79 (1992); DAVID S. MACHIO, LEGAL GUIDE TO WORKING WITH ENVIRONMENTAL CONSULTANTS (1992); JOEL S. MOSKOWITZ, ENVIRONMENTAL LIABILITY AND REAL PROPERTY TRANSACTIONS: LAW AND PRACTICE 77-93 (1989).

[198]42 U.S.C. § 9619.

[199]42 U.S.C. § 9619(c)(2). *See also* Amtreco, Inc. v. O.H. Materials, Inc., 802 F. Supp 443 (M.D. Ga. 1992).

[200]42 U.S.C. § 9619(a)(2).

[201]*Id.* § 9619(c)(5)(1994). The final regulations implementing Superfund Response Action Contractor (RAC) Indemnification were implemented on Jan. 25, 1993. See 58 Fed. Reg. 5972 (1993).

velopment Corp.[202] The court reversed a ruling dismissing a third-party complaint contribution costs under § 9613(f) of CERCLA against James L. Ferry & Sons ("Ferry"), a construction contractor. The dispute arose when Catellus Development Corp.'s ("Catellus") predecessor sold land to the City of Richmond, California ("Richmond"), which then hired Ferry to "excavate and grade a portion of the land for a proposed housing development." Ferry spread some of the displaced soil containing hazardous substances over other parts of the property. Richmond sued Catellus to recover a portion of the cleanup costs. In response, Catellus "filed a third-party complaint against Ferry for contribution under 42 U.S.C. § 9613(f)(1), alleging that Ferry exacerbated the extent of contamination by extracting the contaminated soil from the excavation site and spreading it over uncontaminated areas of the property."

The court reiterated the "well-settled rule that 'operator' liability under Section § 9607(a)(2) only attaches if the defendant had authority to control the cause of the contamination at the time the hazardous substances were released into the environment." The court held that the allegations that Ferry "excavated the tainted soil, moved it away from the excavation site, and spread it over uncontaminated portions of the property," were sufficient to support a claim that a hazardous substance was disposed of. It based this holding on the finding in *Tanglewood* that "the dispersal of contaminated soil during the excavation and grading of a development site" constitutes a disposal and that a disposal can occur during the subsequent movement or dispersal of hazardous substances.[203]

It is important for remediation contractors to note that there is no due diligence defense to liability for contractors like there is for landowners. Thus, even a contractor that used its best efforts to discover any hazardous substances and used state-of-the-art technology in an effort to discover such contamination can be held liable for remedial measures if the contractor failed to discover the contamination and dispose of it in a proper manner. Moreover, since disposal technologies are still in their infancy and waste disposed of in an EPA-approved landfill has an expected life of no more than 100 years, contractors have no way of cutting off liability under CERCLA for future cleanup even where cleanup was undertaken in a legal, permitted, and state-of-the-art manner.

Insurance Carriers

The advent of strict, retroactive, joint, and several liability for the generation, transportation, or disposal of hazardous substances has spawned nationwide litigation and debate over the precise meaning of the words in the nation's various Comprehen-

---

[202]976 F.2d 1338 (9th Cir. 1992).

[203]*Id.* The Court next examined whether the contractor could be found liable under § 9607(a)(3). The court stated "[l]iability for releases under § 9607(a)(3) is not endless; it ends with that party who both owned the hazardous waste and made the crucial decision how it would be disposed of or treated, and by whom." Concluding that plaintiff had "not alleged that Abbott owned any hazardous waste or made any decision on how it would be disposed," the Court held that the plaintiff's claim under (a)(3) was untenable.

sive General Liability's (CGL) policies. The basic language in virtually all CGL policies, which derived from standard insurance industry forms, is identical.[204]

It has been noted that:

> [s]ince Congress passed the Comprehensive Environmental Response, Compensation and Liability Act in 1980, the legislation, commonly known as "Superfund", has led to a long and expensive battle over who is to pay for cleaning up past pollution at industrial sites. . . . Insurers claim that all this has led to a "deep-pocket syndrome", with the firm with the most insurance being singled out to pay the lion's share of a clean-up bill . . . even if it was not the worst offender. If the insurance industry has to pick up much of the tab for pollution, the result could be even more devastating than a big natural catastrophe. To cope with this threat companies have boosted their reserves. A.M. Best estimates that the industry's "survival ratio", which measures the number of years it would take to exhaust reserves based on the present rate of claims, will have risen to 6.9 at the end of the year, up from 5.2 in 1990.[205]

The huge Superfund and toxic tort risks associated with the nation's polluted industrial sites have become essentially uninsurable. The casualty insurance industry, buffeted by the effects of declining interest rates on its investment income, has been stung by the willingness of many courts to impose liability for "gradual" pollution under CGL policies and appalled by Congress's attempt to create huge new liabilities for conduct that was insured (and premium levels gauged) when no such liabilities existed. Thus, led by the London reinsurers, insurers withdrew from the U.S. market for quite a while, refusing to write insurance to cover any form of environmental risk.

In a study published in March 1994, A.M. Best, an insurance rating agency, estimated that if Superfund was not reformed, domestic and foreign insurers in America could end up paying as much as $1.5 trillion in environmental liability claims over the next twenty-five years. A study published in 1992 by the RAND Corporation, concluded that some 88 percent of the cash paid out by insurers on Superfund-related claims in 1986-91 had been spent on defending policyholders and on litigation to decide who should be responsible for cleanups.[206]

PRPs typically pursue their insurance carriers in an attempt to establish coverage under standard CGL or Environmental Impairment Liability (EIL) policies.[207] Carriers vigorously resisted such coverage, arguing that (1) CERCLA cleanup costs are not "damages" under the policies, (2) the "pollution exclusion

---

[204]*See* George Pendygraft, et al., *Who Pays for Environmental Damage: Recent Developments in CERCLA Liability and Insurance Coverage Litigation,* 21 IND. L. REV. 117, 140 (1988).

[205]THE ECONOMIST at 9 (Dec. 3, 1994).

[206]*Id.*

[207]Insurers and insured spend approximately $500 million a year on Superfund litigation involving coverage. *Hearings Before the Subcomm. on Policy Research and Insurance of the House Comm. on Banking, Finance, and Urban Affairs,* 101st Cong., 2d Sess. at 3 (Sept. 27, 1990) (testimony of Dr. Joel S. Hirschhorn).

clause" bars coverage in most cases, and (3) policies predating the enactment of CERCLA do not cover an insured's payments to the government in a CERCLA cost recovery suit. Courts are divided as to whether cleanup costs are compensable under the property damage provision of the standard CGL policies.[208] Courts are also divided as to whether there is coverage even when the insured was ordered to undertake cleanup of the pollution itself.[209]

Coverage under the CGL policy is triggered by an "occurrence"—"an accident, including continuous or repeated exposure to conditions, which results in *bodily injury* or *property damage* neither expected nor intended from the standpoint of the insured[.]"[210] The standard form CGL policy issued between 1970 and 1985 includes the following pollution exclusion clause:

> This insurance does not apply: . . . (f) to *bodily injury* or *property damage* arising out of the discharge, dispersal, release or escape of smoke, vapors, soot, fumes, acids, alkalis, toxic chemicals, liquids or gases, waste materials or other irritants, contaminants or pollutants into or upon land, the atmosphere or any water course or body of water; but this exclusion does not apply if such discharge, dispersal, release or escape is *sudden and accidental.* . . .[211]

Policies issued prior to 1970 generally included no such exclusion and may serve as a basis for finding insurance coverage for pollution predating many of the federal environmental statutes.

The meaning of the pollution exclusion and other limiting phrases in insurance contracts is critical as insurance companies, private parties, the government, users of landfills and waste sites, and property owners ask whether the insurance company must defend relevant lawsuits or indemnify the insured against liability to third persons. Interpretation of insurance contracts is a matter of state law. The nation's courts are hopelessly split; as one judge noted, "[t]he cases swim the reporters like fish in a lake."[212] Indeed, there are different interpretations both be-

---

[208]*See, e.g.,* Avondale Indus., Inc. v. Travelers Indem. Co., 887 F.2d 1200, 1207 (2d Cir. 1989), *cert. denied,* 110 S.Ct. 2588 (1990); Township of Glouchester v. Maryland Cas. Co., 668 F. Supp. 394 (D. N.J. 1987); United States v. Conservation Chem. Co., 653 F. Supp. 152, 184-200 (W.D. Mo. 1986).

[209]*See* Bankers Trust Co. v. Hartford Accident & Indem. Co., 518 F. Supp. 371, 373 (S.D. N.Y. 1981), *vacated on other grounds,* 621 F. Supp. 685 (S.D. N.Y. 1981); Port of Portland v. Water Quality Ins. Syndicate, 549 F. Supp. 233 (D. Ore. 1982), *modified,* 796 F.2d 1188; Chemical Applications Co., Inc. v. The Home Indem. Co., 425 F. Supp. 777 (D. Mass. 1977); Boeing Co. v. Aetna Cas. & Sur. Co., 787 P.2d 507 (Wash. 1990); C.D. Spangler Constr. Co. v. Industrial Crankshaft and Eng'g. Co., 388 S.E.2d 557 (1990); United States Aviex Co. v. Travelers Ins. Co., 336 N.W.2d 838 (Mich. App. 1983).

[210]Just v. Land Reclamation, Ltd., 456 N.W.2d 570, 572 (Wis. 1990).

[211]*Id.* (second emphasis added).

[212]Morton International, Inc. v. General Accident Insurance Co., 629 A.2d 831, 856 (N.J. 1993) For a succinct jurisdiction-by-jurisdiction comparison, *see* ACL Technologies v. Northbrook Property & Casualty, 22 Cal. Rptr. 2d 206, 208-12 (Cal. App. Dist. 1993) (demonstrating that the profusion of conflicting case law renders premature any claim by one side or the other for possession of the "majority rule").

tween states and among state and federal courts within the same state.[213] Some jurisdictions interpret the contract language as ambiguous, interpret it broadly, and generally rule in favor of the insured.[214] Others deem the words unambiguous, interpret them narrowly under the "plain meaning" doctrine, and limit coverage to damages from instantaneous polluting events. This approach usually favors the insurance company.[215] Indeed, although *insure* means "to make sure, certain, or secure,"[216] the only certainty in CGL policies is that of expensive, extended, and exasperating litigation.

Sometimes insurance coverage is found, sometimes it is not. Sometimes an insurance company will defend a CERCLA action, and sometimes it will not. It is clear that there is not enough insurance capacity to cover all environmental problems that may potentially be covered by CGL and other insurance policies. It is this lack of capacity that has created such uncertainty in the insurance market. Allowing insurance coverage for CERCLA cleanups will undoubtedly continue to put some insurance companies out of business; state courts interpreting insurance contracts undoubtedly feel this political pressure.

State, County, and City Governments

The six-hundred (600) member U.S. Conference of Mayors announced "that abandoned industrial sites, dubbed 'brownfields,' are the top environmental problem facing America's cities."[217] Municipalities have a "critical interest in industrial redevelopment."[218] "Public managers are faced with the fiscal reality that their cities' older sections have increased or disproportionate risk of environmental contamination." As one commentator explained:

> Contaminated industrial sites constitute a serious problem for the nation's cities. Once flourishing factories and mills produced goods that improved lives and won wars. In the process, they supported generations of workers and kept towns growing. Their peak is past and their future is bleak. Many are vacant and deteriorating, leaving blanks on the tax rolls and symbolizing decline to nearby residents. . . . In older cities, factories often lie on the river or harbor and in the old inner city core. Restoration of these sites to productive use is a high priority of public officials and community leaders. Abandoning

---

[213]*See generally* Morton, 629 A.2d at 855-871; *see also* S. Hollis & M. Greenlaw, *The CGL Policy and the Pollution Exclusion Clause: Using the Drafting History to Raise the Interpretation Out of the Quagmire,* 23 COLUM. J. L. & SOC. PROBS. 223, 262 (1990).

[214]Those states include the following: Alaska, Colorado, Delaware, Florida, Georgia, Illinois, Missouri, New Jersey, Washington, and Wisconsin.

[215]Those states include the following: California, Iowa, Maine, Massachusetts, Michigan, Minnesota, New York, Oregon, Pennsylvania, and South Carolina. Illinois courts are split.

[216]THE AMERICAN HERITAGE DICTIONARY 667 (2d college ed. 1982).

[217]Randy Lee Loftis, *EPA Targets Toxic Sites in Nation's Urban Areas, Development Gets Boost from Cleanup Plan,* THE DALLAS MORNING NEWS (March 13, 1995).

[218]Reed D. Rubinstein, *Shortening the 10-Foot Pole,* THE CONNECTICUT LAW TRIBUNE (May 15, 1995).

the facilities, transportation links, and other infrastructure simply weakens the community. . . . the road to recovery is littered with obstacles.[219]

Another commentator focused the blame squarely on CERCLA:

[CERCLA] and other environmental regulations, however, cast doubt on the wisdom of municipal ownership of risks associated with fee simple ownership of land. The city is a "potentially responsible party" (i.e., "deep pocket") in the chain of title. Further, research suggests that the perceived risk of redevelopment of contaminated sites may lead to market failure, because investors overvalue the possibility of excessive expense beyond their actual cleanup costs. Hence, there may be a stigma attached to polluted properties beyond actual costs.[220]

These concerns have led many "cities such as Buffalo [to] not even foreclose on abandoned industrial complexes in redevelopment zones because the potential for staggering cleanup costs and liability claims once they take ownership."[221]

"Municipalities and other local governments also can find the financial burden of Superfund liability difficult to carry, whether they incur liability as the 'owners or operators' of municipal landfills at which hazardous waste was disposed of, or as 'generators' or 'transporters' of trash ('municipal solid waste' or 'MSW') sent to a private landfill where it became mixed with hazardous wastes."[222] The federal government does not generally pursue municipalities who sent only MSW to a landfill, but private PRPs very often do sue municipalities.

---

[219]*Id.* at 1. *See also* Johnine J. Brown, *Environmental Justice Conflict Could End with Justice If Brownfields Are Reclaimed,* ILLINOIS LEGAL TIMES (June 1995), who said:

The blight created by brownfields takes many forms. If the sites are inactive, the neighborhood is without jobs. If abandoned, the city forgoes property taxes and may even inherit the problem. Brownfields aren't pretty to look at, and some are thought to contribute to contamination of drinking water or neighborhood health problems. Some become flydumps. Others are taken over by gangs, or are systematically stripped of copper wire, bricks and other valuables by looters so politically correct they recycle from morning to night, mostly night.

[220]Simons, Robert, *How Clean Is Clean? Contaminated Property Development,* APPRAISAL J. (July 1994). *See also* Ellen JoAnne Gerber, *Industrial Property Transfer Liability: Reality versus Necessity,* 40 CLEVELAND STATE L. REV. 177-208 (1992); Tex Ann Reid, Edward M. Clar, Anthony M. Diecidue, and Mark F. Johnson, *Assessing a Municipality's Ability to Pay Superfund Cleanup Costs,* Federal Environmental Restoration Conference and Exhibitions (Washington, D.C.: Federal Environmental Restoration Conference and Exhibitions 1992); Bill Mundy, *The Impact of Hazardous and Toxic Material on Property Value: Revisited,* APPRAISAL J. 463-71 (October 1992); Peter J. Patchin, *Contaminated Properties—Stigma Revisited,* APPRAISAL J. 168-72 (April 1991).

[221]DEBORAH COONEY ET AL., REVIVAL OF CONTAMINATED INDUSTRIAL SITES: CASE STUDIES (Northeast-Midwest Inst. 1992). *See also* Mike Dries, *A Long Road to Urban Redevelopment,* 12 MILWAUKEE BUS. J. 1 (May 13, 1995).

[222]Testimony of Lois Schiffer, assistant attorney general, Environment and Natural Resources Division, Department of Justice, before the Senate Environment Superfund, Waste Control and Risk Assessment Superfund, FDCH Congressional Testimony (April 27, 1995).

Private PRPs often complain that the large volume of the trash contributed by the municipalities severely raises the cost of the remedies at these "codisposal" sites.

For example, in *New Jersey v. Gloucester Envtl. Mgmt. Servs.*[223] one of the generator groups involved in the litigation surrounding the Gloucester Environmental Management Services, Inc. (GEMS), Landfill, a Superfund site ranked twelfth on EPA's NPL, filed a Third-Party Complaint against fifty-two municipalities, seeking contribution from these local municipalities arising from their alleged generation and disposal of hazardous substances at the landfill. Recognizing the complexity of the problem the court said:

> American households are said to dispose of about 1.6 million tons of hazardous waste annually (i.e., an estimated one percent of the 160 million tons of total municipal solid waste). . . . The municipal solid waste, with its hazardous component, is said to present a considerable hazardous potential . . . at certain sites. While the potential for harm from release of hazardous components at a typical municipal landfill may not be great, this potential may be greatly aggravated at a Superfund site in which municipal wastes in high volumes contribute their hazardous components to the hazardous industrial wastes also on site to yield formidable problems of containment and remediation. It is the volume of hazardous substances in municipal waste, and not the isolated or occasional paint can or pesticide from the individual household, that potentially threatens the environment in a material way as envisioned by CERCLA.[224]

Thus, most courts have held that the municipalities could be held liable under CERCLA. As one court explained:

> It is clear from the definition of "person" in 42 U.S.C. § 9601(21) that municipalities are explicitly included as PRPs for purposes of the liability provisions of 42 U.S.C. § 9607(a). . . . Additional evidence of this Congressional intent to consider municipalities as PRPs comes from CERCLA's limited exceptions to potential municipal liability which are found in 42 U.S.C. §§ 9601 (20)(D) and 9607(d)(2). . . . If Congress had the ability to make explicit exemptions from liability in these sections, it had the ability to make exemptions in 42 U.S.C. § 9607(a). The fact that municipalities are "persons" under CERCLA and that no such exceptions were made for municipalities under 42 U.S.C. § 9607(a) is compelling evidence that Congress intended municipalities to be held liable as PRPs under § 9607(a). . . . Communities should recognize that potential liability under CERCLA applies regardless of whether the HHW [household waste] was picked up as a part of a community's routine waste collection service and disposed of in a municipal waste

---

[223]821 F. Supp. 999 (D. N.J. 1993).
[224]*Id.*

landfill . . . or if the HHW was gathered as part of a special collection program and taken to a hazardous waste landfill. . . .[225]

Hence, under CERCLA, if a municipality arranges for the disposal or treatment of hazardous substances, it may be held liable for contribution or response costs under CERCLA if a subsequent release or threatened release requires cleanup efforts. "The concentration of hazardous substances in municipal solid waste—regardless of how low a percentage—is not relevant in deciding whether CERCLA liability is incurred."[226] The fact that municipal solid waste is not specifically mentioned as a hazardous substance does not exempt it from CERCLA's reach.[227] Most courts reached this conclusion despite noting the magnitude of potential liability cities may suffer.[228]

The courts noted that municipalities, like private parties, are not completely without recourse from CERCLA liability. They can bring third-party contribution actions:

Appellant municipalities are not without recourse to avoid inequitable and disproportionate burdens that may arise from their liability as third-party contributors. Courts have the authority to "allocate response costs among liable parties using such equitable factors as the court determines are appropriate." § 9613(f)(1). An array of equitable factors may be considered in this allocation process, including the relative volume and toxicity of the substances for disposal of which the municipalities arranged, the relative cleanup costs incurred as a result of these wastes, the degree of care exercised by each party with respect to the hazardous substances, and the financial resources of the parties involved. Consequently, the amount of liability imposed will not necessarily be a function solely of the total volume of municipal waste disposed of in the landfills, but rather will be a function of the extent to which municipal dumping of hazardous substances both engendered the necessity, and contributed to the costs, of cleanup.[229]

This judicial recommendation for municipal protection is only partial. As one court recognized:

This court emphasizes the limited nature of its holding. *We are not holding that municipalities will be held equally culpable with other PRPs. . . .* That was not the issue presented to us, and even CERCLA suggests that munici-

---

[225]*Id.*

[226]*Id.*

[227]*Id. See also* B.F. Goodrich Co. v. Murtha, 958 F.2d 1192 (2nd Cir. 1992).

[228]*Gloucester,* 821 F. Supp. at 999 (citing Interim Municipal Settlement Policy, 54 Fed. Reg. 51,071 [1989]).

[229]*Gloucester,* 821 F. Supp. at 999.

palities should not be held equally culpable due to the relatively low toxicity level of MSW. . . . Section 113(f)(1) of CERCLA . . . gives courts the discretion to resolve contribution actions according to "such equitable factors as the court determines are appropriate." . . . In enacting CERCLA, members of Congress repeatedly emphasized the environmental degradation and harm caused by industrial hazardous substances, . . . and nothing in this opinion displaces to municipalities any part of the burden that must be predominantly shouldered by the industrial generators and haulers of such wastes. But nothing in the statute or legislative history excludes municipalities from bearing an appropriate share of liability at a Superfund site. *We are merely holding that a municipality that has generated or arranged for the disposal of municipal solid wastes at a facility may be liable under CERCLA § 107(a)(3) for an equitable share of responsibility* upon a third-party contribution claim under CERCLA § 113(f). . . .[230]

Under current law, there is no way for a municipality to protect itself from the risk of extensive litigation draining its already scarce resources only to find ultimately that the municipalities' fair share in a site is zero dollars.[231] Although no damages must be paid, the cost of litigation and the depletion of legal resources has caused many municipalities great consternation when addressing redevelopment issues in the context of brownfield sites.

---

[230]*Id.* (emphasis added).

[231]*See generally* Meske, *The Solid Waste Dilemma: Municipal Liability and Household Hazardous Waste Management,* 23 ENVTL. L. REP. (ELI) 355 (1993), *citing* U.S. EPA, Proceedings of the Fourth National Conference on Household Hazardous Waste Management 4 (1989); Ferrey, *The Toxic Time Bomb: Municipal Liability for the Cleanup of Hazardous Waste,* 57 GEO. WASH. L. REV. 197, 202 (1988). *See also* Note, *A Proposed Scheme of Municipal Waste Generator Liability,* 100 YALE L. J. 805-06 (1990).

CHAPTER 3

State Statutes Designed to Aid
Brownfields Redevelopment

CERCLA Equivalents
RCRA Equivalents
State UST Laws
OPA Equivalents
Specialized Stakeholder Protection Statutes under State Law
State Voluntary Cleanup Laws

State Brownfields Statutes

The existence of potentially contaminated and abandoned property is not a new problem in many metropolitan areas, especially older central cities and suburbs. Federal and state laws governing the treatment of these sites may require remediation (cleanup) of property before redevelopment and can contribute to uncertain liability for property owners or users. As a result of these and other factors, reuse of these sites can be hindered. While more attention has been focused on brownfields in the past few years, information on the extent of the problem and the level of contamination at these sites is limited. Estimates range from tens of thousands to 450,000 brownfields sites in the country. In addition, their condition may vary from zero, low, or moderate contamination to extremely hazardous conditions, while many sites have still not been evaluated.

Unquestionably, it is "[t]he states . . . [that] have taken the lead in encouraging redevelopment of brownfields properties."[1] While state policies to address brownfields sites vary considerably, the most common approaches are:

state Superfund programs, and stakeholder adaptations to them,
environmental lien laws,
property transfer laws, and
voluntary cleanup programs.

Each includes a process for site assessment and remediation, with state Superfund and property transfer laws operating on an enforcement-driven basis. Each

---

[1]Stephen C. Jones, *Unless Congress Authorizes the EPA to Grant Developers Releases from Liability, New Inner-City Cleanup Programs May Be of Limited Value*, THE NATIONAL LAW JOURNAL (May 15, 1995).

addresses a different state on the brownfields continuum, as the following table indicates:

Stages of Brownfield Evolution		State Legal Responses
Stage 1: Strong real estate market	Private sector absorbs costs	State and federal environmental regulations add costs but do not deter project
Stage 2: Brownfield traps	Business owner afraid to sell, lease, refinance, or expand	State transfer triggers statutes State voluntary cleanup programs State amnesty programs
Stage 3: Mothballed property	Property sits idle and dirty	No current law dealing with this problem—the reluctant seller
Stage 4: Tax-deliquent properties	Government must clean up	CERCLA and state equivalents Environmental lien laws State nuisance statutes RCRA

In general, the evolution of state laws reflect a movement from state or federally funded public works style cleanup programs to smaller sites with little to no state oversight. Early laws dealt primarily with Stage 4 brownfield projects—the most polluted sites. Second- and third-generation laws, the transaction triggered laws and voluntary cleanup programs, address Stage 2 brownfields. These Stage 2 sites are generally not within the regulatory radar of state or federal environmental agencies and tend to have only mild to slightly severe (but not chronic) contamination.

Voluntary programs are particularly popular because they allow private parties to initiate cleanups and avoid some of the cost and delays associated with state Superfund or other litigation driven programs. In particular, many state-run voluntary programs offer added incentives to private parties to participate, including:

technical assistance and flexibility in cleanup standards;
liability assurances, such as a "covenant not to sue" or "no further action" letter from the state; and
financial support in the form of grants and low-interest loans.

Although states have developed a variety of means to address brownfields, experience is still limited. As more cleanup and redevelopment activity occurs, especially through voluntary programs, additional information will become available about the effectiveness of state programs that may also help identify solutions to a number of unresolved issues. While state brownfield programs unquestionably substantially enhance industrial development and lending activities, they have not alleviated the problem of brownfields paralysis. The releases and covenants offered to date have often had "reopeners" or "reservations of rights" with respect to site conditions not known at the time the settlement was framed. In addition, the releases offered typically relate to state and federal Superfund prosecution only.

Settling parties may remain exposed to prosecution under other state and federal environmental statutes (such as RCRA or OPA) as well as to common law theories of liability. Moreover, there is no preclusion from third-party or private-party actions by predecessors, successors, and adjoining landowners (including toxic tort and diminution of property value claims). Finally, the enforceability of informal pronouncements by federal and state agencies is questionable.

CERCLA Equivalents

Most states have enacted laws similar to CERCLA that impose liability for cleanup expenditures on landowners and others. State Superfund programs were created in most instances to address sites that are not considered hazardous enough to be placed on the NPL but that a state believes may pose a significant health risk and warrant remediation. Although state Superfunds were developed to clean up hazardous waste sites, they are not specifically designed to address brownfields and the many issues involved with the reuse of these sites. Approximately forty-five states operate their own Superfund programs in the United States. Many of them include authorities and capabilities similar to the federal Superfund program.

While there is some variation among the programs, state Superfund or cleanup programs are generally characterized by the following features:

procedures for emergency response actions and permanent remediation of environmental and human health risk;
provisions for a cleanup fund or other financing mechanisms to support program activities;
enforcement authority to identify and compel responsible parties to pay for site assessment and cleanup;
authorized state agency with staff charged with responsibility for oversight of remediation activities and provisions for public participation in the remediation process.

Many voluntary cleanup and business transfer programs are run as amendments to the state equivalent of CERCLA or RCRA (or both), while others are broader than the scope of these state statutes. The legal genesis of these adaptations lies in the brownfields continuum. CERCLA and the state equivalents were designed to enable states and the federal government to clean up stage 4, abandoned contaminated sites. As these chronicly contaminated sites were cleaned up or scheduled for cleanup, many parties began looking for creative legal means to cover other less-contaminated sites. Thus, many private parties (not governmental authorities) began to use the CERCLA and state equivalent statutes to force environmental cleanup in stage 3 sites. These stage 3 sites are generally less contaminated and historically were not of concern to regulatory agencies. Private use of the state and federal cleanup laws stretched the definition of what type of sites should be the subject of cleanup litigation. State and federal agencies began to see their enforcement discretion invaded by private parties. Sites for which the environmental agencies had little or no concern were now the subject of expensive private litigation.

Business transfer laws were developed as an adaptation of state CERCLA equivalents to force parties that are likely to have contamination to deal with the issues in the open before transfer. The laws required parties to disclose contamination to the potential parties to a property transfer and in many cases to the state agency as well. These laws attempted to force parties to deal with shifting environmental risks before a transaction was closed, rather than litigating it later under the state or federal CERCLA laws.

When the business transfer laws failed to stem the tide of environmental litigation brought by private parties, many states began to develop voluntary clean up laws. These laws took an even more affirmative step in attempting to reduce the ability of private parties to set aside the enforcement discretion of environmental agencies. The laws allow private parties to clean up contaminated sites under varying degrees of state oversight and then use the legal certificate awarded by the state as a defense to any further environmental cleanup liability brought by private parties after the fact.

State Environmental Lien Laws

At present, almost all states have enacted statutory schemes that parallel CERCLA and authorize state environmental agencies to clean up hazardous waste sites and seek reimbursement for expenditures. Twenty-one states now allow the imposition of environmental liens, many of which exceed the parameters of the CERCLA lien by attaching not only to the real property subject to cleanup, but also to revenues, personal property, and real property not related to the contaminated site.[2] These liens enable the government to ensure recovery for cleanup costs by attaching a lien on the property of the owner of the contaminated property subject to cleanup.

The general format of most of these environmental lien statutes follows CERCLA section 107(l): the lien arises upon the state's expenditure of cleanup funds, and it becomes immediately effective when notice of the lien is filed in the county where the property is located.[3] Other than two statutes that authorize liens only to secure payment of judicially determined debts,[4] only one state lien provides the property owner with prior notice and an opportunity to challenge the validity of the lien in an administrative hearing.[5]

Several states authorize environmental liens with "superpriority" status. Unlike the CERCLA lien, which assumes priority only over subsequently filed liens on the same property, superliens assume priority over all other existing, as well as

---

[2]*See, e.g.,* N.H. Rev. Stat. Ann. § 147-B:2, 147-B:10.

[3]*See, e.g.,* Ark. Code Ann. § 8-7-41, 8-7-51.

[4]*See* Md. Code Ann., Health Envtl. § 7-266; Mont. Code Ann. § 75-10-720.

[5]*See* Minn. Stat. Ann. § 514.671-4.

subsequent, liens, encumbrances, and security interests on the property.[6] The superliens may attach to the hazardous waste site, as well as to all other property owned by the liable party, including noncontaminated realty, personal property, and business revenues.[7] The liens have superpriority only as they apply to the realty, personal, and business revenues of the site itself. As to all other attachable property or revenues, the liens have priority only over subsequent encumbrances.[8]

The environmental lien, and particularly the superlien, impacts all aspects of property ownership and transactions. For example, financing the purchase of real or personal property is risky where the lender does not know that the property is contaminated and subject to a superior environmental lien.[9] This is particularly problematic where the lien statutes do not provide a period of time in which the state must recognize that a claim for cleanup costs exists and file the lien. In fact, in most jurisdictions the lien may be filed many years after the pollution occurred or was cleaned up.[10]

Also, the existence of a superlien affects a secured lender's ability to ensure repayment of the loan. Foreclosing on a lien is ineffective if a superior environmental lien is large enough.[11] Furthermore, traditional business expectations, such as in bankruptcy proceedings, may be negated by the superlien. Where the environmental lien takes priority and depletes the estate in its entirety, the claims of perfected, secured creditors, ordinarily paid first, are worthless.[12]

Real estate developers, investors, lenders, and title insurers have all become more cautious in property transactions as a result of the environmental lien.[13] The title insurance industry, for example, has responded to the existence of superliens by revising standard contract forms, expanding title searches, and receiving from the EPA periodic lists of properties on which CERCLA liens have been filed.[14]

In light of the environmental lien's potential to cause irreparable harm to those holding interests in the attached property, it is not surprising that questions have been raised as to its constitutionality.[15] In *Reardon v. United States*,[16] the first

---

[6]*See, e.g.,* ME. REV. STAT. ANN. tit. 38, sec. 1371.

[7]*See* ME. REV. STAT. ANN. tit. 38, § 1371; MASS. GEN. LAWS ch. 21E, § 13; N.H. REV. STAT. ANN. § 147-B:2, 147-B:10; N.J. STAT. ANN. § 58:10-23.11f(f).

[8]*See id.*

[9]Johnine J. Brown, *Superfunds and Superliens: Super Problems for Secured Lenders,* TOXICS L. REP. (BNA) (March 16, 1988), *reprinted in Impact of Environmental Regulations on Business Transactions,* 445, 454 (1988).

[10]Richard L. Epling, *Environmental Liens in Bankruptcy,* 44 BUS. LAW. 85, 87 (1988).

[11]*See* Brown, *supra,* note 9 at 545.

[12]Beth Anne Smith, Comment, *State "Superlien" Statutes: An Attempt to Resolve the Conflict between the Bankruptcy Code and Environmental Law,* 59 TEMPLE L.Q. 981, 1009 (1986).

[13]Norman R. Newman, *How to Counsel the Land Developer on Superfund and Superliens,* 34 PRAC. LAW, No. 7, at 13, 22-26 (1988).

[14]*See, e.g.,* ALTA Endorsement Form 8.1 (3/27/87) Environmental Protection Lien; see generally Robert S. Bozarth, *Environmental Liens and Title Insurance,* 23 U. RICH. L. REV. 305, 314 (1989).

[15]Reardon v. United States, 947 F.2d at 1517-18 (due process challenge to EPA lien imposed under CERCLA § 107 should be evaluated under two-part analysis: [1] whether statute authorizes the taking of a "significant property interest"; and [2] if so, what process is due under the circumstances).

[16]947 F.2d 1509 (1st Cir. 1991).

appellate court to consider the issue declared that the CERCLA lien provision violates the due process clause of the Fifth Amendment because it fails to require that the property owner receive notice and a hearing prior to the filing of the lien.

The *Reardon* court issued two significant holdings regarding the CERCLA lien. First, the court held that Section 113(h) of CERCLA did not bar judicial review of the Reardons' claims that the lien statute is unconstitutional. The court interpreted Section 113(h) as preventing review of challenges to the EPA's administration of CERCLA—that is, selection of the proper response action.[17] Accordingly, the court found that Section 113(h) does not foreclose review of constitutional challenges to CERCLA itself, which do not involve the merits of any particular removal or remedial action.[18]

Second, the court declared that the CERCLA lien statute violates the due process clause of the Fifth Amendment, because it fails to provide the property owner with preattachment notice and a hearing.[19] In reaching this conclusion, the court relied primarily on *Connecticut v. Doehr*[20] in which the Supreme Court held that even the temporary or partial impairment of property rights that a lien entails is sufficient to merit due process protection.[21] Because the CERCLA lien clouds title, limits alienability, and affects current and potential mortgages, the lien deprives the owner of a significant property interest within the meaning of the due process clause.[22]

Although no state environmental lien provision has been challenged on due process grounds, the *Reardon* court's rationale could readily be applied to a constitutional analysis of the majority of state lien statutes. As a result, numerous state legislatures may be faced with the inevitability of constitutionally mandated lien reform.

The *Reardon* decision relied heavily on the rationale of *Connecticut v. Doehr*,[23] which did not involve an environmental lien statute. The *Reardon* court applied the same standard of review to determine the constitutionality of an environmental lien statute as that of other attachment liens. Under the *Doehr* and *Reardon* analyses, an environmental lien statute, in order to pass constitutional muster, must provide at least notice and a preattachment hearing, unless an immediate postattachment hearing is provided in the face of exigent circumstances.[24]

---

[17]*Id.* at 1514.

[18]*Id.*

[19]*Id.* at 1523.

[20]111 S. Ct. 2105 (1991).

[21]*C.f.,* Reardon v. United States, 947 F.2d at 1525-28 (dissent, argued that Congress never considered whether or how CERCLA § 113(h) would apply to a lien challenge. Congress's reasoning in enacting § 113[h] does not apply to PRPs whose property is encumbered with a lien).

[22]*Reardon,* 947 F.2d at 1518.

[23]111 S. Ct. 2105, 115 L. Ed. 2d 1 (1991).

[24]Note that EPA's filing of a lien can be considered enforcement activity because, in filing a lien to secure payment of a liability under CERCLA § 107, the government is seeking to enforce the liability provision. Reardon v. United States, 947 F.2d at 1512-13. But note that Judge Cyr, dissenting, disagreed with the characterization of a lien as "enforcement activity."

As noted, environmental liens can cause many practical problems for the property owner, lender, developer, or creditor.[25] There is a serious question, however, whether the alleged due process protections required by *Reardon* will alleviate these problems. Remember, a title search will not reveal the potential existence of the superlien, even though the property may have been subject for many years to an ongoing governmental investigation and hazardous substance cleanup, during which a lien could be imposed at any time. The presence of hazardous material on land, and the concomitant possibility of governmental liens, does not constitute a title defect covered by a title insurance policy.[26] This problem is somewhat alleviated in states that require the government to record notice as soon as it designates the property as a contaminated site or begins to expend funds to investigate or clean up the property.[27]

The states that have enacted regular lien statutes are as follows: Alaska, Arkansas, Illinois, Indiana, Iowa, Kentucky, Maryland, Michigan, Minnesota, Montana, Ohio, Oregon, Pennsylvania, Tennessee, and Texas. Like the CERCLA lien, these state liens have no superpriority status. They have priority only over subsequently filed encumbrances.

Each state's environmental lien law must be examined to determine what property is subject to attachment and what procedural or appellate rights are provided to the property owner. For example, only Minnesota provides for a preattachment hearing at the state environmental protection agency to challenge the state's intended action of filing a lien.[28] Other states may set time limits within which the government may file a lien to secure payment of cleanup costs.[29]

The state statutes also vary in their approaches to provide administrative or judicial review of the filing of an environmental lien. The most severe restriction of postattachment review is imposed in Michigan and Pennsylvania, which absolutely prohibit judicial review of agency action prior to the state initiating an enforcement or cost recovery lawsuit.[30]

Other states have no specific provisions within their CERCLA counterparts regarding judicial review of administrative action. In these jurisdictions, challenges to environmental liens must be asserted in accordance with the states'

---

[25]Reardon v. United States, 947 F.2d at 1514 (If the expanded definitions of *removal* and *remedial* to include related "enforcement activity" encompass suits to recover attorneys' fees and other cleanup costs, then the definitions also logically include the imposition of liens to ensure complete recovery) (Judge Cyr, dissenting, disagreed with characterization of a lien as "enforcement activity").

[26]*See* Lick Mill Creek Apartments v. Chicago Title Ins. Co., 283 Cal. Rptr. 231 (Cal. Ct. App. 1991); Chicago Title Ins. Co. v. Kumar, 506 N.E.2d 154, 156 (Mass. App. Ct. 1987).

[27]*See, e.g.,* TENN. CODE ANN. § 68-46-209.

[28]*See* MINN. STAT. ANN. §§ 115B.01-.37.

[29]*See* ARK. CODE ANN. §§ 8-7-417, -516 (within 30 days of the last cleanup act on the property); TENN. CODE ANN. § 68-212-209 (within one year of cleanup completion); IOWA CODE ANN. § 455B.396 (West 1990) (within 120 days after costs are incurred).

[30]*See* MICH. STAT. ANN. § 13.32(16a); 35 PA. CONS. STAT. § 6020.509.

administrative procedure acts, which generally allow judicial review only of final agency action and after the exhaustion of administrative remedies.[31] Whether the filing of a lien constitutes final and thus reviewable agency action must be determined according to each state's laws.

In contrast, several states specifically provide the opportunity for a post-attachment administrative hearing to challenge state agency action and for immediate judicial review following the administrative hearing.[32]

Although most states must bring a civil action to obtain reimbursement, several states authorize environmental liens with "superpriority" status. As to all other attachable property or revenues, the liens have priority only over subsequent encumbrances.[33] These provisions provide that any expenditures made pursuant to a state's Superfund statute constitute a first-priority lien on the real property of the person responsible for the contamination problem. In some states, however, tax liens have priority over the superlien.[34] The effect of these superlien statutes on commercial lending and real estate transactions can be far reaching. Liability is limited in some states to containment and/or cleanup costs,[35] but in other states liability is imposed for damages to natural resources and even for increases in property value resulting from the cleanup.[36] Although these statutes typically have recording provisions, some states have inadequate recording requirements.[37] Subsequent mortgagees or purchasers, therefore, may enter transactions involving real estate without actual notice of the lien.

The superlien states are Connecticut,[38] Louisiana,[39] Maine,[40] Massachusetts,[41] New Hampshire,[42] and New Jersey.[43] The New Hampshire superlien is perhaps the most onerous from a property owner's perspective.[44] The New Hampshire environmental lien attaches upon the business revenues and all real and personal property of the party liable under the state's environmental cleanup statute. The lien has superpriority status as to "the real property on which the hazardous waste . . . is located" and as to "the business revenues generated from the facility

---

[31]*See, e.g.,* ARK. CODE ANN. 8-7-417-516; ILL. REV. STAT. ch. 111-1/2, § 1021.3.

[32]*See* ALA. STAT. §§ 44.62.560, 46.03.820, 46.08.140; CONN. GEN. STAT. ANN. §§ 22a-436-437.

[33]*See, e.g.,* MASS. GEN. LAWS ch. 21E, § 13.

[34]*See* ARK. CODE ANN. §§ 8-7-417, 8-7-516; TENN. CODE ANN. § 68-46-209(a).

[35]*E.g.,* N.H. REV. STAT. ANN. § 147-B:10(I).

[36]ARK. CODE ANN. § 8-7-417; TENN. CODE ANN. § 68-46-209(a), (b).

[37]*E.g.,* TENN. CODE ANN. § 68-46-209(a), (b) (a lien may be recorded up to one year after the state completes cleanup).

[38]CONN. GEN. STAT. § 22a-452a.

[39]LA. REV. STAT. ANN. § 30:2281.

[40]ME. REV. STAT. ANN. tit. 38, § 1371.

[41]MASS. GEN. LAWS ch. 21E, § 13.

[42]N.H. REV. STAT. ANN. § 147-B:10.

[43]*See* Kessler v. Tarrats, 476 A.2d at 326 (the lien was a proper exercise of police power and served a legitimate public purpose; it also withstood the contracts clause of the U.S. Constitution, which provides that "No state shall . . . pass any . . . Law impairing the Obligation of Contracts," and did not violate the secured creditor's private property rights).

[44]N.H. REV. STAT. ANN. § 197-B:10.

on which hazardous waste . . . is located."[45] As to residential real property and all other real or personal property and business revenues, the lien has no priority over previously perfected encumbrances.[46]

Not surprisingly, the trend toward superliens has faltered. Since 1988, Arkansas and Tennessee deleted superlien provisions from their cleanup statutes and efforts to create environmental superliens in Kansas, New York, and Pennsylvania were defeated.[47] Currently, twenty-one states authorize liens to secure cleanup costs incurred by state environmental protection agencies.[48] The general format of the majority of these statutes parallels the CERCLA lien statute: the lien arises upon the state's expenditure of cleanup funds, and it becomes immediately effective when notice of the lien is filed in the county where the property is located.

Perhaps as a consequence of the constitutional cloud placed on the CERCLA lien, a new trend has arisen in the environmental lien arena. States are expanding their jurisdictions to either conduct a cleanup or force others to do so under state nuisance principles. These statutes are often broader in scope than the CERCLA-based environmental lien statutes.

Business Transfer Statutes

Another significant trend predating legislation labeled as "brownfields" legislation is the emergence of state property transfer legislation. This important development was initiated by the enactment, in late 1983, of the New Jersey Environmental Cleanup Responsibility Act (ECRA).[49] The Connecticut Transfer Act subsequently was enacted in 1985, the Illinois Responsible Property Transfer Act in 1988, and the Indiana Responsible Property Transfer Law in 1989. As these laws gained national notoriety, legislation—modeled on ECRA—was proposed in numerous jurisdictions.

The Connecticut Transfer of Hazardous Waste Establishment Act[50] also ties environmental cleanups to business transactions involving the transfers of real property. The intent of the Act is to protect purchasers by compelling full disclosure of the environmental conditions of properties before transfer. Unlike the New Jersey ECRA, the Connecticut Act does not require the removal of any hazardous waste from the property and does not permit the avoidance of the transaction if the

---

[45]*Id.*

[46]*See* N.H. REV. STAT. ANN. §§ 147-B:2, :10-b (1990). *See also* CONN. GEN. STAT. ANN. § 22a-452a; LA. REV. STAT. ANN. § 30:2281; ME. REV. STAT. ANN. tit. 38, § 1371; MASS. GEN. LAWS ch. 21E, § 13; N.J. STAT. ANN. § 58:10-23.11f(f).

[47]Robert S. Bozarth, *Environmental Liens and Title Insurance,* 23 U. RICH. L. REV. 305, 324 (Spring 1989).

[48]The states that currently have environmental lien statutes are Alaska, Connecticut, Illinois, Indiana, Iowa, Kentucky, Louisiana, Maine, Maryland, Massachusetts, Michigan, Minnesota, Montana, New Hampshire, New Jersey, Ohio, Oregon, Pennsylvania, and Texas. The states with superlien statutes are Connecticut, Louisiana, Maine, Massachusetts, New Hampshire, and New Jersey.

[49]N.J. STAT. ANN. § 13:1K-6 *et seq.*

[50]CONN. GEN. STAT. §§ 22a-134-22a-134e.

Act is violated. The Connecticut Act permits the transferee of the property to recover damages from the transferor. Additionally, the transferor is strictly liable for all cleanup and removal costs and for all direct damages resulting from a violation.

Another ECRA-inspired law is the Illinois Responsible Property Transfer Act of 1988.[51] Like the Connecticut Transfer of Hazardous Waste Establishment Act,[52] the Illinois Act is intended to inform parties involved in certain real property transactions of existing environmental liabilities associated with the property, as well as previous use and the environmental condition of the property. The Illinois Act imposes stiff civil penalties of up to $1,000 for each day a violation continues. Moreover, if a person knowingly gives false information in the disclosure statement, the person is fined $10,000 for each day the violation continues.

The concept of legally requiring disclosure of environmental concerns as a condition precedent to closing a real estate transaction grew.[53] Additionally, the following states have statutes requiring approval of state environmental officials of transfers: Colorado,[54] Iowa,[55] Missouri,[56] and New York.[57]

Moreover, certain states enacted statutes requiring approval by state environmental officials. For example, under Colorado law, "[T]he board of county commissioners or the governing body of the municipality" must approve a "substantial change in the ownership of a hazardous waste disposal site" before it becomes effective.[58] In Iowa, the director of the Department of Natural Resources must approve any substantial change in the way "in which a hazardous waste or hazardous substance disposal site . . . is used" or any sale of such property.[59] If the director believes that a violation has, or may, occur, the director may seek an injunction to prevent the violation and a civil penalty of up to $1,000 per day for each day the violation continues.[60] Similarly, in Missouri, the director of the Department of Natural Resources must approve any substantial change in the way "in which an abandoned or uncontrolled hazardous waste disposal site . . . is used."[61] If the director believes that a violation has occurred, or may occur, the director may seek a civil injunction to prevent such violation and a civil penalty of up to $1,000 per day for each day the violation continues.[62] Under New York law, a person may not "substantially change" the way "in which an inactive hazardous waste disposal site, for which the commissioner of health [has issued] a declaration" without

---

[51]ILL. ANN. STAT. ch. 30, para. 901.
[52]CONN. GEN. STAT. §§ 22a-134-22a-134e.
[53]W.VA. CODE § 20-5E-20(a).
[54]COLO. REV. STAT. ANN. § 25-15-206(1).
[55]IOWA CODE ANN. § 445B.430 (1), (2).
[56]MO. ANN. STAT. § 260.465(1).
[57]N.Y. ENVTL. CONSERV. LAW § 27-1317.
[58]COLO. REV. STAT. ANN. § 25-15-206(1).
[59]IOWA CODE ANN. § 455B.430(1), (2).
[60]Id. § 455B.430(4).
[61]MO. ANN. STAT. § 260.465(1).
[62]Id. § 260.465(4).

approval by the commissioner.[63] The commissioner will disapprove a change of use if the new use will hamper substantially a proposed, pending, or completed remedial action at the site or will harm the environment or health.[64]

Based upon the success of these business transfer statutes, many states began to expand the scope of environmental disclosure legislation to include disclosure of asbestos, radon, and underground storage tanks as well as the existence of hazardous wastes or hazardous substances on the property. In addition, certain states began requiring deed restrictions to be placed on transferred property that was cleaned up to less than a residential use (or an "edible dirt") standard.

Property transfer laws are, by definition, an indirect method for identifying and initiating cleanup activities. Property transfer provisions exist in the states as laws, regulations, or policies that make the transfer of real property, or ownership or control of such property, contingent on the discovery, identification, investigation, cleanup, or disclosure of the existence of contamination. These provisions vary across the states. Some simply require disclosure of the environmental condition of a site. Others require a more advanced level of site investigation. A few states require complete cleanup before a transfer can occur. To date, approximately twenty-four states have some form of deed restriction or property transfer requirements.

State Role in the RCRA State/Federal Partnership

Over forty states have legislation in place similar to and often exceeding the dictates of the federal RCRA program. Thus, a state program that exceeds the federal requirements regulating waste management and underground storage tanks exists in most states. These standards are often higher than the federal standards in that they require greater cleanup levels.

RCRA § 3004[65] requires that the owner of a hazardous waste facility provide notification to the general public to alert a potential purchaser that the property was previously used to manage hazardous waste.[66] The notice must be placed in a deed or other document that would allow discovery during a typical title search. Penalties for failure to provide this type of notice may be assessed up to $25,000.

---

[63]N.Y. ENVTL. CONSERV. LAW § 27-1317.

[64]*Id.*

[65]42 U.S.C. § 6924; 40 C.F.R. § 264.1199(b)(1).

[66]*See* ALA. ADMIN. CODE r. 14-5.07; ALASKA ADMIN. CODE tit. 18, § 62.410; ARIZ. COMP. ADMIN. R. & REG. 18-8-264; Reg. Ark. Dep. of Pol'n Ctrl. & Ecol. § 15; FLA. ADMIN. CODE ANN. r. 391-3-11-.10; Idaho Dept. Health & Welfare, Div. of Envtl. Rules & Reg., § 01.5232; Me. Dep. Env't Prot., Rules for Haz. Waste Mgmt., ch. 855, § 80(H); MD. REGS. CODE tit. 10, § 51.05.07(j)(1); Miss. Dep't. Nat. Res. Haz. Waste Mgmt. Regs. pt. 264; Mont. Admin. Rules § 16.44.702; NEB. ADMIN. RULES & REG. tit. 128, ch. 21001; NEV. ADMIN. CODE ch. 444, § 444.9025; N.M. Haz. Waste Mgmt. Reg., r.501; N.D. ADMIN. CODE § 33-24-05-68; OHIO ADMIN. CODE § 3745-55-19; Okl. Rules & Reg. for Indus. Waste Mgmt. § 210; S.D. ADMIN. R. § 74:28:25:01; TENN. COMP. R. & REG. ch. 1200, § 1-11-.05(7)(j); TEX. ADMIN. CODE tit. 31, § 335.125; UTAH CODE ANN. § 6606(5); Va. Haz. Waste Mgmt. Reg. § 9.6(j); WISC. ADMIN. CODE, Sec. Nat. Res. § 181.42(9)(i). Hawaii and Vermont have no implementing regulations.

Since the RCRA program is delegable to the states, many states have enacted restrictions similar to RCRA § 3004 in their respective statutes governing hazardous waste management. These statutes must be at least as stringent as the federal rule, but some go further.[67]

The following states require that some form of general notice be placed on the public record or in the chain of title to alert prospective purchasers and the public of a site's status as formerly managing hazardous waste: Alaska,[68] Arizona,[69] Arkansas,[70] California,[71] Connecticut,[72] Colorado,[73] Delaware,[74] Florida,[75] Georgia, Hawaii,[76] Idaho,[77] Illinois,[78] Indiana,[79] Iowa,[80] Kansas,[81] Louisiana,[82] Maryland,[83] Michigan,[84] Minnesota,[85] Mississippi,[86] Missouri,[87] Montana,[88] Nebraska,[89] Nevada,[90] New Mexico,[91] North Carolina,[92] North Dakota,[93] Ohio,[94] Oklahoma,[95] Oregon,[96] Pennsylvania,[97] Rhode Island,[98] South Carolina,[99] South

---

[67]Note that the federal notice requirement is only applicable in states that have not had the RCRA program delegated to them.

[68]ALA. STAT. § 46.03.299; ALA. ADMIN. CODE tit. 18, § 62.410.

[69]ARIZ. REV. STAT. ANN. § 49-922; ARIZ. COMP. ADMIN. R. & REG. § 18-8-264.

[70]ARK. CODE ANN. § 82-4204; Reg. Ark Dep. of Pollution Control & Ecology, § 15.

[71]CAL. HEALTH & SAFETY CODE § 25245; CAL. CODE REG. § 264.119.

[72]CONN. GEN. STAT. § 22a-116(d); CONN. AGENCIES REGS. § 22a-449(c)-29(k).

[73]COLO. REV. STAT. § 25-15-208; 6 Colo. Code Reg. § 264.119.

[74]DEL. CODE ANN. tit. § 6305; Regs. Governing Haz. Waste § 264.120.

[75]FL. STAT. ANN. § 403.704; FL. ADMIN. CODE ANN. r. 391-3-11-10.

[76]HAW. REV. STAT. § 342.

[77]IDAHO CODE § 39-4405; Idaho Dept. Health & Welfare, Div. of Envt. Rules & Reg. § 01.5232.

[78]ILL. REV. STAT., ch. 111 1/2, para. 1021 (n) and para 1039(g); ILL. ADMIN. CODE tit. 35, § 724.219.

[79]IND. CODE §13-7-8.5-5.14; IND. ADMIN. CODE tit. 320, r. 4.1-21-9.

[80]IOWA CODE ANN. § 455 B.481; IOWA ADMIN. CODE r. 141 (455B).

[81]KAN. STAT. ANN. § 65-3431; KAN. ADMIN. REGS. § 28-31-8(c).

[82]LA. REV. STAT. ANN. § 1136; LA. ADMIN. CODE tit. 33, § 3525.

[83]MD. CODE ANN., HEALTH-ENVTL. § 7-208; MD. REGS. CODE tit. 10, § 51.05.07(J)(1).

[84]MICH. COMP. LAWS ANN. § 115B.16; MICH. ADMIN. CODE r 299.9613.

[85]MINN. STAT. ANN. § 115B.16; MINN. R. 7045.0496.

[86]MISS. CODE ANN. § 17.17.27(1); Miss. Dep't. Nat. Res. Haz. Waste Mgmt. Regs. part 264.

[87]MO. ANN. STAT. § 260.370; MO. CODE REGS. tit. 10, § 25-7.264(G).

[88]MONT. CODE ANN. § 75-10-405; MONT. ADMIN. R. 16.44.702.

[89]NEB. ADMIN. R. & REG. tit. 128, ch. 21001.

[90]NEV. REV. STAT. § 459.490; NEV. ADMIN. CODE ch. 444, § 444.9025.

[91]N.M. STAT. ANN. § 74-4-4; N.M. Haz. Waste. Mgmt. Reg., r. 501.

[92]N.C. GEN. STAT. § 130A-294; N.C. ADMIN. CODE tit.10, r. 10F.0032(h).

[93]N.D. CENT. CODE § 23-20.30-03; N.D. ADMIN. CODE § 33-24-05-68.

[94]OHIO REV. CODE ANN. § 3734.12(D)(8); OHIO ADMIN. CODE § 3745-55-19.

[95]OKLA. STAT. ANN. tit. 63, § 1-2004; Okla. Rules & Reg. for Industrial Waste Mgmt. § 210.

[96]OR. REV. STAT. § 466.015; OR. ADMIN. R. 340-104-001.

[97]35 PA. CONS. STAT. § 6020.512; 25 PA. CODE § 75.265(0)(20).

[98]R.I. GEN. LAWS § 23-19.1-6; Rules & Reg. for Haz. Waste Gen., Treatment, Storage, & Disposal, § 9.16.

[99]S.C. CODE ANN. § 44-56-30; S.C. CODE REG. 61-79.264.120.

Dakota,[100] Tennessee,[101] Texas,[102] Utah,[103] Virginia,[104] Washington,[105] West Virginia,[106] and Wisconsin.[107]

State UST Laws

Underground storage tanks (USTs), which are normally utilized to hold petroleum and other fuels at commercial gas stations and trucking fleet facilities, is another area that has been interceded by state regulations. The regulation and cleanup of USTs was first promulgated by the federal government in 1984.[108] Congress passed what now is subtitle I of the Resource Conservation and Recovery Act (RCRA) in order to afford some mitigation and control over what was considered the pervasive problem of leaking storage tanks whose substances leached into the groundwater.[109]

One of the primary designs of this program was to allow for state authorization and implementation of the UST program. As such, most actions against UST owners originate from state administrative law. States have, for the most part, adopted the standards established by the federal law, instead of establishing higher thresholds.[110] In addition, most states have established underground storage tank funding programs to finance the cleanup of UST sites and to compensate victims.[111] Normally, the cleanup fund programs are financed by a tax implemented on fuel sale within the state, apportioning the costs of remediation to all consumers of petroleum.

Unfortunately, the costs of cleanup are normally disproportionately expensive, due to the tendency of remediation contractors performing overly extensive

---

[100]S.D. CODIFIED LAWS. ANN. § 44-56-30; S.D. ADMIN. RULES 74:28:25:01.

[101]TENN. CODE ANN. § 68-46-107; TENN. COMP. R. & REGS. ch. 1200-1-11-.05(7)(j).

[102]TEX. REV. CIV. STAT. ANN. art. 4477-7, § 4(c); TEX. ADMIN. CODE tit. 31, § 335.125.

[103]UTAH CODE ANN. § 26-14-6; Utah Stat. Ann. tit. 10, § 6606(5).

[104]VA. CODE ANN. § 10-266; Va. Haz. Waste Mgmt. Reg. § 9.6(J).

[105]WASH. REV. CODE § 70.105.130; WASH. ADMIN. CODE § 173-303-610(10).

[106]W.VA. CODE § 20-5E-6; W. Va. Reg. tit. 47, § 35-15.4.

[107]WISC. STAT. ANN. § 144.62; WISC. ADMIN. CODE, § NAT. RES. 181.42(9)(i).

[108]*See* Heidi Brieger, *LUST and the Common Law: A Marriage of Necessity*, 13 B.C. ENVTL. AFF. L. REV. 521 (1986). Federal regulation of USTs came to fruition in the 1970s, predating the promulgation of RCRA. *Id.* at 523.

[109]42 U.S.C. §§ 6991-6991I (1998). The statute provided for the regulation of petroleum fuels as well as CERCLA listed hazardous substances. *Id.* § 6991(2). However, the statute excluded from coverage hazardous wastes as defined under RCRA. *Id.*

[110]*See, e.g.,* CAL. HEALTH & SAFETY CODE § 25299; MICH. COMP. LAWS ANN. § 299.701; MASS. GEN. LAWS ch. 21E, §§ 1-18.

[111]*See* California Underground Storage Tank Cleanup Fund, CAL. HEALTH & SAFETY CODE § 25299.50; Michigan Underground Storage Tank Financial Assurance Act, MICH. COMP. LAWS ANN. § 299.801 *et seq.*

and unnecessary remediation in order to avoid subsequent liability of their own.[112] In order to control these remediation and recovery costs expended by the state, the use of deductibles has been sought as a remedy for forcing entities that are financially capable of bearing some of the costly burdens to assume a share of the cleanup.[113] Other cost control options might include state oversight regarding the type of permits the UST owner must have before the owner can seek cost recovery for cleanup.[114]

But, despite these measures to lessen the burden on the states, most state-supported UST recovery funds are drying up. For example, the Michigan cleanup fund (MUSTFA)[115] has consistently overextended its resources since 1992.[116] In addition, EPA has given little guidance as to what should be done when state funding programs exceed their ability to cover the costs of remediation, and thus go broke.[117]

OPA Equivalents

Similarly, a majority of states have legislation in place similar to or exceeding the standards of the federal OPA. Thus, a state program (often exceeding the federal requirements regulating oil spills and cleanups) exists in most states. Again, these standards are often higher than the federal standards in that they typically require greater cleanup levels. They allow for state as well as federal oversight into the cleanup process. In addition, many of the OPA equivalents place disclosure and reporting requirements not only on property owners and operators, but some states such as Virginia require any person with notice of a spill to report it. Failure to report can result in civil and criminal sanctions for those with knowledge of the spill.

Specialized Stakeholder Protection Statutes: Adaptations to the CERCLA Model

Even prior to the recent trend of state voluntary cleanup legislation specifically addressing brownfield issues, many states enacted legislation amending the state equivalent of RCRA, CERCLA, or OPA to address the needs of various brown-

---

[112]See Amber, *Fund in Crisis as Cleanups Linger,* CRAIN'S GRAND RAPIDS BUS. J., Nov. 23, 1992, § 1, at 1.

[113]One such state is Michigan, which requires a $10,000 deductible to be carried by the property owner as a form of self-insurance over storage tank leakages. *See* MICH. COMP. LAWS ANN. § 299.809.

[114]See *Lloyd Properties*, 1993 WL 42259 (administrative hearing of the California Water Resources Board, January 21, 1993).

[115]MICH. COMP. LAWS ANN. § 299.801.

[116]See Carolyn Claerhout, *Financial Difficulties Shake the Michigan Underground Storage Tank Fund,* MICH. LAW WEEKLY, Nov. 30, 1992, at 22.

[117]See Laurel B. Calkins, *Environmental Fund Tapped Out; Contractors Fold, Lay Off Workers,* CRAIN'S HOUSTON BUS. J., May 17, 1993 §1, at 1.

field stakeholders. Legislation protecting lenders began to be enacted at the state level in 1986, when the Maryland legislature amended its CERCLA equivalent[118] to legislatively veto a decision by the Federal District Court for Maryland, which found potential lender liability on motion for summary judgment.[119] The trend accelerated following the Eleventh Circuits decision in *United States v. Fleet Factors Corp.*[120] The lending community's concern over CERCLA was heightened due to dicta in *Fleet Factors* that declared that the unexercised ability to control the affairs of the debtor was sufficient to void the secured creditor exemption.[121] Of course, such legislation is ineffective against any federal CERCLA actions, public or private.

Again identifying potential pitfalls in the federal CERCLA statute, many states have also enacted legislation protecting trustees from owner or operator liability under the state equivalent of CERCLA.[122] These statutes too were largely the result of political efforts on the state levels to show disagreement with the federal judiciary's interpretation of trustee liability under CERCLA.

State Voluntary Cleanup Laws

Approximately twenty-one states encourage the reuse and redevelopment of contaminated industrial property through enacting brownfields restoration and voluntary cleanup legislation. Many additional states are debating brownfields restoration or voluntary cleanup legislation. Depending upon the jurisdiction involved, volunteers satisfying program requirements may be shielded from private third-party actions and citizen suits under state law.

These state voluntary cleanup programs are new and have developed to replace the old, enforcement-liability-driven systems of the state Superfund and property transfer laws. Fifteen states have announced brownfield economic redevelopment programs during the last two years; seven states have enacted such legislation within the past year alone. Many of these programs apply not just to prospective purchasers but also to site owners and operators who report site contamination and agree to enter into a voluntary remediation program.

---

[118]MD. CODE ANN., HEALTH ENVTL. § 7-201.

[119]United States v. Maryland Bank & Trust Co., 632 F. Supp. 573 (D. Md. 1986).

[120]901 F.2d 1550 (11th Cir. 1990), *cert. denied*, 498 U.S. 104 (1991).

[121]*See, e.g.,* CAL. CIV. PROC. CODE § 726.5; COLO. REV. STAT. § 13-20-703; DEL. CODE ANN. tit. 7, § 9105(c)(3); HAW. REV. STAT. ch. 415, § 5/22.2; KEN. REV. STAT. ANN. § 224.01-400(26)-(27); ME. REV. STAT. ANN. tit. 38, §§ 1362 & 1367-A; MICH. COMP. LAWS ANN. § 299.612a; MINN. STAT. ANN. § 115B.03; MO. ANN. STAT. § 427.031; OR. REV. STAT. § 465.255; S.D. CODIFIED LAWS ANN. § 34A-15-3 through 5.

[122]*See, e.g.,* ALA. CODE § 19-3-11; CONN. GEN. STAT. § 45a-234(39); IND. CODE § 13-7-8.7(e) & (f); KEN. REV. STAT. ANN. § 244.01-400(27); MICH. COMP. LAWS ANN. § 299.612 a(10)-(12); N.C. GEN. STAT. § 32.27(8.1); OR. REV. STAT. § 465.440; TENN. CODE ANN. § 35-50-110(32); VA. CODE ANN. § 26-7.4.

State	VCP	Statutory Authority (where program is operated under agency enforcement discretion)
Arizona	Water Quality/Assurance Revolving Fund (WQARF) Voluntary Program; UST Voluntary Program; Voluntary Environmental Mitigation Use Restriction (VEMUR) Voluntary Program	
Arkansas	Arkansas Brownfield Program (1995)	Remedial Action Trust Fund (RATFA) Brownfields Redevelopment Amendment
California	Voluntary Cleanup Program (1993); Expedited Remedial Action Program (ERAP) (1994); Unified Agency Review Law (1993); Private Site Management Program Act (1995); Prospective Purchaser Policy (1996)	
Colorado	Voluntary Cleanup and Redevelopment Act (1994)	
Connecticut	Urban Sites Remedial Program; 1995 Amendments to Connecticut's Transfer Act; 1995 Voluntary Site Remediation Programs	
Delaware	Voluntary Cleanup Program (1995); Blue Collar Jobs Credit; Targeted Areas Credit	Hazardous Substances Cleanup Act (HSCA)
Illinois	Voluntary Cleanup Program (1989) (the 1989 Program); Site Remediation Program Created by 1995 Legislation (the 1995 Program)	
Indiana	Voluntary Remediation of Hazardous Substances and Petroleum Act (Voluntary Remediation Act) (1993)	
Kentucky	Senate Bill 219 (1996)	
Maine	Voluntary Response Action Program (1993)	
Maryland	Brownfields Voluntary Cleanup and Revitalization Programs (SB 205 and HB 5)	
Massachusetts	Massachusetts Contingency Plan (1990)	Chapter 21E
Michigan	1995 Amendments to the Michigan Environmental Response Act (MERA); Leaking Underground Storage Tank (LUST) Act	
Minnesota	Voluntary Investigation and Cleanup Program (VIC)	1988 Amendment to Minnesota Environmental Response and Liability Act (MERLA); Land Recycling Act of 1992

(*continued*)

Missouri	Brownfields Initiative (1995); Supplements the Missouri Voluntary Remediation Program	
Nebraska	Remedial Action Plan Monitoring Act (1995)	
New Jersey	Voluntary Cleanup Program; Environmental Opportunity Act (1996); Landfill Reclamation Act (1995); Hazardous Discharge Site Remediation Fund (1993)	Procedures for Department Oversight of Contaminated Sites (Oversight Rules), promulgated under the New Jersey Spill Compensation and Control Act (Spill Act)
New York	Voluntary Cleanup Program (1994)	
Ohio	Voluntary Real Estate Reuse and Cleanup Program (Voluntary Action Program) (1994 and 1996 Amendments)	
Oregon	Recycled Lands Act of 1995	
Pennsylvania	Land Recycling and Environmental Remediation Standards (Act 2); Economic Development Agency, Fiduciary, and Lender Environmental Protection Act (Act 3); Industrial Sites Environmental Assessment Act (Act 4)	
Rhode Island	Industrial Property Remediation and Reuse Act (Reuse Act) (1995)	
Tennessee	Voluntary Cleanup Oversight and Assistance Plan (VOAP) (1994 and 1995 amendments)	
Texas	Voluntary Cleanup Program (1995)	Hazardous Substances Mitigation Act
Utah	Voluntary Agreement Program (1989 and 1997 amendments)	(HSMA)
Vermont	Redevelopment of Contaminated Properties Program (RCPP)	Hazardous Waste Management Act (HWMA)
Virginia	Voluntary Remediation Program (VRP) (1995)	
Washington	Independent Remedial Action Program (IRAP)	Model Toxics Control Act (MTCA)
West Virginia	Voluntary Remediation and Redevelopment Act (1996)	
Wisconsin	Land Recycling Act (1994)	

State voluntary cleanup laws are primarily aimed at protecting potential purchasers of used industrial property, developers, and commercial lenders. As such, any remedial activity conducted under these programs is designed to occur with minimal regulatory oversight.[123]

---

[123]For a more detailed discussion of brownfield cleanup and reuse, *see* Office of Technology Assessment, *State of the States on Brownfields Programs for Cleanup and Reuse of Contaminated Sites* (June 1995).

Generally, state brownfields restoration and voluntary cleanup programs:

limit liability of nonresponsible property owners for the voluntary cleanup of hazardous substances and/or petroleum site contamination;[124]

limit prospective purchasers' and developers' liability when purchasing, reusing, or redeveloping brownfields;[125]

protect lenders accepting industrial property as security for commercial financing from liability;[126]

protect trustees from assuming liability when forced to administer estates suffering from site contamination.[127]

Typically, state agencies will provide a release and covenant not to sue to parties who are prepared to make some predefined investment in site remediation activities. Alternatively, some states have been willing to issue letters stating that no action will be required at a specific site. In some states, the programs have been implemented by statute. In other states, the programs have no legislative authority and are simply the result of policy initiatives.

Most programs, provide that a state agency will issue a formal covenant not to sue that becomes effective once agreed-upon site remediation activities or payment obligations have been completed. Even without a formal program, some states have demonstrated a willingness to work with prospective purchasers and investors to provide comfort in the form of a written letter.

Through the use of specific numeric cleanup standards or site-specific risk assessment techniques, most state programs require assurances that reuse or redevelopment of brownfields will safeguard the public health, safety, welfare, and environment. Other programs provide the volunteer with a choice between cleaning up to:

background levels;

cleanup levels based upon numerical equations, with default values based upon contamination and land use; and

site-specific risk assessment.

Property owners satisfying program requirements generally qualify for Certificates of Completion or No Further Action (NFA) letters issued by the states. Cer-

---

[124]See COLO. REV. STAT. ANN. § 25-16-310(1) (West 1994); IND. CODE ANN. § 13-7-8.9-18 (West 1994); MICH. COMP. LAWS ANN. § 299.614(a) (West 1994); MINN. STAT. ANN. § 115.175 [subdiv. 1] (West 1994); OHIO REV. CODE ANN. § 3746.12 (Baldwin 1994).

[125]Id.

[126]See IND. CODE ANN. § 13-7-8.9-18[2][A] (West 1994) (pursuant to § 18(2)(A), liability protection afforded to parties undertaking voluntary remedial activity may be legally transferred to third parties, such as lenders); MINN. STAT. ANN. § 115B.175 (subdiv. 6)(2) (West 1994); OHIO REV. CODE ANN. § 3746.25(A)(1) (Baldwin 1994).

[127]See OHIO REV. CODE ANN. § 3746.27(A) (Baldwin 1994); see also IND. CODE ANN. § 13-7-8.9-18(2)(A) (West 1994).

tificates of Completion or NFA letters are submitted to state environmental regulators in lieu of the issuance of a Covenant Not to Sue to the "volunteer(s)."

Several programs provide economic incentives for property owners to voluntarily clean up, reuse, and redevelop brownfields sites. The most obvious incentive applicable to all programs is increased land values. The remediation of hazardous substances can, however, be a very expensive proposition. For small and midsized businesses lacking the financial capacity to underwrite environmental assessments and voluntary cleanups, some states have created Voluntary Cleanup Funds or other economic incentives to assist qualified voluntary cleanup program participants. For example, Illinois provides "volunteers" with a 25 percent tax credit for the costs of remediation. In the state of Ohio, "volunteers" receive a 10 percent tax abatement for the increase in land values, with an additional 10-year tax abatement from local municipalities for qualifying properties.

Certain programs create an express duty on the part of the state to petition EPA on behalf of the volunteers to: (1) withhold from listing a brownfields site on either CERCLIS or the NPL and (2) petition EPA to waive pursuing federal enforcement proceedings against participating parties who have successfully completed the voluntary cleanup program. Similarly, under certain state cleanup programs, satisfactory completion of voluntary cleanup programs may protect property owners from actions brought by third parties and citizen groups for further site remediation. Other voluntary cleanup programs do not preclude participants from seeking cost recovery from the parties responsible for on-site contamination due to the presence of hazardous substances and/or petroleum.

Some brownfields sites may be ineligible for participation in state voluntary cleanup programs. For instance, in Ohio, the existence of groundwater contamination will *currently* bar a property owner from participating in voluntary cleanups, although ultimately sites with groundwater contamination will be included in the program as ultimately promulgated. Involvement in ongoing state or federal environmental enforcement actions or site listing on the NPL generally prevents a property owner from engaging in a state voluntary cleanup program. Finally, most programs apply only to property owners or other volunteers who are not responsible for existing site contamination. Under most state brownfields restoration and voluntary cleanup programs, parties owning or operating the property during the period of site contamination may not elect to remediate the site.

Following is a region by region description of each of the brownfields statutes developed by the states to encourage the recycling of land.

CHAPTER 4

U.S. EPA Region I:
Connecticut, Maine, Massachusetts, New Hampshire,
Rhode Island, and Vermont

Introduction

The states in Region I, the northeast corridor, were among the earliest to develop laws to aid in the cleanup of brownfields. As such, the state laws in Region I reflect less sophisticated legal mechanisms than in some other states (such as Pennsylvania), which enacted laws later and were able to benefit from the lessons learned in Region I.

All the states in Region I have enacted CERCLA-like statutes. The states in Region I, however, adapted these CERCLA clones to incorporate environmental liens that exceeded the federal authority under CERCLA. Despite constitutional deficiencies now recognized, most Region I states have some form of superlien in place.

Most of the Region I states also have transfer-triggered disclosure laws in place, and most have developed voluntary cleanup programs. The enactment of voluntary cleanup programs developed as a result of lessons learned in connection with the transaction-triggered disclosure laws. Specifically, many entities wanted to be able to voluntary clean up properties, with state oversight and approval, before engaging in a property transfer. Since cleanup was inevitable under the transfer laws, many companies began to recognize that it would be cheaper and more expedient to clean up in advance of sale. By so doing, the seller had to work only with the state environmental agency and would not need to also negotiate with a would-be purchaser.

The following table describes the more recent programs developed in Region I to encourage voluntary cleanup:

States	Brownfields Financing Programs	VCP with Written Assurances	Incentives to Attract Private Development
Connecticut	X	X	X
Maine		X	
Massachusetts	X	X	X
New Hampshire		X	X
Rhode Island		X	X
Vermont		X	

Connecticut

Voluntary Cleanup Program

In 1994, the Connecticut General Assembly added a voluntary cleanup measure to its already existing Hazardous Waste Establishment Transfer Act and Hazardous Waste Site Remediation Act.[1] In essence, the Urban Sites Remedial Action Program operates in conjunction with the Connecticut 1985 business transfer statute. A $30 million bond fund is included in the law to help encourage site assessment and cleanup. The law is administered by the Connecticut Department of Environmental Protection's (DEP) Water Bureau in conjunction with the Department of Economic Development (DED). The benefits of the program[2] include:

Social and Economic	Health Benefits
1) Returning abandoned and underutilized sites to productive use;	1) Minimizing or eliminating public exposure to pollutants on specific brownfields sites;
2) Creating the potential for new tax revenue to the state and municipality;	2) Cleaning up sites that otherwise would not be cleaned up for decades; and
3) Utilizing existing infrastructure and thereby minimizing the expansion of new infrastructure into undeveloped areas;	3) Minimizing the environmental impacts associated with industrial sprawl into clean suburban areas.
4) Helping to reverse urban decay and revitalizing cities;	
5) Creating short-term construction and longer term permanent manufacturing jobs.	

In conducting the state review, Connecticut divided site cleanup into three different categories:

Type	Description
I	Receive expedited review and approval of voluntary cleanup proposals.
II	Involve more state participation in the development of cleanup proposals; DEP may conduct the site investigation and draft the remedial action plan (RAP). If DED classifies a site as economically significant, then DEP can hire contractors to complete the RAP.
III	Can be purchased by the state for cleanup (assuming cleanup costs do not exceed $15 million). The state then remediates and leases the land. Lease payments include reimbursement for cleanup costs.

In administering the type I program, the amendments provide for an owner of property identified on the inventory of hazardous waste disposal sites[3] who

---

[1]P.A. No. 95-183, approved June 30, 1995; repealing CONN. GEN. STAT. § 22a-134, and substituting CONN. GEN. STAT. §§ 22a-134a to 22a-134d.

[2]DEP Fact Sheet, Connecticut's Urban Sites Remedial Action Program, reprinted in KEY ISSUES IN U.S. EPA REGION I (Oct. 2, 1996, Boston, Massachusetts).

[3]This inventory is maintained pursuant to CONN. GEN. STAT. § 22a-133c.

wishes to perform voluntary cleanup of the property to submit to the commissioner of environmental protection an environmental condition assessment form.[4] The commissioner shall then notify the owner within thirty days whether a review and approval of any such remedial action will be necessary.[5]

If the commissioner determines that prior approval is not necessary, the property owner is required to submit a statement of proposed actions for investigation and remediating their property and a schedule of implementation for such activities.[6] This statement must be submitted within ninety days of the commissioner's notification.[7] However, the commissioner may require the owner to submit copies of technical plans and reports related to the cleanup activities.[8] In addition, the property owner has the duty to notify the commissioner of any modifications to his proposed schedule.[9]

The commissioner may also determine that formal review and approval of the site remediation is necessary. In such a case, the property owner has the obligation to submit to the commissioner a proposed schedule for: (1) investigating and remediating the property; and (2) all technical plans, technical reports, and progress reports related to the investigation and remediation.[10]

If approved, the property owner must submit further technical plans, technical reports, and progress reports to the commissioner in accordance with the approved schedule and perform all actions identified in the approved technical plans, reports, and progress reports in accordance with the approved schedule.[11] If modifications are made to the original plans, they must be approved in writing by the commissioner.[12]

Connecticut utilizes a "licensed environmental professional"[13] to verify that parcels that have petitioned for voluntary cleanup meet the standards that were proposed to, and approved by, the commissioner.[14] A copy of such verification must be submitted to the commissioner at the completion of the remediation work.[15]

For purpose of the law, the term *environmental professional* is defined in the statute as:

> a person who is qualified by reason of his knowledge . . . to engage in activities associated with the investigation and remediation of pollution and

---

[4]S.B. No. 1189 § 3(a).
[5]*Id.*
[6]*Id.* § 3(b).
[7]*Id.*
[8]*Id.*
[9]*Id.*
[10]*Id.* § 3(c).
[11]*Id.*
[12]*Id.*
[13]*See* Environmental Program Fact Sheet, reprinted in KEY ISSUES IN U.S. EPA REGION I (Oct. 2, 1996, Boston, Massachusetts).
[14]*Id.* § 3(b), (c).
[15]*Id.*

sources of pollution including the rendering or offering to render to clients professional services in connection with the investigation and remediation of pollution and sources of pollution.[16]

The program is licensed by the State Board of Examiners of Environmental Professionals.[17]

At the completion of the environmental remediation, and verification of such work, the property owner may petition the commissioner for a Form II statement of remediation.[18] This document provides that any discharge, spill, uncontrolled loss, seepage, or filtration of hazardous waste that had occurred on the property has been properly remediated by the property owner and that this remediation has been approved by the state and verified by a certified environmental professional.[19]

The fee for this environmental remediation assessment is $2,000, which must be paid at the time that the initial assessment form is submitted to the state.[20]

Connecticut regulations allow, depending on the facts and circumstances, use of environmental land use restrictions as part of a brownfields cleanup plan. These restrictions must be recorded in the municipal land records. Examples of land use restrictions include: (1) a requirement that the property may not be used for residential purpose, or (2) that a certain building or impermeable cap not be disturbed. Restriction of the land is left to the discretion of the property owner, but is binding upon successors and assigns.

A person electing to record an environmental land use restriction must obtain a detailed survey of the property that is subject to the restriction. Restrictions need not apply to the entire parcel; in fact, more often restrictions apply only to a portion of the property. The proposed land use restriction must be submitted to the commissioner prior to recording. Where the remediation is voluntary, the restriction must be reviewed by the licensed environmental professional rather than the commissioner. All subordination agreements associated with a land use restriction must, however, be reviewed by the commissioner. The landowner should obtain relevant agreements from any lienholders or persons who otherwise might have an interest in the property. These agreements must irrevocably subordinate the lien or other interest to the environmental land use restriction. All land use restrictions must give opportunity for notice and comment by the public before the restriction is recorded.

---

[16]*Id.* § 4(a).

[17]*See* Environmental Program Fact Sheet, reprinted in KEY ISSUES IN U.S. EPA REGION I (Oct. 2, 1996, Boston, Massachusetts).

[18]*Id.* § 3(d).

[19]*Id.* § 1(12), 3(d).

[20]*Id.* § 3(e).

Connecticut Transfer of Hazardous Waste Establishment Act

Early Connecticut law tied environmental cleanups to business transactions involving property transfers. Basically, the Connecticut Transfer of Hazardous Waste Establishment Act[21] is intended to protect purchasers by compelling full disclosure of the environmental conditions of properties before transfer. Since the underlying environmental risks exist independently under CERCLA and other environmental laws, the practical effect of the Act is to bring into focus environmental considerations that should be routinely addressed in all transactions.

It is this focus that led those in business to begin a voluntary cleanup program that would allow those with contaminated property to correct problems before transferring property. The later legislation, thus, allowed certain property owners to potentially bypass or at least expedite the effect of the earlier law.

The Connecticut Transfer of Hazardous Waste Establishment Act applies to the "transfer of [an] establishment," defined as follows:

> the transfer of any operations which involve the generation, recycling, reclamation, reuse, transportation, treatment, storage, handling or disposal of hazardous waste, or any other transaction or proceeding through which an establishment undergoes a change in ownership, including, but not limited to, sale of stock in the form of a statutory merger or consolidation, sale of the controlling share of the assets, the conveyance of real property, change of corporate identity or financial reorganization, but excluding corporation reorganization not substantially affecting the ownership of the establishment.[22]

Thus, this broad term includes the sale of stock through a merger or consolidation, the disposition of the controlling share of assets, a sale of real property, a change in corporate identity, and a financial reorganization significantly affecting the ownership of the establishment.

Unlike other state laws,[23] the definition of *establishment* is not tied to the SIC system. Instead, the state includes "any establishment [that since] May 1, 1967, generated more than one hundred kilograms of hazardous waste per month or [that] recycled, reclaimed, reused, stored, handled, treated, transported or disposed of hazardous waste generated by another" entity.[24]

The Act defines *hazardous waste* as "any waste material, except by-product material, source material or special nuclear material, . . . which may pose a present or potential hazard to human health or environment when improperly disposed of, treated, stored, transported, or otherwise managed."[25] The definition excludes

---

[21]CONN. GEN. STAT. ANN. §§ 22a-134 to 22a-134e.

[22]*Id.* § 22a-134(1).

[23]See New Jersey ISRA.

[24]*Id.* § 22a-134(3).

[25]*Id.* § 22a-115(1).

sewage.[26] Thus, the Act applies to, for example, a sale of stock substantially affecting the ownership of a business engaged in the transportation or handling of hazardous waste since May 1967.

The Connecticut Act does not specifically authorize the voiding of the underlying transaction in the event of noncompliance with the Act. The Connecticut Act also differs in several other significant respects from those of New Jersey. Specifically, the Connecticut Act:

(1) does not apply to closures or termination of leases;
(2) deals only with hazardous wastes, not hazardous substances or petroleum products;
(3) excludes low-volume generators directly from its coverage; and
(4) has relatively simple reporting requirements.

When applicable, the Act also requires the transferor to submit a "negative declaration" to both the transferee and the Connecticut commissioner of environmental protection. A "negative declaration" is:

a written declaration on a form prescribed by the commissioner stating (1) that there has been no discharge, spillage, uncontrolled loss, seepage or filtration of hazardous waste on-site, or that any such discharge, spillage, uncontrolled loss, seepage or filtration has been cleaned up in accordance to procedures approved by the commissioner or determined by him to pose no threat to human health or safety or the environment which would warrant containment and removal or other mitigation measures and (2) that any hazardous waste which remains on-site is being managed in accordance with [the Act] and regulations. . . .[27]

The Connecticut Act does not require that all hazardous substances or wastes be removed from the property. All that is necessary in order to make a negative declaration is that the seller provide assurances that any hazardous waste is being properly managed.[28]

If the transferor is unable to make a negative declaration, the transferor, before a transfer, must certify to the Connecticut commissioner of environmental protection that, "to the extent necessary to minimize or mitigate a threat to human health or the environment," the transferor "will contain, remove, or otherwise mitigate" the effects of any hazardous wastes according to a schedule and procedures approved by the commissioner and embodied in an administrative order, stipulated judgment, or consent agreement.[29]

---

[26]CONN. GEN. STAT. ANN. § 22a-134(4).

[27]Id. § 22a-134(2), (5).

[28]Id. § 22a-134(5).

[29]Id. § 22a-134a(c).

If the transferor violates the Act, the transferee may recover damages from the transferor.[30] Furthermore, the transferor is strictly liable for all cleanup and removal costs and for all direct damages resulting from the violation.[31]

If a person knowingly gives false information or does not comply with the Act, the person is civilly liable for up to $100,000.[32] In contrast to the New Jersey laws, the Connecticut Act does not empower the commissioner to void the transaction in the event of noncompliance with the Act.

Specialized Stakeholder Laws—Trustee

Under Connecticut law, a fiduciary has the power to "take any action necessary to deal with or prevent problems created by environmental hazards, including, but not limited to, conducting assessments, taking any remedial action to contain, clean up or remove environmental hazards and expending estate or trust assets to accomplish any such action."[33]

Environmental Lien

Connecticut is one of a minority of states that enacted a superlien.[34] Connecticut laws provide that, after June 3, 1985, any amount paid by the commissioner of environmental protection to contain and remove or mitigate the effects of a spill will be a lien against the real estate on which the spill occurred or from which it emanated. The law carves out an exception where the environmental lien against real estate has been transferred "in accordance with the provisions of Sections 22a-134 to 22a-134d, inclusive, shall not have priority over any previous transfer or encumbrance."[35]

To be effective, a certificate of lien must be filed in the land records of each town in which the real estate is located, describing the real estate, the amount of the lien, and the name of the owner as grantor. In addition, the commissioner must mail a copy of the certificate to such persons and to all other persons of record holding an interest in such real estate over which the commissioner's lien is entitled to priority.

Upon presentation of a certificate of lien, the town clerk must endorse it with his identification and the date and time of receipt and record it. Generally, the lien will "take precedence over all transfers and encumbrances recorded on or after June 3, 1985, in any manner affecting such interest in such real estate or any part of it on which the spill occurred or from which the spill emanated, or real estate which has been included, within the preceding three years, in the property description of such real estate and is contiguous to such real estate."[36]

---

[30]CONN. GEN. STAT. ANN. § 22a-134b.
[31]*Id.*
[32]*Id.* § 22a-134d.
[33]*Id.* § 45a-234(39).
[34]CONN. GEN. STAT. § 22a-452a.
[35]CONN. GEN. STAT. § 22a-452a.
[36]*Id.* § 22a-452a(c).

The superlien does not apply to real estate that is exclusively residential. In all other real estate, the lien will take precedence over any transfer or encumbrance recorded after the commissioner files with the town clerk notice of intent to file a lien on the land records in the town in which the real estate is located.

When a recorded environmental lien has been paid or reduced, the commissioner, upon request of any interested party, must issue a certificate discharging or partially discharging such lien, which certificate must be recorded in the same office in which the lien was recorded. The town clerk must note the recording of the certificate of discharge upon the original notice of lien.[37]

Any action for the foreclosure of the lien must be brought by the attorney general in the name of the state in the superior court for the judicial district in which the property subject to the lien is situated. If such property is located in two or more judicial districts, the action may be brought in the superior court for any one such judicial district. The court may limit the time for redemption or order the sale of such property or make such other or further decree as it judges equitable.

RCRA

Connecticut, like other states, has a state statute requiring that some form of general notice be placed on the public record or in the chain of title to alert prospective purchasers and the public of a site's status as formerly managing hazardous waste, as defined under RCRA.[38]

In addition, the Connecticut Transfer Act[39] requires transferors of facilities that produce more than 100 kilograms each month of hazardous waste or which deal with hazardous wastes generated by other entities, to certify to the transferee that either:

(1) there has been no discharge of hazardous waste at the property or
(2) if a discharge existed, it has been cleaned up.

The certification is to be filed with the state, and if the transferor cannot certify to one of the above two items, either the transferor or transferee must certify that it will clean up the site.[40]

---

[37]Any action for reduction or discharge of such lien or any appeal therefrom shall be in accordance with the provisions of sections 49-35a to 49-35c, inclusive, except that the forms prescribed in section 49-35a shall be modified as the court deems appropriate.

[38]CONN. GEN. STAT. § 22a-116(d); CONN. AGENCIES REGS. § 22a-449(c)-29(k); COLO. REV. STAT. § 25-15-2086; Colo. Code Reg. § 264.119.

[39]CONN. GEN. STAT. ANN. § 22a-454(b).

[40]See I. Leo Motiuk et al., Environmental Transfer Law Update: New Jersey and the Nation, TOX-ICS L. REP. (BNA) 487, 488 (1992).

Note also that the governor of Connecticut signed the Landowner Liability Act, Public Act. No. 93-375, on June 30, 1993, effective as of that date, which establishes an innocent landowner defense to avoid liability for spills or discharges on property. New Jersey's ISRA provides for a similar innocent purchaser defense.

Maine

Voluntary Cleanup Program

In 1993, Maine established its Voluntary Response Action Program (VRAP)[41] and its Controlled Sites Program. The VRAP program was modeled after Minnesota's Voluntary Investigation and Cleanup (VIC) legislation.[42] The Controlled Sites Program is triggered when voluntary cleanup sites are not completed in a satisfactory manner. To date, at least forty-one sites have completed site cleanup successfully. In addition, approximately fifty sites are currently being processed. The sites "admitted into the program have displayed a wide-range of contaminants."[43] These include: coal tar wastes; metals; PAHs; PCBs; solvents and gasoline, fuel oil, and other petroleum products.

The VRAP Program is a technical assistance as well as a voluntary response action program.[44] It consists of four basic stages:[45]

Stage	Brief Description
1	A thorough environmental investigation is completed by the applicant's consultant, resulting in a remedial action work plan.
2	The VRAP program approves of the work plan and issues a "no action assurance" letter to the applicant, outlining protections the applicant will receive if the remedial actions are satisfactorily completed.
3	The applicant's consultant completes the remedial actions outlines in the work plan and demonstrates that those are satisfactorily completed.
4	The VRAP program issues a "Commissioner's Certificate of Completion," which outlines the facts, conclusions, conditions, and assurances associated with the remedial actions at the site.

"Remedial goals at sites are developed on a site specific basis, taking into account the proposed future use of the property, the risks posed to real receptors, and the institutional controls which may be implemented to effectively reduce risks posed by the site."[46] Depending on the level of cleanup undertaken, the VRAP may exempt individuals who initiate site investigation and cleanup from liability

---

[41]ME. REV. STAT. ANN. tit. 38, § 343-E.

[42]Maine Voluntary Remedial Action Plan (VRAP) Program, reprinted in KEY ISSUES IN U.S. EPA REGION I (Oct. 2, 1996, Boston, Massachusetts).

[43]Maine Voluntary Remedial Action Plan (VRAP) Program, reprinted in KEY ISSUES IN U.S. EPA REGION I (Oct. 2, 1996, Boston, Massachusetts).

[44]DEP Issue Profile: The Voluntary Clean-up Program (Sept. 1995), reprinted in KEY ISSUES IN U.S. EPA REGION I (Oct. 2, 1996, Boston, Massachusetts).

[45]Maine Voluntary Remedial Action Plan (VRAP) Program, reprinted in KEY ISSUES IN U.S. EPA REGION I (Oct. 2, 1996, Boston, Massachusetts).

[46]Maine Voluntary Remedial Action Plan (VRAP) Program, reprinted in KEY ISSUES IN U.S. EPA REGION I (Oct. 2, 1996, Boston, Massachusetts).

under Maine's environmental laws.[47] It also provides a "no-action" guarantee from the Maine Department of Environmental Protection.[48] Participants in the VRAP pay an initial oversight fee to DEP, and may incur additional costs for DEP's services.

Maine DEP has yet to create all regulations for VRAP. Thus, to date, any site may be eligible for cleanup, so long as the cleanup is consistent with federal environmental laws.[49] DEP currently decides program eligibility on a site-by-site basis. Parties are responsible for payment of state oversight costs at a rate not to exceed $50 per hour. In addition, parties entering into the program must submit a $500 nonrefundable registration fee.[50]

As with other voluntary cleanup programs, a proposed action plan for either partial or complete cleanup must be submitted to DEP.[51] The plan must detail the investigation of the site, the results of the investigation, and an analysis of the findings.[52] Implementation of the complete plan begins once it is approved by DEP. "On occasion, an applicant will have a site specific risk-assessment performed following the Maine Department of Environmental Protection's Guidance Manual for Human Health Risk Assessments at Hazardous Substance Sites."[53]

DEP provides a "no-action" letter upon the approval of the plan, in which DEP assures the parties that it will not take legal action if the plan and any additional site requirements are followed.[54] Once the site is cleaned per the plan's requirements and meets DEP satisfaction, DEP issues a certificate of completion to the program applicant.[55] If the owner or operator who caused contamination at the site did not apply to the program, then that person will not receive any liability protection.[56]

Notice and Reporting

Maine law provides that "any environmental professional who determines discharges of hazardous waste, special waste, hazardous substances, or petroleum products are present at a site must report such information to the client, if, in their judgment, the discharges will require removal or remediation to prevent significant

---

[47]*Id.* § 343-E (1)(2)(6)(8)(9); James T. Kilbreth & Juliet T. Browne, Maine (unpublished manuscript, on file with author).

[48]ME. REV. STAT. ANN. § 343-E(9).

[49]*Id.* at 3-4; ME. REV. STAT. ANN. § 343-E(1).

[50]*See* Charles Bartsch & Elizabeth Collaton, COMING CLEAN FOR ECONOMIC DEVELOPMENT at 74.

[51]ME. REV. STAT. ANN. § 343-E(1),(2),(4).

[52]*Id.* § 343-E(4).

[53]Maine Voluntary Remedial Action Plan (VRAP) Program, reprinted in KEY ISSUES IN U.S. EPA REGION I (Oct. 2, 1996, Boston, Massachusetts).

[54]*Id.* § 343-E(9).

[55]*Id.* § 343-E(5).

[56]DEP Issue Profile: The Voluntary Clean-up Program (Sept. 1995), reprinted in KEY ISSUES IN U.S. EPA REGION I (Oct. 2, 1996, Boston, Massachusetts).

threats to public health or the environment."[57] If the client is not the owner or operator of the site, "that client must report such findings to the owner/operator who, in turn, must disclose the information to the department within a "reasonable" time period. Generally, DEP considered ten working days from the time the person receives the analytical information to be a reasonable time. In addition, parties such as real estate brokers may have obligations pursuant to their professional codes of conduct to report the release.

Specialized Stakeholder Laws—Lender

Maine was one of the states that amended its state equivalent of CERCLA to include some specialized lender liability protection. In Maine, the liability of fiduciaries and lenders is also defined by statute. A person may not be deemed a responsible party under state law if that person is "a fiduciary, as defined in section 1362, subsection 1-D, but that exclusion does not apply to an estate or trust of which the site is a part"[58] or "a lender, as defined in section 1362, subsection 1-B, who, without participating in management of a site, holds indicia of ownership primarily to protect a security interest in the site."[59] Participating in management is

---

[57]DEP Issue Profile: The Voluntary Clean-up Program (Sept. 1995), reprinted in KEY ISSUES IN U.S. EPA REGION I (Oct. 2, 1996, Boston, Massachusetts).

[58]Under the Act, *fiduciary* means:

a person (a) who is acting in any of the following capacities: an executor or administrator as defined in section one of chapter one hundred ninety-seven, including a voluntary executor or a voluntary administrator; a guardian; a conservator; a trustee under a will or inter vivos instrument creating a trust under which the trustee takes title to, or otherwise controls or manages, property for the purpose of protecting or conserving such property under the ordinary rules applied in the courts of the commonwealth; a court-appointed receiver; a trustee appointed in proceedings under federal bankruptcy laws; an assignee or a trustee acting under an assignment made for the benefit of creditors pursuant to sections forty through forty-two of chapter two hundred and three; or a trustee, pursuant to an indenture agreement or similar financing agreement, for debt securities, certificates of interest of participation in any such debt securities, or any successor thereto, and (b) who holds legal title to, controls, or manages, directly or indirectly, any site or vessel as a fiduciary for purposes of administering an estate or trust of which such site or vessel is a part.

[59]The statute says further:

"Indicia of ownership primarily to protect a security interest," only those interests in real or personal property typically acquired and held as security or collateral for payment or performance of an obligation. Such interests shall include, without limitation, a mortgage, deed of trust, lien, security interest, assignment, pledge, or other right or encumbrance against real or personal property, including those security interests which have a contingent interest component, which are furnished by the owner thereof to assure repayment of a financial obligation; and contractual participation rights in such interest; provided, that the contract conferring such rights confers no other interest in the site or vessel.

Such interests shall not include the following:—those to protect any interest in property owned or held for investment purposes; a lease or a consignment which may not be considered a secured transaction under applicable principles of commercial law; interests of a person acting as a trustee defined as executing decision-making control over the borrower's management of oil or hazardous materials,[60] while the borrower is in possession of the facility or exercising control over substantially all of the operational aspects of the borrower's enterprise. Participation in management does not include the following:

(1) conducting or requiring site assessments of the property;[61]
(2) engaging in periodic or regular monitoring of the business;
(3) financing conditioned on compliance with environmental laws;
(4) providing general business or financial advice, excluding management of hazardous materials and oil;
(5) providing general advice with respect to site management;
(6) policing the security interest or loan;
(7) engaging in workout activities prior to foreclosure; or
(8) participating in foreclosure proceedings.

The following must be considered in determining whether a secured lender is "acting diligently to sell or otherwise divest" or as "evidence of diligent efforts to sell or divest":

(1) use of the property during the period;
(2) market conditions;
(3) marketability of the site; or
(4) legal constraints on the sale or divestment.[62]

---

of property or of a business; and any interest other than one created as a bona fide security interest in real or personal property.

"Offshore oil facility," any oil facility of any kind located in, on, or under any submerged land within the jurisdiction of the commonwealth including, without limitation, the territorial sea; provided, however, that it shall not include a vessel.

[60]A *hazardous material* is defined as:

material including, but not limited to, any material, in whatever form, which, because of its quantity, concentration, chemical, corrosive, flammable, reactive, toxic, infectious or radioactive characteristics, either separately or in combination with any substance or substances, constitutes a present or potential threat to human health, safety, welfare, or to the environment, when improperly stored, treated, transported, disposed of, used, or otherwise managed. The term shall not include oil. The term shall also include all those substances which are included under 42 U.S.C. § 9601(14), but it is not limited to those substances.

[61]The Act defines *assess* and *assessment* as:

such investigations, monitoring, surveys, testing, and other information gathering activities to identify: (1) the existence, source, nature and extent of a release or threat of release of oil or hazardous materials; (2) the extent of danger to the public health, safety, welfare and the environment; and (3) those persons liable under section five. The term shall also include, without limitation, studies, services and investigations to plan, manage and direct assessment, containment and removal actions, to determine and recover the costs thereof, and to otherwise accomplish the purposes of this chapter.

[62]ME. REV. STAT. ANN. tit. 38, § 342-B.

Liability will not necessarily be imposed on a lender that holds the property for longer than the five-year period if the lender works diligently to dispose of the property. The exemption for lenders will not apply in three circumstances. First, lenders are not exempt if "the secured lender causes, contributes to or exacerbates the discharge, release or threat of release." Second, the exemption will be lost if "the secured lender participates in management of the site prior to acquiring ownership of the site. Finally, the secured lender will not be protected by the exemption if, after acquiring ownership of the site and upon obtaining knowledge of a release or threat of release, the secured lender does not:

(1) notify the department within a reasonable time after obtaining knowledge of a release or threat of release;
(2) provide reasonable access to the department and its authorized representatives so that necessary response actions may be conducted;
(3) undertake reasonable steps to control access and prevent imminent threats to public health and the environment; and
(4) act diligently to sell or otherwise divest the property within a limited time period of up to five years from the earlier of the lender's possession or ownership.

There is a rebuttable presumption that the second lender is acting diligently to sell or otherwise divest the property during the first eighteen months after taking possession. The secured lender must demonstrate by a preponderance of the evidence diligent efforts to sell or divest the property during the next forty-two months.

When a lender has ownership or possession of a site pursuant to a security interest in the site, the term *owner* or *operator* means a person who owned or operated the site immediately prior to that secured lender obtaining ownership or possession of the site.[63]

---

[63]*Owner* or *operator* is defined under the Act as:

(a)(1) in the case of a vessel, any person owning, operating or chartering by demise such vessel, (2) in the case of a site, any person owning or operating such site, (3) in the case of an abandoned site, any person who owned, operated, or otherwise controlled activities at such site, vessel, onshore oil facility, offshore oil facility, deepwater port, or pipeline, any person who owned, operated or otherwise controlled activities at such site immediately prior to such abandonment, except that, in the case of an onshore oil facility or offshore oil facility, this definition shall not include an agency or political subdivision of the federal government or the commonwealth, or any interstate body, that owned an onshore oil facility or offshore oil facility and that, as the owner, transferred possession and right to operate the onshore oil facility or offshore oil facility to another person by lease, assignment, or permit, immediately prior to such abandonment, (4) in the case of an onshore oil facility, other than a pipeline, any person owning or operating the onshore oil facility, except that this definition shall not include an agency or political subdivision of the federal government or the commonwealth, or any interstate body, that owns an onshore oil facility and that, as the owner, transfers possession and right to operate the onshore oil facility to another person by lease, assignment, or permit, (5) in the case of an offshore oil facility, other than a pipeline or a deepwater port licensed under the U.S. Deepwater Port Act of 1974, the lessee or permittee of the area in which the offshore oil facility is located or the holder of a right of use and easement granted under an applicable law

Maine has also enacted protections for persons who voluntarily undertake cleanup at a contaminated site.[64] Note that the Maine law includes petroleum spills in its statute and so its coverage is more stringent than the federal laws pertaining to site cleanup.

Under the Maine law, a person may not be deemed a responsible party if the person investigates the discharge, release, or threatened release and undertakes and completes response actions to remove or remedy all known discharges, releases, and threatened releases at an identified area of real property in accordance with a voluntary response action plan approved by the state commissioner. In addition, the state may approve a voluntary response action plan that does not require removal or remedy of all discharges, releases, and threatened releases at an identified area of real property conditioned based on consideration of the following:

(1) if reuse or development of the property is proposed, the voluntary response action plan provides for all response actions required to carry out the proposed reuse or development in a manner that protects public health and the environment;

(2) the response actions and the activities associated with any reuse or development proposed for the property will not cause, contribute, or exacerbate discharges, releases, or threatened releases that are not required to be removed or remedied under the voluntary response action plan and will not interfere with or substantially increase the cost of response actions to address the remaining discharges, releases, or threatened releases; and

(3) the owner of the property that is the subject of the partial voluntary response action plan agrees to cooperate with the commissioner, the requestor, or the commissioner's authorized representatives to avoid any action that interferes with the response actions.

---

of the commonwealth or the U.S. Outer Continental Shelf Lands Act, for the area in which the offshore oil facility is located if such holder is a different person from lessee or permittee; provided, however, that this definition shall not include an agency or political subdivision of the federal government or the commonwealth, or any interstate body, that owns an offshore oil facility and that, as the owner, transfers possession and right to operate the offshore oil facility to another person by lease, assignment, or permit, (6) in the case of a deepwater port licensed under the U.S. Deepwater Port Act of 1974, the licensee, (7) in the case of a pipeline, any person owning or operating the pipeline, (8) when a fiduciary who is not an owner or operator pursuant to this definition has title or control or management of a site or vessel, the grantor or settlor of the estate or trust in question, to the extent the assets of the estate or trust are insufficient to pay for liability pursuant to this chapter, and (9) when a secured lender who is not an owner or operator pursuant to this definition has ownership or possession of a site or vessel, any person who owned or operated such site or vessel immediately prior to such secured lender obtaining ownership or possession of such site or vessel. The term shall include any estate or trust of which the site or vessel is a part. The term shall not include the commonwealth to the extent the commonwealth holds or held any right, title, or interest in a site or vessel solely for the purpose of implementing or enforcing the commonwealth's rights or responsibilities pursuant to this chapter, unless the commonwealth caused or contributed to the release or threat of release.

[64]ME. REV. STAT. ANN. tit. 38, § 343-E.

The commissioner may condition the protection from liability on the requestor's agreement to any or all of the following terms:

(1) to provide access to the property to the commissioner and the commissioner's authorized representatives;
(2) to allow the commissioner or the commissioner's authorized representatives to undertake activities at the property including placement of borings, wells, equipment, and structures on the property; and
(3) to the extent the requestor has title to the property, to grant easements or other interests in the property to the department for any of the above purposes.

All agreements must apply to and be binding upon the successors and assigns of the owner. However, no protections will be afforded to a successor or assign of the person who obtained approval if that successor or assign had knowledge that the approval was obtained by fraud or intentional misrepresentation or by knowingly failing to disclose material information. To the extent the requestor has title to the property, the requestor must also record the agreement or a memorandum approved by the commissioner that summarizes the agreement in the registry of deeds for the county where the property is located.

In addition to the person who undertakes and completes a voluntary response action pursuant to an approved voluntary response action plan, the liability protection also includes "a person providing financing to the person who undertakes and completes the response actions or who acquires or develops the identified property" and a "lender or fiduciary . . . who arranges for the undertaking and completion of response actions." Persons who acquire or develop the identified property and arrange for the undertaking and completion of response actions are similarly provided protection, as are successors or assigns of a person to whom the liability protection applies.[65]

The protection from liability does not apply to: (1) a person who causes, contributes, or exacerbates a discharge, release, or threatened release that was not remedied under an approved voluntary response action plan; (2) a partial voluntary response action plan that does not require removal or remediation of all known releases; or (3) a person who obtains approval of a voluntary response action plan by fraud or intentional misrepresentation, or by knowingly failing to disclose material information and their successors and assigns. The state will issue "no-action" letters concerning applicability where parties are unclear whether they are protected or not.

---

[65]The law also contains a catchall protection for:

A person acting in compliance with a voluntary response action program approved by the commissioner who, while implementing the voluntary response action plan and exercising due care in implementation, causes, contributes or exacerbates a discharge or release, provided that the discharge or release is removed or remediated to the satisfaction of the commissioner.

Specialized Stakeholder Laws—Trustee

Maine was one of the states that amended its state equivalent of CERCLA to include some specialized trustee liability protection. In Maine, a fiduciary may not be deemed a responsible party and that person is not subject to department orders or other enforcement proceedings, liable or otherwise responsible for discharges, releases, or threats of releases of a hazardous substance, hazardous waste, hazardous matter, special waste, pollutant or contaminant, or a petroleum product or by-product. The exclusion does not apply to an estate or trust of which the site is a part.[66]

The exclusion also does not apply if the fiduciary causes, contributes to, or exacerbates the discharge, release, or threat of release. Nor will the exclusion apply if, after acquiring title to or commencing control or management of the site, the fiduciary does not: (1) notify the department within a reasonable time after obtaining knowledge of a release or threat of release; (2) provide reasonable access to the site to the department and its authorized representatives so that necessary response actions may be conducted; and (3) undertake reasonable steps to control access and prevent imminent threats to public health and the environment.

Notwithstanding the exemption from liability, a fiduciary may be named as a party in an administrative enforcement proceeding or civil action brought by the state for purposes of requiring the submission of information or documents relating to an uncontrolled hazardous substance site, for purposes of proceeding against the assets of the estate or trust for reimbursement, fines, or penalties, or for purposes of compelling the expenditure of assets of the estate or trust by the fiduciary to abate, clean up, or mitigate threats or hazards posed by a discharge or release, or to comply with state environmental laws and regulations or the terms of a department order of enforcement proceeding. The statute does not, however, require the fiduciary to expend his own funds or to make the fiduciary personally liable for compliance pursuant to an order or enforcement proceeding.[67]

A fiduciary who arranges for the undertaking and completion of response actions will, like the person who undertakes and completes a voluntary response action pursuant to an approved voluntary response action plan, be exempt from liability under the state Superfund equivalent.

---

[66]Section 342-B defines "assets of the estate or trust" as:

assets of the estate or trust of which the site is a part; assets that subsequent to knowledge of the release are placed by the fiduciary or the grantor in an estate or trust over which the fiduciary has control if the grantor is or was an owner or operator of the release site at the time of the transfer; and assets that are transferred by the fiduciary upon or subsequent to knowledge of the release for less than full and fair consideration, to the extent of the amount that the fair market value exceeded the consideration received by the estate or trust.

[67]Exceptions are enumerated in section 568, subsection 4, paragraph B or section 1365, subsection 6.

Environmental Lien

Maine[68] is one of a minority of states that enacted a superlien.[69] Under Maine law, all costs incurred by the state for the abatement, cleanup, or mitigation of hazards posed by an uncontrolled hazardous substance site and all interest and penalties will be a lien against the real estate of the responsible party.

The priority of the lien will be dependent upon the facts and circumstances of the case. Any lien filed on real estate that encompasses an uncontrolled hazardous substance site has precedence over all encumbrances on the real estate, including liens of the state or any political subdivision, recorded after July 7, 1987.[70] Any lien filed on any other real estate of the party responsible for the uncontrolled hazardous substance site will have precedence over all transfers and encumbrances filed after the date that the lien is filed with the Registry of Deeds.

Notice of the lien is accomplished by the mailing, by certified mail, of a certificate of lien signed by the commissioner of environmental protection to all those persons of record holding an interest in the real estate over which the commissioner's lien is entitled to priority. In addition, a certificate may, but need not, be filed for record in the office of the clerk of any municipality in which the real estate is situated. The lien will not, however, be effective until it is filed with the Registry of Deeds for the county in which the real estate is located. The lien must include a description of the real estate, the amount of the lien, and the name of the owner as grantor.

The lien does not apply to: (1) a unit of real estate that consists primarily of real estate used or under construction as single or (2) multifamily housing at the time the lien is recorded or to property owned by a political subdivision except for the real estate that encompasses an uncontrolled hazardous substance site and that is owned by a political subdivision.

When the lien has been paid or reduced, the commissioner, upon request by any person of record holding interest in the real estate that is the subject of the lien, must issue a certificate discharging or partially discharging the lien. The certificate must be recorded in the registry in which the lien was recorded. Any action of the foreclosure of the lien must be brought by the state attorney general in the superior court for the judicial district in which the property subject to the lien is situated.[71]

---

[68]ME. REV. STAT. ANN. tit. 38, § 1371.

[69]The other states include: Connecticut, Louisiana, Maine, Massachusetts, New Hampshire, and New Jersey.

[70]The term *real estate* is defined as including "all real estate of a responsible party that has been included in the property description of the affected real estate within the 3-year period preceding the date of filing of the lien or on or after July 7, 1987, whichever period is shorter."

[71]ME. REV. STAT. ANN. tit. 38, § 1371.

RCRA

Maine, like other states, has a state statute requiring that some form of general notice be placed on the public record or in the chain of title to alert prospective purchasers and the public of a site's status as formerly managing hazardous waste.[72]

Massachusetts

"In 1990, seven years after the Massachusetts Superfund Law (M.G.L. c.21E) was enacted, less than one quarter of the 4,200+ confirmed and suspected hazardous waste sites were being assessed or cleaned up, and the backlog of sites requiring attention by the Department of Environmental Protection (DEP) was growing rapidly."[73] Many concerns were identified:

> regulatory barriers were preventing sites from being assessed and cleaned up;
> DEP was not able to devote its resources to finding and cleaning up the most serious waste sites;
> there was a lack of clear standards and guidelines defining when and how sites should be cleaned up.

Accordingly, in July 1992, Governor Weld signed amendments to chapter 21E. These changes redesigned the program to provide new incentives for private parties to undertake cleanups. Specifically, "the liability of secured lenders and fiduciaries was clarified to ensure that, as long as they meet certain requirements, they will not be held responsible for cleanup costs."[74] In addition, "the rights of liable parties and others who respond to releases were clarified to make it easier to get other liability parties to contribute to response actions."[75] "A dispute resolution mechanism" was provided for private parties to obtain participation from other PRPs.[76] Finally, private parties conducting response actions were given the right "to gain access to contaminated property for assessment and cleanup work."[77]

In so privatizing smaller cleanup operations, the state could concentrate its limited resources on larger sites.[78] Key features of the pioneering Massachusetts program include:

---

[72]ME. REV. STAT. ANN. tit. 38, § 1303-A; Maine Dep. Env't Prot., Rules for Haz. Waste Mgmt., ch. 855 § 80(H).

[73]Cleaning Up Contaminated Property: The Massachusetts Waste Site Cleanup Program, http://www.magnet.sts . . . o/dep/bwsc/progdes.htm (7/7/97 12:51:29).

[74]*Id.*

[75]*Id.*

[76]*Id.*

[77]*Id.*

[78]*Id. See also* Two Years Later: How the New 21E Program Is Measuring Up, reprinted in, American Bar Association 4th Annual Fall Meeting (Oct. 4, 1996, Boston, Massachusetts).

reliance on licensed site professionals (LSPs);
opportunities for quicker cleanups;
DEP oversight commensurate with the level of contamination and health risk;
DEP audits of 20 percent of the cleanup actions completed by LSPs.

Voluntary Cleanup Program

"Massachusetts has made significant progress toward addressing the problems posed by brownfield sites. Working cooperatively, the state's Executive Office of Environmental Affairs, Department of Economic Development and Department of Environmental Protection, along with the Department of Housing and Community Development, the Massachusetts Office of Business Development and the Attorney General, have developed a number of tools for cleaning up and redeveloping contaminated property in urban and industrial areas."[79]

In 1992, Massachusetts began its voluntary cleanup program. It consists of various parts:

Streamlined Waste Site Cleanup Program;
Clean Site Initiative;
Massachusetts Economic Development Incentive Program; and
  outreach and technical assistance.[80]

The Massachusetts Clean Sites Initiative encourages redevelopment of contaminated properties by providing a covenant not to sue from the Commonwealth of Massachusetts upon successful completion of the cleanup. The project is limited for use in Economic Target Areas (ETAs), which are identified by the Massachusetts Economic Assistance Coordinating Council. Exceptions may, however, be made if property not included in an ETA would provide an "exceptional economic development opportunity."[81]

To be eligible, a site must be:	
Located in one of the 30 areas defined as economic targets; or	Deemed by the Executive Office of Business Development (EOBD) as a development opportunity.

Applicants must be prospective purchasers of the property. Current owners and other PRPs are not eligible for participation in the Massachusetts program regardless of where the site is located. "A property that is not contaminated, or that

---

[79]Massachusetts Brownfields Strategy (3/13/97) at 2.

[80]*Id.* at 2-4.

[81]Abelson et al., Massachusetts, at 19 (unpublished manuscript, on file with author). For a further discussion on Massachusetts cleanups, see Terry J. Tondro, *Reclaiming Brownfields to Save Greenfields: Shifting the Environmental Risks of Acquiring and Reusing Contaminated Land,* 27 CONN. L. REV. 789 (1995).

has already been cleaned up to meet DEP's requirements, is not eligible for the Covenant."[82] The state explained:

> Covenants between the Commonwealth of Massachusetts and prospective buyers or tenants will encourage *cleanup and redevelopment of contaminated property* in areas targeted by the Commonwealth for economic development, by ending liability for releases that occurred in the past once cleanup is complete.[83]

The Clean Sites Initiative is run under the auspices of the Department of Environmental Protection, the Executive Office of Economic Affairs, and the Office of the Attorney General. A program applicant must certify that he is:

> a bona fide prospective purchaser or prospective tenant who has no legal responsibility for cleanup;
> not currently subject to pending administrative or judicial enforcement actions for compliance with environmental laws and regulations (or the applicant must disclose all such actions and must have established an approved compliance schedule); and
> willing and able to ensure that the site is cleaned up in accordance with DEP standards.

In exchange for a cleanup commitment, participants in the program receive:

> assurances that once cleanup is complete, they will be relieved of further liability for past contamination of the site;
> protection from liability for damage to natural resources on and around the site; and
> the ability to transfer the covenant to future property owners when the site is sold.

There are two situations in which a covenant can be reopened:

Before cleanup is completed:	If DEP finds that response actions have not been conducted in substantial and material compliance with the MCP.
After cleanup is completed:	If DEP finds response actions did not meet the standard of care in effect at the time they were undertaken.

To date, at least forty-four Covenants Not to Sue have been granted to prospective purchasers or tenants of contaminated sites.[84] "The program has

---

[82]Clean Sites Initiative establishes covenants to ensure cleanups, prevent future state lawsuits at hazardous waste sites (Feb. 9, 1995), http://www.magnet.state.ma.us/dep.

[83]*Id.*

[84]Massachusetts Brownfields Strategy (3/13/97) at 3.

provided an important ancillary benefit: a number of previously unknown contaminated sites have been reported to DEP, since properly notifying the agency of the presence of oil or hazardous materials is a prerequisite for participation."[85]

The state also has a Waste Site Cleanup program, which is designed for sites with "sub-standard development potential."[86] The program is designed to "encourage faster assessment and cleanup of contaminated sites."[87] Enacted as 1992 amendments to Chapter 21E, the state equivalent of CERCLA, the law "gave the private sector both more responsibility for cleanups and greater flexibility to get them done."[88] The rules for reporting assessing and cleaning up releases of contaminants were substantially revamped under the new law and took effect on October 1, 1993. According to the state:

> It is easier today for property owners to restore contaminated sites for development or sale, and for the state to focus its limited resources on the most serious environmental threats. DEP no longer directly oversees assessment and cleanup at most sites. Instead, property owners hire private environmental professionals licensed by an independent state board (known as Licensed Site Professionals or LSPs) to evaluate site conditions and oversee response actions. DEP audits the results at a portion of all sites each year to ensure adherence to state cleanup standards.[89]

The new rules include:

> clear release notification thresholds that screen out problems not likely to pose significant risks to public health or the environment;
> opportunities and incentives for cleaning up small problems quickly and reducing risks;
> generic cleanup standards for the most common contaminants, eliminating the need for detailed risk assessment and uncertainty about "how clean is clean enough";
> consideration of future land use so that sites intended for commercial or industrial development, for example, do not have to be restored to cleaner conditions required for residential development—resulting in considerable cost savings; and
> clear endpoints to the process—known as Response Action Outcomes (RAOs)—that provide citizens, lenders, prospective buyers and tenant with definitive information on what has been done at a site and why.[90]

---

[85]*Id.* at 3.

[86]*See* Charles Bartsch & Elizabeth Collaton, COMING CLEAN FOR ECONOMIC DEVELOPMENT at 75.

[87]Massachusetts Brownfields Strategy (3/13/97) at 2.

[88]*Id.* at 2.

[89]*Id.*

[90]*Id. See also* http://www.magnet.stat . . . o/dep/bwsc/progdes/htm (7/02/97 12:51:30).

As with other voluntary cleanup programs, a party's application is reviewed, and if approved, ultimately a covenant not to sue is issued. Once the application is filed, DEP and EOBD review it to determine site eligibility. They also look to see if the applicant has any outstanding environmental debts. The goal is to return applications within thirty days; although this is nearly impossible where outstanding debts exist.

The state ranks sites under both the Clean Sites Initiative and Waste Site Cleanup Program, as is depicted in the following table:

Tier	
IA	Encompasses the most serious and complex types of corrective action; Superfund-style cleanup; The Licensed Site Professional submits Phase I and Phase II site assessments for approval from DEP (this may include risk assessment and feasibility studies); remedial work plan is then developed and approved by DEP before actual remediation begins; upon completion of the remediation plan, the LSP submits a final report (which may include samples) to DEP; if approval is granted, written assurances for the owner may be given.
IB	Much more limited state oversight; LSPs are required to submit DEP permit requests before they can undertake site remediation. Once the permit is granted, the LSP initiate remediation without DEP oversight.
IC	Much more limited state oversight; LSPs are required to submit DEP permit requests before they can undertake site remediation. Once the permit is granted, the LSP initiate remediation without DEP oversight.
II	Can be managed by a LSP without DEP approvals or permits.

Cleanup commences, and when finished, the party responsible for the cleanup must prepare a Response Action Outcome (RAO) statement for the DEP. Cleanup standards used by the state require that the applicant demonstrate that there is "no significant risk" to human health or the environment. The following risk reduction levels apply:

Risk Reduction Level	For
10^{-6}	the individual
10^{-5}	the aggregate cancer-causing contaminants
Hazard waste index level of 1	non-cancer-causing contaminants

These levels may be achieved in three ways:

Method	Description
Method 1	Makes use of numerical standards derived for 107 of the most common contaminants
Method 2	Uses method 1 framework and then allows for site-specific adjustments
Method 3	Allows for a full risk assessment of the site to enable a site-specific standard to be created for the property

DEP will audit 20 percent of sites that go through the voluntary program. Affected citizens have the right, under the law, to request public participation plans from the LSP hired to remediate the site.

Upon receipt of the RAO statement, the DEP issues a certificate of completion. The certificate merely reflects DEP's receipt of the RAO statement. DEP does not actually approve any action conducted. The agency does not evaluate the cleanup. Nor does DEP protect against future site audits.[91] To the contrary, the program allows for rather liberal reopeners. These include:

if the AG finds that the response actions taken at the site did not meet the relevant standards at the time the action was undertaken; or

new releases occurred at the site after the cleanup;

releases not detected by the site assessment or addressed by the response action are later found; or

when contamination found in surrounding properties is determined to stem from the applicant's site.

Of course, the covenant will also be voided if the AG determines that it was obtained through fraud.

In the first three years of operation, the state of Massachusetts licensed 439 individuals as qualified LSPs. More than 6,200 spill response and other risk reduction actions were taken by the state. Approximately 5,500 site assessments and cleanups received LSP approvals to get out of the Massachusetts equivalent of CERCLA, the Massachusetts Contingency Plan (MCP) system. Over one-fifth of these sites had "languished for years under the old rules." More than 380 audits of private sector response actions conducted by LSPs were audited by DEP, with a 78 percent "pass" rate. Finally, more than 550 sites were required to insert institutional controls, such as deed notices or deed restrictions in the title to the site in exchange for less expensive cleanups.[92]

Financing

In 1993, Massachusetts enacted the Massachusetts Economic Development Incentive Program. This program was designed to "stimulate job creation in blighted areas, attract new businesses, encourage existing businesses to expand and increase overall economic development readiness."[93] The program is administered by the Massachusetts Office of Business Development and the Department of Housing and Community Development.

Properties within a designated Economic Opportunity Area as designated by the Economic Assistance Coordinating Council are eligible for the following benefits:

5 percent state investment tax credit;

10 percent abandoned building tax deduction;

---

[91] Abelson et al., Massachusetts, at 21.

[92] Massachusetts Brownfields Strategy (3/13/97) at 2.

[93] *Id.*

priority for state capital funding; and

municipal tax benefits (including special tax assessments or tax increment financing).

These benefits are often obtained in conjunction with a Covenant Not to Sue under the Clean Sites Initiative.[94]

Transaction Triggered Statute

Massachusetts previously enacted a statute legally requiring disclosure of environmental concerns as a condition precedent to closing a real estate transaction. The Massachusetts[95] statute governs the transfer of hazardous waste sites. It requires notice and disclosure.[96] Under the statute, no transfer or lease of land upon which or in which hazardous waste has been disposed may occur "until notice of such disposal is recorded in the registry of deeds" or in the land court for the district where the property is located.[97] The Act also limits the use of such land to a hazardous waste disposal facility until notice of such disposal is recorded.[98]

Like other states, Massachusetts has a state statute enacted along with the RCRA program requiring that some form of general notice be placed on the public record or in the chain of title to alert prospective purchasers and the public of a site's status as formerly managing hazardous waste.[99] The Massachusetts Hazardous Waste Management Act's (MHWMA)[100] provisions apply to land on which hazardous waste has been disposed.[101] If the owner conveys, leases, or devotes the property to any use other than as a facility for disposal, the Act's requirements are triggered.[102] Once triggered, the owner must record a notice that the land has been used for disposal of hazardous waste in the Registry of Deeds.[103] If the land is already registered, the notice must be recorded in the registry section of the Land Court for the district prior to conveyance.[104] Violation of MHWMA might result in civil and criminal penalties.[105]

No transfer or lease of land upon which or in which hazardous waste has been disposed may occur "until notice of such disposal is recorded in the registry of

---

[94]Massachusetts Brownfields Update (Autumn 1995), reprinted in KEY ISSUES IN U.S. EPA REGION I (Oct. 2, 1996, Boston, Massachusetts).

[95]MASS. ANN. LAWS ch. 21c, § 7.

[96]The other states include: Illinois, Minnesota, Missouri, New York, Pennsylvania, and West Virginia.

[97]MASS. GEN. LAWS ch. 21C, § 7.

[98]Id.

[99]MASS. GEN. LAWS ANN., ch. 21C, § 4(g); MASS. REGS. CODE tit. 310, § 30.595.

[100]Id.

[101]Id. at § 7.

[102]Id.

[103]Id.

[104]Id.

[105]Id. at § 9.

deeds" or in the land court for the district where the property is located.[106] The Act also limits the use of such land to a hazardous waste disposal facility until notice of such disposal is recorded.[107]

Specialized Stakeholder Laws—Lender

Massachusetts was one of the states that amended its state equivalent of CERCLA to include some specialized lender liability protection.

A secured lender will not be deemed an owner or operator under Massachusetts law if the secured lender did not cause or contribute to the release or threat of release or causes the release or threat of release to become worse than it otherwise would have been. In addition, the secured lender must have, before acquiring ownership or possession of the site or vessel, not participated in the management of the site or vessel. For purposes of this definition, participation in the management of the site or vessel will include substantially divesting from the borrower or any other person possession or control over those aspects of the operations involving the management of oil or hazardous materials. Participation in management will not include doing any combination of one or more of the following:

(1) requiring or conducting assessments of the site or vessel or any portion thereof;

(2) engaging in the regular or periodic monitoring of the business conducted at or on the site or vessel;

(3) making the provision or continuation of financing conditional on covenants, representations, or warranties concerning any combination of one or more of the following: the proper handling, use, storage, transport, or disposal of oil or hazardous materials at or from the site or vessel; the timely and proper response to releases or threats of release of oil or hazardous materials at or from the site or vessel; the maintenance of the site or vessel in compliance with this chapter and other applicable laws or regulations; or the periodic submission of information concerning oil or hazardous materials at or from the site or vessel;

(4) providing periodic advice, information, guidance, or direction concerning the general business and financial aspects of a borrower's operations, excluding advice, guidance, or direction concerning those aspects of the borrower's operations involving the management of oil or hazardous materials . . . ;

(5) providing general information concerning a borrower's obligations pursuant to this chapter and the Massachusetts Contingency Plan and information on how the borrower may identify and select a waste site cleanup

---

[106]MASS. GEN. LAWS ch. 21C, § 7.
[107]*Id.*

professional license . . . and general information concerning a borrower's obligations pursuant to other federal, state or local laws concerning the transportation, storage, treatment, and disposal of hazardous waste or hazardous materials; and

(6) engaging in financial workouts, restructuring, or refinancing of the borrower's obligation, or undertaking activities to protect or preserve the value of the security interest in a site or vessel excluding activities involving those aspects of the borrower's operations involving the management of oil or hazardous materials.

After acquiring ownership or possession of the site or vessel, and upon obtaining knowledge of a release or threat of release of oil or hazardous material, the secured lender must notify the department immediately upon obtaining knowledge of a release or threat of release for which notification is required. The secured lender must also provide reasonable access to the site or vessel to employees, agents, and contractors of the department to conduct response actions, and to other persons intending to conduct necessary response actions. In addition, the secured lender must undertake reasonable steps to: (1) prevent the exposure of persons to oil or hazardous materials by fencing or otherwise preventing access to the site or vessel, and (2) contain[108] the further release or threat of release of oil or hazardous materials from a structure or container. If the secured lender elects to voluntarily undertake a response action or portion of a response action at the site or vessel, the secured lender conducts such response action or portion of a response action in compliance with the law. Finally, the secured lender must act diligently to sell or otherwise divest itself of ownership or possession of the site or vessel.

Whether the secured lender is acting or has acted diligently to sell or otherwise divest itself of ownership or possession of the site or vessel will be determined as follows. During the first eighteen months after the secured lender first acquired ownership or possession of the site or vessel, whichever occurs earlier, there will be a presumption that the secured lender is acting diligently to sell or otherwise divest itself of ownership or possession of the site or vessel. This presumption may be rebutted by a preponderance of the evidence.

If the secured lender has not divested itself of ownership or possession of the vessel or site after the expiration of the eighteen-month period, then during the next forty-two months, the burden of proof will be on the secured lender to demonstrate by a preponderance of the evidence that the secured lender is acting diligently to

---

[108]The act defines *contain* or *containment* as:

actions taken in response to a release or threat of release of oil or hazardous material into the environment to prevent or minimize such release so that it does not migrate or otherwise cause or threaten substantial danger to present or future public health, safety, welfare or the environment. The term shall also include security measures, including, without limitation, the building of fences for the purpose of limiting and restricting access to a site or vessel where there has been a release or there is a threat of a release of oil or hazardous materials.

sell or otherwise divest itself of ownership or possession of the site or vessel. In determining whether or not the secured lender is acting diligently to sell or otherwise divest itself of ownership or possession of the site or vessel, the following factors will be considered:

(1) the use or uses to which the site or vessel was put or is being put during the period in question;

(2) market conditions;

(3) the extent of contamination of the site or vessel and the impacts of such contamination on marketability of the site or vessel;

(4) the applicability of, and compliance by such person with, federal and state requirements relevant to sale or divestment of property in which such person holds or formerly held a security interest; and

(5) legal constraints on sale or divestment of ownership or possession.

If the secured lender has not divested itself of ownership or possession of the site or vessel within five years after the secured lender first acquired ownership or possession, whichever occurred earlier, and can establish by a preponderance of the evidence that it satisfies its burden of proof that it acted to dispose of the property, then no liability will be imposed. A secured lender who can establish by a preponderance of the evidence that it satisfies the conditions required to show diligent efforts to dispose of the property, but who has not divested itself of ownership or possession of the site or vessel within five years after the secured lender first acquired ownership or possession, and whose property has been the site of a release of oil or hazardous material for which the department has incurred costs for assessment, containment, or removal will be liable to the commonwealth only to the extent of the value of the property following the department's assessment, containment, and response actions, less the total amount of costs reasonably paid by said person for carrying out assessment, containment, and response actions in compliance with the Massachusetts Contingency Plan and all other applicable requirements of state law.

The Massachusetts law protects secured lenders from liability as an owner or an operator, but yields no protections if the secured lender acted as an arranger[109] or transporter under the state hazardous waste laws. In addition, the exclusion from owner or operator liability applies only with respect to releases and threats of release that first begin to occur before such secured lender acquires ownership or possession of the site or vessel. A secured lender will be deemed an owner or operator with respect to any release or threat of release that first begins to occur at or from a site or vessel during the time that such secured lender has ownership or possession of it for any purpose. A secured lender will be deemed an owner or

---

[109]*See* Ashland Oil v. Sanford Products Corp., 810 F. Supp. 1057 (D. Minn. 1993) (court rejected argument that lender was liable as an arranger because it facilitated the sale of assets from one debtor to another).

operator of an abandoned site if the secured lender owned, operated, or held ownership or possession of the site immediately prior to the abandonment.[110]

For purpose of this law, *secured lender* is defined as "a person who holds indicia of ownership in a site or vessel primarily to protect that person's security interest in said site or vessel." The definition also includes:

(1) two persons when one holds indicia of ownership in a site or vessel primarily to protect the other person's security interest in that site or vessel if the person holding the indicia of ownership is (a) wholly owned by the person holding the security interest, or (b) an affiliate of the person holding the security interest and both are wholly owned, directly or indirectly, by the same person; and

(2) persons who hold contractual participation rights in a security interest, and any of the following that hold indicia of ownership in a site or vessel primarily to protect that security interest: a wholly owned subsidiary of any such person; an affiliate of any such person if both are wholly owned, directly or indirectly, by the same person; and any entity formed among such persons, subsidiaries, or affiliates.[111]

---

[110]Under the statute:

"Statement of claim" or "statement," an instrument signed by the commissioner, describing a particular site or sites or vessel or vessels and naming the person or persons then deemed by the commissioner to be liable under the provisions of this chapter with respect to each such site or vessel and their residential addresses, to the extent known to the commissioner, and declaring a lien upon the property of such person or persons for the payment of amounts due or to become due from such person or persons to the commonwealth under the provisions of this chapter; provided, however, that neither failure to state any such address nor the designation of an incorrect address shall invalidate such statement; and provided further, that successive statements, naming other persons so deemed to be liable, may be issued; and provided further, that if the property in question is owned or possessed by a fiduciary or a secured lender, who is not liable under this chapter with respect to the site or vessel in question, the mention of the fiduciary or the secured lender in the statement as a person who owns or possesses the site or vessel shall not constitute a finding or evidence that such person is liable under this chapter, and the lien shall only be on the property in question and not upon all property of such fiduciary or secured lender.

[111]In addition, a fiduciary will not be deemed an owner or operator if all of the following requirements are met:

(1) No act of the fiduciary causes or contributes to the release or threat of release or causes the release or threat of release to become worse than it otherwise would have been.

(2) After acquiring title to or commencing control or management of the site or vessel, the fiduciary satisfies all of the following conditions:

(A) the fiduciary notifies the department immediately upon obtaining knowledge of a release or threat of release for which notification is required pursuant to, and in compliance with, section seven or regulations promulgated pursuant thereto,

(B) the fiduciary provides reasonable access to the site or vessel to employees, agents, and contractors of the department to conduct response actions, and to other persons intending to conduct necessary response actions,

Other protections accorded lenders in Massachusetts[112] include the right to be notified by the state when the department takes or arranges for such response actions it deems necessary. Prior to undertaking any response action, the department must notify the owner or operator of the site or vessel or a fiduciary or secured lender that has title to or possession of a site or vessel of its intent to take such action. Notice is not, however, required when the department does not know the identity or location of the owner or operator or of a fiduciary or secured lender

---

    (C) the fiduciary notifies the department immediately upon obtaining knowledge or reasonable basis for believing that there are not sufficient assets available in the estate or trust to satisfy one or more of the conditions set forth in subclause (D) of clause (2) of this subsection of this definition,

    (D) upon obtaining knowledge of a release or threat of release of oil or hazardous material, and to the extent that assets are available in the estate or trust,

        (i) the fiduciary undertakes reasonable steps to (a) prevent the exposure of persons to oil or hazardous materials by fencing or otherwise preventing access to the site or vessel, and (b) contain the further release or threat of release of oil or hazardous materials from a structure or container, and

        (ii) if there is significant evidence of an imminent hazard to public health, safety, welfare, or the environment from oil or hazardous materials at or from the site or vessel, the fiduciary takes action to control the potential for health damage, human exposure, safety hazards, and environmental harm through appropriate short term measures,

    (E) if the fiduciary elects to voluntarily undertake a response action or portion of a response action at the site or vessel, the fiduciary conducts such response action or portion of a response action in compliance with the requirements of this chapter and the Massachusetts Contingency Plan, and

    (F) in the case of a fiduciary who is acting pursuant to an assignment made for the benefit of creditors, the fiduciary notifies the department in writing of the assignment immediately upon his acceptance thereof.

A fiduciary who takes any action referred to in clauses (b) (2) (A) through (D) of this definition shall not be deemed an owner or operator solely because said fiduciary took such action.

Nothing in this chapter shall preclude claims against a fiduciary solely in the fiduciary's representative capacity.

Nothing in clause (b) (2) (D) of this definition shall require a fiduciary to utilize any funds other than the assets of the estate or trust to satisfy the conditions set forth in clause (b) (2) (D) of this definition, provided that all of the conditions of clause (b) of this definition are met.

For purposes of this chapter, the assets of the estate or trust shall include: (i) any assets in the estate or trust of which the site or vessel is a part (hereafter "the estate or trust"); (ii) any assets that are, at or subsequent to the time the fiduciary obtained knowledge of a release or threat of release, placed in any other estate or trust by action of or on behalf of a settlor or grantor of the estate or trust of which the site or vessel is a part when such settlor or grantor is or was an owner or operator of the site or vessel, if such other estate or trust is or was controlled or managed by the fiduciary of the estate or trust of which the site or vessel is a part; and (iii) any assets that are at or subsequent to the time the fiduciary obtained knowledge of a release or threat of release, transferred by the fiduciary out of the estate or trust of which the site or vessel is a part for less than full and fair consideration, as determined by the fiduciary in good faith.

Nothing in this definition shall affect the liability, responsibilities, or rights of a grantor or beneficiary of an estate or trust, or of any person other than a fiduciary acting solely in his representative capacity, pursuant to this chapter or any other law.

[112]MASS. ANN. LAWS ch. 21E, 4.

that has title to or possession of a site or vessel, or when because of an emergency or other circumstances, the giving of notice would be impractical.[113]

Specialized Stakeholder Laws—Trustee

Massachusetts was one of the states that amended its state equivalent of CERCLA to include some specialized trustee liability protection.

---

[113]Another concern of lenders is, of course, environmental lien statutes. The Massachusetts environmental lien statute provides:

Any liability to the commonwealth under this chapter shall constitute a debt to the commonwealth. Any such debt, together with interest thereon at the rate of twelve percent per annum from the date such debt becomes due, shall constitute a lien on all property owned by persons liable under this chapter when a statement of claim naming such persons is recorded, registered or filed. If a fiduciary or a secured lender has title to or possession of the property, and if the fiduciary or secured lender is not a person liable under this chapter when a statement of claim is recorded, registered or filed, such debt, together with interest thereon at the rate of twelve percent per annum from the date such debt becomes due, shall constitute a lien on the property in question when a statement of claim describing such property is duly recorded, registered or filed. If the site described in such statement comprises real property, such lien shall be effective when duly recorded and indexed in the grantor index in the registry of deeds or registered in the registry district of the land court for the county or district wherein the land lies so as to affect its title, and describes the land by metes and bounds or by reference to a recorded or registered plan showing its boundaries. In addition, such lien shall be effective with respect to such other real property owned by such person when notice thereof is duly recorded and indexed in the registry of deeds or registered in the registry district of the land court for the county or district wherein any such other land lies. If the site described in such statement is personal property, whether tangible or intangible, the statement shall be filed in accordance with the provisions of section 9-401 of chapter one hundred and six. Any such statement shall be sufficient if executed or approved by the commissioner of the department. Any lien recorded, registered or filed pursuant to this section shall have priority over any encumbrance theretofore recorded, registered or filed with respect to any site, other than real property the greater part of which is devoted to single or multi-family housing, described in such statement of claim, but as to other real property shall be subject to encumbrances or other interests recorded, registered or filed prior to the recording, registration or filing of such statement, and as to all other personal property shall be subject to the priority rules of said chapter one hundred and six. Such lien shall continue in force with respect to any particular real or personal property until a release of the lien signed by the commissioner is recorded, registered or filed in the place where the statement of claim as to such property affected by the lien was recorded, registered or filed. In addition to discretionary releases of liens, the commissioner shall forthwith issue such a release in any case where the debt for which such lien attached, together with interest and costs thereon, has been paid or legally abated. If no action to enforce or foreclose the lien is brought by the deadline prescribed in subsection (1) of section eleven A, the lien shall be dissolved after said deadline. This section shall not apply to any property, real or personal, tangible or intangible, any money, fees, charges, revenues or otherwise, owned payable to or by, held in trust by or for, or otherwise owned, operated or managed by the Massachusetts Municipal Wholesale Electric Company established pursuant to chapter seven hundred and seventy-five of the acts of nineteen hundred and seventy-five, Massachusetts municipal light departments organized under chapter one hundred and sixty-four or any other special law, or with respect to any property real or personal whatsoever of municipal light departments administered pursuant to chapters forty-four and one hundred and sixty-four A. Notwithstanding the foregoing, the aforesaid Massachusetts Municipal Wholesale Electric Company and municipal light departments shall use their authority as provided by applicable statutes to assess, contain or remove any such oil or hazardous

Under Massachusetts law, a fiduciary will not be deemed an owner or operator if no act of the fiduciary causes or contributes to the release or threat of release or causes the release or threat of release to become worse than it otherwise would have been. Nor will a fiduciary be held liable if, after acquiring title to or commencing control or management of the site or vessel, the fiduciary:

Step	Discussion
(1)	notifies the department immediately upon obtaining knowledge of a release or threat of release for which notification is required;
(2)	provides reasonable access to the site or vessel to employees, agents, and contractors of the department to conduct response actions, and to other persons intending to conduct necessary response actions; and
(3)	notifies the department immediately upon obtaining knowledge or reasonable basis for believing that there are not sufficient assets available in the estate or trust to undertake reasonable steps to: (a) prevent the exposure of persons to oil or hazardous materials by fencing or otherwise preventing access to the site or vessel, and (b) contain the further release or threat of release of oil or hazardous materials from a structure or container.

material release for which they are responsible under this chapter. The provisions of this section shall apply to any site or vessel which has been the subject of a response action and which is owned or possessed by a fiduciary or secured lender, except that nothing in this section shall be deemed to allow the commonwealth to take any action otherwise authorized by this section with respect to any property while it is owned or possessed by a secured lender except to (1) record, register or file a lien or release of a lien as provided in this section, (2) in the case of real property, foreclose upon a lien and subsequently sell the property in question in accordance with the procedures set forth in chapter two hundred and forty-four, and (3) in the case of personal property, take possession and sell, lease or otherwise dispose of the secured property in accordance with the procedures for the disposition of collateral set forth in part five of section nine of chapter one hundred and six. If the property is sold for less than the amount of the lien, a secured lender who meets the requirements of clause (b) of the definition of owner or operator in section two shall not be deemed an owner or operator of the site or vessel in question and shall not be liable to the commonwealth for the deficiency.

Mass. Ann. Laws ch. 21E, § 13 (1993). *See also* Acme Laundry Co. v. Secretary of Envtl. Affairs, 575 N.E.2d 1086 (Mass. 1991) (liability for release of hazardous materials is created when commonwealth engages in information-gathering activities relative to assessment, containment, and removal of releases, and if costs are incurred in these activities, lien may be placed on all properties of person responsible for release); Chicago Title Ins. Co. v. Kumar, 506 N.E.2d 154 (Mass. App. Ct. 1987) (owner of property in which there had been release of hazardous material is liable to commonwealth for cleanup of site, after commonwealth has expended funds on owner's behalf, at which point commonwealth can attach lien by recording statement of claim in appropriate registry of deeds); Chicago Title Ins. Co. v. Kumar, 506 N.E.2d 154 (Mass. App. Ct. 1987) (release of hazardous material on insured property at time of conveyance of title without recording of notice of existence of hazardous waste, and power of commonwealth to impose future lien on property for cleanup expense, are not defects in or lien or encumbrance on title within meaning of title insurance policy and do not make title unmarketable). *See generally* Mendler, Massachusetts Conveyancers' Handbook with Forms §§ 6:1-6:10, Title Insurance; Mendler, Massachusetts Conveyancers' Handbook with Forms, §§ 13:1-13:9, Environmental Controls; Frederic and Baird, *Introduction: A Brief Overview of Federal and Massachusetts Environmental Laws*, 75 Mass. L. Rev. 91, (1990); Holt and Stearns, *A Survey of Recent Environmental Decisions in Massachusetts and the First Circuit*, 75 Mass. L. Rev. 103 (1990).

If there is significant evidence of an imminent hazard to public health, safety, welfare, or the environment from oil or hazardous materials at or from the site or vessel, the fiduciary takes action to control the potential for health damage, human exposure, safety hazards, and environmental harm through appropriate short-term measures. If the fiduciary elects to voluntarily undertake a response action or portion of a response action at the site or vessel, the fiduciary conducts such response action or portion of a response action in compliance with the requirements of this chapter and the Massachusetts Contingency Plan.

In the case of a fiduciary who is acting pursuant to an assignment made for the benefit of creditors, the fiduciary must notify the department in writing of the assignment immediately upon his acceptance thereof.

A fiduciary who undertakes activities to prevent the exposure of persons to oil or hazardous materials by fencing or otherwise preventing access to the site or vessel or acts to contain the further release or threat of release of oil or hazardous materials from a structure or container will not be deemed an owner or operator solely because said fiduciary took such action.

The law does not protect the fiduciary from or preclude claims against a fiduciary solely in the fiduciary's representative capacity. Neither does the statute require a fiduciary to utilize any funds other than the assets of the estate or trust to maintain the safety and cleanup of the site.

The statute defines assets of the estate or trust as including:

(1) any assets in the estate or trust of which the site or vessel is a part (hereafter "the estate or trust");

(2) any assets that are, at or subsequent to the time the fiduciary obtained knowledge of a release or threat of release, placed in any other estate or trust by action of or on behalf of a settlor or grantor of the estate or trust of which the site or vessel is a part when such settlor or grantor is or was an owner or operator of the site or vessel, if such other estate or trust is or was controlled or managed by the fiduciary of the estate or trust of which the site or vessel is a part; and

(3) any assets that are at or subsequent to the time the fiduciary obtained knowledge of a release or threat of release, transferred by the fiduciary out of the estate or trust of which the site or vessel is a part for less than full and fair consideration, as determined by the fiduciary in good faith.

The law dictates that the definition will not "affect the liability, responsibilities, or rights of a grantor or beneficiary of an estate or trust, or of any person other than a fiduciary acting solely in his representative capacity, pursuant to this chapter or any other law."[114]

---

[114]The law states:

"Statement of claim" or "statement," an instrument signed by the commissioner, describing a particular site or sites or vessel or vessels and naming the person or persons then deemed by the commissioner to be liable under the provisions of this chapter with respect to each such site or vessel

"When a fiduciary[115] who is not an owner or operator pursuant to this definition has title or control or management of a site or vessel, the grantor or settlor of the estate or trust in question, to the extent the assets of the estate or trust are insufficient to pay for liability." The term shall include any estate or trust of which the site or vessel is a part.

Massachusetts law also protects fiduciaries by requiring that, prior to undertaking any response action, the department must notify the owner or operator of the site or vessel or a fiduciary or secured lender that has title to or possession of a site or vessel of its intent to take such action.[116] The notice is not required when the department does not know the identity or location of the owner or operator or of a fiduciary or secured lender that has title to or possession of a site or vessel, or when because of an emergency or other circumstances, the giving of such notice would be impractical.[117]

---

and their residential addresses, to the extent known to the commissioner, and declaring a lien upon the property of such person or persons for the payment of amounts due or to become due from such person or persons to the commonwealth under the provisions of this chapter; provided, however, that neither failure to state any such address nor the designation of an incorrect address shall invalidate such statement; and provided further, that successive statements, naming other persons so deemed to be liable, may be issued; and provided further, that if the property in question is owned or possessed by a fiduciary or a secured lender, who is not liable under this chapter with respect to the site or vessel in question, the mention of the fiduciary or the secured lender in the statement as a person who owns or possesses the site or vessel shall not constitute a finding or evidence that such person is liable under this chapter, and the lien shall only be on the property in question and not upon all property of such fiduciary or secured lender.

[115]*Fiduciary* is defined as a person: (a) who is acting in any of the following capacities: an executor or administrator as defined in section one of chapter one hundred ninety-seven, including a voluntary executor or a voluntary administrator; a guardian; a conservator; a trustee under a will or inter vivos instrument creating a trust under which the trustee takes title to, or otherwise controls or manages, property for the purpose of protecting or conserving such property under the ordinary rules applied in the courts of the commonwealth; a court-appointed receiver; a trustee appointed in proceedings under federal bankruptcy laws; an assignee or a trustee acting under an assignment made for the benefit of creditors pursuant to sections forty through forty-two of chapter two hundred and three; or a trustee, pursuant to an indenture agreement or similar financing agreement, for debt securities, certificates of interest of participation in any such debt securities, or any successor thereto, and (b) who holds legal title to, controls, or manages, directly or indirectly, any site or vessel as a fiduciary for purposes of administering an estate or trust of which such site or vessel is a part. MASS. ANN. LAWS ch. 21E, § 2.

[116]MASS. ANN. LAWS ch. 21E, § 4.

[117]*See* Griffith v. New England Tel. and Telegraph Co., 610 N.E.2d 944 (Mass. 1993), *superseding*, 585 N.E.2d 751 (1992) ("owner or operator" of site liable for costs of contamination cleanup is present owner or operator); Acme Laundry Co. v. Secretary of Envtl. Affairs, 575 N.E.2d 1086 (Mass. 1991) (commonwealth can recover costs for investigation, planning, monitoring, and supervision of cleanup activities of party responsible for release of hazardous materials); Sanitoy, Inc. v. Ilco Unican Corp., 602 N.E.2d 193 (Mass. 1992) (ALM GL c 21E § 15 authorizes trial judge to award attorney's fees and expert witness fees in private party suit for reimbursement brought pursuant to § 4 of Massachusetts Oil and Hazardous Material Release Prevention Act); Sanitoy, Inc. v. Ilco Unican Corp., 602 N.E.2d 193 (Mass. 1992) (where plaintiff recovered 30 percent of its response costs from the defendant, who was responsible for 30 percent of contamination of plaintiff's land, judge erred in awarding to plaintiff only 30 percent of its reasonable attorney's fees and expert witness fees, since plaintiff was entitled to

Environmental Lien

Under Massachusetts law,[118] any liability to the state under its equivalent of CER-CLA constitutes a debt to the commonwealth.[119] Any such debt, together with interest thereon at the rate of 12 percent per annum from the date such debt becomes due, will constitute a lien on all property owned by persons liable under this chapter when a statement of claim naming such persons is recorded, registered, or filed.

recover full amount of its reasonable attorney's fees and expert witness fees); Sanitoy, Inc. v. Ilco Unican Corp., 602 N.E.2d 193 (Mass. 1992) (while ALM GL c 21E § 4, authorizing private right of action for reimbursement for costs of assessment, containment, and removal of contamination, does not provide for recovery of attorney's fees or expert witness fees, § 15 does provide for recovery of such fees); Oliveira v. Pereira, 605 N.E.2d 287 (Mass. 1992) (action for reimbursement of reasonable costs of assessment, containment, and removal of oil or hazardous material from contaminated site sounds in tort, and three-year statute of limitations under ALM GL c 260 § 2A applies; proper date of accrual of action for reimbursement of costs of assessment, containment, and removal is date costs were paid, not when plaintiff knew property was contaminated); Griffith v. New England Tel. and Telegraph Co., 610 N.E.2d 944, *superseding*, 585 N.E.2d 751 (Mass. App. Ct. 1992) (ALM GL c 21E § 4 creates private right of action whereby any person who undertakes containment and removal of oil or hazardous material can recover cleanup costs from person responsible for contamination; former lessee of property, which had stored gasoline and fuel oil in three underground tanks was not "owner or operator" under ALM GL c 21E § 2 and thus was not liable for costs of contamination cleanup); Garweth Corp. v. Boston Edison Co., 613 N.E.2d 92 (Mass. 1993) (there is no cause of action under ALM GL c 21E § 5 permitting recovery for economic loss not directly resulting from environmental damage; consequently, corporation whose property was not contaminated or damaged by release of oil had no right to recover for economic loss in nature of delay in completion of contract with third party whose property was contaminated by release of oil); Garweth Corp. v. Boston Edison Co., 613 N.E.2d 92 (Mass. 1993) (plaintiff has private right of action to enforce purpose of Massachusetts Oil and Hazardous Material Release Prevention Act); Sheehy v. Lipton Indus., Inc., 507 N.E.2d 781 (Mass. App. Ct. 1987) (owner of land not required to wait until Department of Environmental Quality Engineering has approved correction of hazardous material problem or sought damages before asserting claim under GL c 21E against prior owner for reimbursement for costs of assessment, containment, and removal; GL c 21E § 4 creates private right of action to enforce provisions of Massachusetts Oil and Hazardous Material Release Prevention Act; summary judgment should not have been entered for seller of land, where buyer stated in affidavit that he expended substantial funds for assessment and removal of hazardous material from land); Griffith v. New England Tel. & Tel. Co., 585 N.E.2d 751, *modified,* Mass. App. LEXIS 430 (Mass. App. 1992) and *review denied,* 412 Mass. 1104 (parties to commercial real estate transaction may not by contract between them allocate payment of liability imposed by ALM GL c 21E for release of hazardous waste); Bortone v. LeConti, 1990 Mass. App. Div. 159 (in contract action to recover deposit paid by plaintiffs pursuant to agreement for purchase and sale of real estate, where seller delivered to buyers document entitled "site report relative to hazardous material" signed by registered, professional engineer, site report provided by seller did not constitute "certification" that property in question was not in violation of ALM GL c 21E as required by parties' agreement; site report stated no conclusion or express assurance on question of threat of release of hazardous materials into environment, which is principal concern of ALM GL c 21E); Hays v. Mobil Oil Corp., 736 F. Supp. 387 (D. Mass. 1990) (estate of deceased gas station owner could not recover indemnity or contribution from oil company with which owner had retail dealer contract under either Massachusetts Oil and Hazardous Material Release Prevention Act [ALM GL c 21E §§ 1 *et seq.*] or unfair trade practices provisions [ALM GL c 93A § 1 *et seq.*] where contract required owner to indemnify oil company against all losses and claims for property damage and oil company's refusal to pay and its commencement of litigation

If a fiduciary or a secured lender has title to or possession of the property, and if the fiduciary or secured lender is not a person liable when a statement of claim is recorded, registered, or filed, such debt, together with interest thereon at the rate of 12 percent per annum from the date such debt becomes due, will constitute a lien on the property in question when a statement of claim describing such property is duly recorded, registered, or filed.

If the site comprises real property, the lien will be effective when duly recorded and indexed in the grantor index in the registry of deeds or registered in the registry district of the land court for the county or district wherein the land lies. The document must affect its title and describes the land by metes and bounds or by reference to a recorded or registered plan showing its boundaries. In addition, the lien will be effective with respect to such other real property owned by such person when notice thereof is duly recorded and indexed in the registry of deeds or registered in the registry district of the land court for the county or district wherein any such other land lies. Thus, Massachusetts[120] is one of a minority of states that enacted a superlien.[121]

If the site is personal property, whether tangible or intangible, the statement will be filed in the UCC records. Any such statement will be sufficient if executed or approved by the commissioner of the department. Any lien recorded, registered, or filed pursuant to this section will have priority over any encumbrance theretofore recorded, registered, or filed with respect to any site, other than real property the greater part of which is devoted to single or multifamily housing, described in such statement of claim, but as to other real property will be subject to encumbrances or other interests recorded, registered, or filed prior to the recording, registration or filing of such statement, and as to all other personal property will be subject to the priority rules of the UCC.

The lien will continue in force with respect to any particular real or personal property until a release of the lien signed by the commissioner is recorded,

---

did not rise to level of rascality or unfairness necessary to stage deceptive trade practices claim); Wellesley Hills Realty Trust v. Mobil Oil Corp., 747 F. Supp. 93 (D. Mass. 1990) (in action against oil corporation for contamination of real property, plaintiff stated claim under ALM GL c 21E since plaintiff alleged not only that corporation owned site, but that it stored oil in underground tank, that this tank sprang leak, and that corporation did not clean up spillage; these factual allegations supported plaintiff's allegation that corporation "caused" release of oil); Dedham Water Co. v. Cumberland Farms Dairy, Inc., 770 F. Supp. 41 (D. Mass. 1991), 33 Env't. Rep. Cas. (BNA) 1607, 21 Envtl. L. Rep. (Envtl. L. Inst.) 21332, *aff'd,* 1992 U.S. App. LEXIS 19256 (1st Cir.) (waterworks companies may not recover from nearby dairy cost of building and operating water purification plant to remove pollutants from waterworks' well field pursuant to Massachusetts Oil and Hazardous Material Release Prevention and Response Act [ALM GL c 21E §§ 4, 5], where dairy had released volatile organic chemicals onto its property, because actual pollution of well field was not caused by releases from dairy and while dairy's discharges potentially threatened well field purification plant was not built in response to that threat).

[118]Mass. Gen. Laws ch. 21E, § 13 (1993).

[119]*Id.*

[120]Mass. Gen. Laws ch. 21E, § 13.

[121]The other states include: Connecticut, Louisiana, Maine, Massachusetts, New Hampshire, and New Jersey.

registered, or filed in the place where the statement of claim as to such property affected by the lien was recorded, registered, or filed. In addition to discretionary releases of liens, the commissioner will forthwith issue such a release in any case where the debt for which such lien attached, together with interest and costs thereon, has been paid or legally abated. If no action to enforce or foreclose the lien is brought by the deadline prescribed, the lien will be dissolved after said deadline.

The environmental lien does not apply to any property, real or personal, tangible or intangible, any money, fees, charges, revenues or otherwise, owned payable to or by, held in trust by or for, or otherwise owned, operated or managed by the Massachusetts Municipal Wholesale Electric Company established pursuant to state law, Massachusetts municipal light departments organized under state law, or with respect to any property real or personal whatsoever of municipal light departments administered pursuant to state law. Nevertheless, the aforesaid Massachusetts Municipal Wholesale Electric Company and municipal light departments are required by law to use their authority as provided by applicable statutes to assess, contain, or remove any such oil or hazardous material release for which they are responsible under this chapter.

The lien law applies to any site or vessel that has been the subject of a response action and that is owned or possessed by a fiduciary or secured lender, except that nothing in the lien law allows the commonwealth to take any action otherwise authorized by this section with respect to any property while it is owned or possessed by a secured lender except to: (1) record, register or file a lien or release of a lien, (2) in the case of real property, foreclose upon a lien and subsequently sell the property in question in accordance with the procedures set forth in chapter two hundred and forty-four, and (3) in the case of personal property, take possession and sell, lease or otherwise dispose of the secured property in accordance with the procedures for the disposition of collateral. If the property is sold for less than the amount of the lien, a secured lender who meets the requirements of the secured creditor exemption, as defined by state law, will not be deemed an owner or operator of the site or vessel in question and will not be liable to the commonwealth for the deficiency.

New Hampshire

Voluntary Cleanup Program

New Hampshire's Site Management Program began as an informal program, operated under the state's enforcement discretion, in 1992. In 1996, the state legislature codified the existing "brownfields program."[122] The state claims it had at least 300 sites that entered the program even before the state enacted its most recent "brownfields" legislation.

---

[122]N. H. REV. STAT. ANN. § 147-F:1 (1996).

Although the program was purported to be quite effective even in the absence of legislation, the new law sought to (1) "give incentives to parties interested in the redevelopment of contaminated properties" and (2) further "expedite the voluntary cleanup" process.[123]

Application for participation[124]

|

DES eligibility determination within thirty days[125]

|

Site investigation report and remedial action plan developed[126]

|

DES Notice of Approved Remedial Action Plan[127]

|

DOJ Covenant Not to Sue

|

Implement remedial action

|

DES issues Certificate of Completion

The program is run by the Department of Environmental Services (DES).[128] Any site is eligible[129] unless it is listed on the NPL, categorized as a LUST site, or contains a landfill. Any person who did not cause the existing contamination on the property may apply for the program.[130] These specifically include:

(1) prospective purchasers;
(2) current property owners if they did not cause or contribute to the site contamination (even if the existing owners would otherwise be considered a PRP);
(3) secured creditors or mortgage holders; or
(4) municipalities owed real estate taxes on the property.

The formal application has recently been codified by statute. Interested persons seek an eligibility determination from the state. To do so, the party must submit a certificate under oath including all the data deemed by the state to be

---

[123]*Id.*

[124]*See* § 147-F:9 (1996) (application for participation in Covenant Not to Sue program).

[125]§ 147-F:10 (1996) (eligibility determination with respect to Covenant Not to Sue).

[126]§ 147-F:11 (1996) (site investigation and remedial action plan).

[127]§ 147-F:12 (1996).

[128]*Id.* § 147-F:2 (1996).

[129]The term "eligible person" is defined under the act in § 147-F:3 (1996).

[130]*See id.* § 147-F:4 (1996). *See also* Environmental Fact Sheet, New Hampshire's "Brownfields" Program, the Voluntary Clean-up Program (Sept. 1995), reprinted in KEY ISSUES IN U.S. EPA REGION I (Oct. 2, 1996, Boston, Massachusetts).

pertinent in verifying eligibility.[131] DES then makes an eligibility determination. For purposes of this initial inquiry, at least the following must be submitted:[132]

(1) a signed, completed application form (provided by DES);
(2) all supporting information required as part of the application package;
(3) an environmental site assessment report (this may also include submittal of an initial characterization report or site investigation and/or remedial action plan for sites that are further along in the investigation and cleanup process);
(4) a nonrefundable application fee of $500.[133]

After receiving the application package, DES will provide a completeness determination within ten days. If the application is found to be complete, DES will provide a written notice of its eligibility determination within thirty days. After an eligibility determination is made, the applicant must make an additional nonrefundable program participation fee of $3,000.[134] The following chart briefly describes the process begun with eligibility determination:

Stage	Description
Eligibility Determination	DES makes eligibility determination within thirty days of finding application complete.
Remedial Action Plan (RAP)	After a work plan is approved by DES, the eligible person will perform necessary investigations and data analysis. If there is no contamination, then a NFA may be issued. If contamination is found a RAP must be developed.
Notice of Approved RAP	RAP describes the proposed actions to clean up the site. RAP is submitted to DES for approval. DES issues a Notice of Approved Remedial Action Plan, which must be recorded in the registry of deeds by the eligible person. DOJ issues a Covenant Not to Sue, which may contain conditions relative to the required actions at the site.
Written Assurances	Upon completion of active site cleanup and DES approval of a completion report, DES will issue a certificate of completion. This may include use restrictions, environmental monitoring requirements, and routine maintenance requirements.[135] This certificate will also be recorded in the county deed of registry.[136]

Very often DES requests that the party prepare an investigative work plan. For this purpose, DES requires the parties to use the standard promulgated for the

---

[131]*Id.* at § 147-F:4 (III).
[132]*Id.*
[133]§ 147-F:14 (1996) (fee schedule).
[134]*Id.*
[135]§ 147-F:15 (1996) (restrictions on future property use).
[136]*Id.*

state's Groundwater Protection program. For parties seeking site-specific standards, DES requires submittal of risk assessment and feasibility studies. The risk assessments procedures must then be approved by the Department of Health and Human Services. Once the remedial plan is approved, then the party and DES will enter into a formal agreement or an administrative order (AO). Complex agreements generally require entering into an AO, which is approved by the state attorney general's office.

Once the remedial plan is approved, the remedial work begins. Upon completion, a formal report must be delivered to DES for approval. If the cleanup is acceptable, written assurances in the form a "certificate of no further action" letter may be provided. The state retains the right to engage in oversight activities during remediation—including staff fieldwork.

The chart below lists the documents that the state of New Hampshire developed for issuance in the brownfields redevelopment process before the voluntary cleanup law was enacted. All these documents are required to be recorded in the land records in the county in which the property is located.

Documents To Be Recorded	Issued By	Purpose
Notice of Approved RAP	DES	Approval of remedial action plan
Certificate of Completion	DES	Approval of site remediation completion
Covenant Not to Sue[137]	DOJ	Written assurances

The program operated under the new law seems to favor the Covenant Not to Sue program.

The liability protection afforded by law provides that "an eligible person is not liable for the remediation of additional contamination or increased environmental harm caused by pre-remedial or site investigation activities, unless attributable to negligence or reckless conduct by the eligible person";[138] however, it applies only to state law. Nevertheless, the law dictates that the assurances provided are transferable, although the conditions for transfer may vary dependent upon the nature of site cleanup at the time of transfer.[139] If the eligible person cannot complete the site cleanup, then the Covenant Not to Sue issued when the RAP was approved provides protection from liability as long as the site is stabilized to the satisfaction of DES and the site is not in worse condition than it was before remediation was begun.

Participants in the program can withdraw from the program at any time.[140] To do so, they must file a notice of withdrawal, stabilize the site, and pay all outstanding fees accrued by the state.[141] The law expressly states that withdrawal

---

[137]*See* §§ 147-F:6 (Covenant Not to Sue) and 147-F:7 (liability protection before covenant issues).
[138]*Id.*

[139]§ 147-F:17 (1996) (sale or transfer of property in program).

[140]§ 147-F:8 (1996) (withdrawal from the program).
[141]*Id.*

from the program will not be considered inconsistent with or interference with the approved remedial action plan. Nor will withdrawal necessarily void the covenant.[142]

The new legislation gives the DES broad authority to operate the voluntary cleanup program. Accordingly, the law requires a great deal of rule-making[143] and enforcement[144] power on the part of DES to clarify and evolve the program.

Expedited Site Assessment Reviews[145]

New Hampshire enacted a program in 1993 to "provide New Hampshire property owners, businesses, industries, and financial institutions with the ability to receive an accelerated DES review when timely critical real estate transfers or financial transactions are dependant on receipt of comments on site assessment reports from DES."[146] To apply for an expedited review, an applicant must include the following:

(1) a transmittal letter requesting the review (this should include a description of the transaction);
(2) an environmental site assessment report;
(3) a copy of the current municipal records indicating the current equalized assessed valuation of the property (land and buildings);
(4) a check made out to Treasurer, State of New Hampshire, based on the fee schedule below:

Equalized Assessed Valuation	Fee
$0 to $250,000	$1,200
$250,001 to $500,000	$1,500
$500,001 to $1,000,000	$2,500
greater than $1,000,000	$5,000

(5) a statement indicating if any person(s) may seek reimbursement from any of the state's petroleum reimbursements funds.

If the sites are eligible for expedited review, the program requires DES to review environmental site assessment reports and the comments to these reports within sixty days from the date when a complete expedited site assessment review package was submitted. The following table indicates the sites that are eligible for expedited review:

---

[142]*Id.*
[143]§ 147-F:18 (1996).
[144]§ 147-F:19 (1996).
[145]Environmental Fact Sheet, Expedited Environmental Site Assessment Reviews (Feb. 1, 1997), reprinted in KEY ISSUES IN U.S. EPA REGION I (Oct. 2, 1996, Boston, Massachusetts).
[146]*Id.* (emphasis in original).

Expedited Reviews Available for These Types of Sites	Explanation/Caveat
Sites for which there is a duty to report	Such sites include: sites with groundwater contamination above ambient groundwater quality standard; and sites where there has been a discharge of wastes or petroleum.
Sites eligible for reimbursement from the state petroleum reimbursement funds	These sites require prior approval of work scopes to ensure full reimbursement.
Sites where regulatory opinions are required by parties other than DES	To complete the real estate or financial transaction

Environmental Lien

New Hampshire[147] is one of a minority of states that has enacted a superlien.[148] Under New Hampshire law, the division of waste management is granted a lien upon the business revenues and all real and personal property of any person subject to liability under the state mini-Superfund statute.[149] In order for the lien created to be valid and effective against the real property of a liable person, the division of waste management must record the notice of lien in the registry of deeds for each county in which such person owns or holds an interest in real property. Upon its recording in a registry of deeds, the notice of lien will be effective against all real property of the person located within such county. In order for the lien to be perfected and valid against the business revenues and personal property, tangible and intangible, of the liable person, the division of waste management must record the notice of lien with the office of the secretary of state in which financing statements are filed.[150] The division of waste management must also file separate notices of lien forms for each person subject to liability. There is no charge for filing a notice of lien. The fee for discharging a notice of lien is, however, borne by the person identified in the notice of lien.

The priority of the lien, as in other states' statutes, is also dependent on the facts and circumstance. With the exception of residential property, the lien will constitute a first-priority lien against real property on which the hazardous waste or hazardous material is located prior to all encumbrances, whether of record or

---

[147]N.H. Rev. Stat. Ann. § 147-B:10.

[149]The other states include: Connecticut, Louisiana, Maine, Massachusetts, New Hampshire, and New Jersey.

[149]Section 147-B:10. *Compare* N.H. Rev. Stat. Ann. § 147-A:3 Dep't of Health & Welfare, Div. Pub. Health Haz. Waste Rules, § 1905.08 (RCRA disclosure requirements).

[150]For purposes of this section, the applicable filings are those made pursuant to N.H. Rev. Stat. Ann. § 382-A:9-401.

inchoate, when the notice of lien is recorded in the registry of deeds for the county in which such real property is located and the notice of lien identifies the record owner of such real property. The first-lien priority includes the business revenues generated from the facility on which hazardous waste or hazardous material is located and personal property located at the facility on which hazardous waste or hazardous material is located and will also be effective when the notice of lien is filed with the secretary of state and the notice of lien identifies the owner of such personal property.

As to all other property (including residential), whether real, personal, or business revenues, the notice of lien will constitute a lien that is effective as of the date and time of recording or filing, without priority on antecedent encumbrances of record when the notice of lien is properly recorded in the appropriate registry of deeds or filed with the secretary of state.

Notwithstanding the lien priorities created, a holder who, either voluntarily or in conjunction with others (including the state and federal government), undertakes cleanup activities or expends funds on other response or remedial costs, will have a lien of equal rank and priority with the state's lien "to the extent of moneys expended for remediation." The division of waste management will also send a copy of the notice of lien to the person identified in the notice of lien at the address set forth in the notice of lien by certified mail, return receipt requested, postage prepaid.

RCRA

New Hampshire, like other states, has a state statute requiring that some form of general notice be placed on the public record or in the chain of title to alert prospective purchasers and the public of a site's status as formerly managing hazardous waste.[151]

Rhode Island

Voluntary Cleanup Program

Rhodes Island's Voluntary Cleanup program was established in 1993 and revised in 1995 by the Industrial Property Remediation and Reuse Act.[152] The major goals of the program are to:

identify, investigate, and clean up contaminated sites; and
bring these sites back to beneficial reuse in the community.

---

[151]N.H. REV. STAT. ANN. § 147-A:3 Dep't of Health & Welfare, Div. Pub. Health Haz. Waste Rules, § 1905.08.
[152]*See* Charles Bartsch & Elizabeth Collaton, COMING CLEAN FOR ECONOMIC DEVELOPMENT at 78-80.

Under the Rhode Island program, responsible parties, voluntary parties, and prospective purchasers are all eligible for the program, provided the site does not pose an imminent threat to human health and the environment. The program was not established and is not under the jurisdiction of the Department of Environmental Management (DEM or RIDEM), under RCRA, LUST, or its superfund program. Instead, it is an independent program.

The law defines those who may participate in the program as one of three categories of persons:[153]

Responsible Parties	Volunteers	Bona Fide Prospective Purchasers
Current owner or operators of the site; past owners or operators of the site; persons who arranged for the disposal at the site; persons who brought hazardous materials, which subsequently release to the environment at the site.	Persons who are not RPs who undertake a site assessment or investigation and provide the results to DEM; persons who are not RPs and undertake and complete a cleanup at a site according to a remedial action work plan approved by DEM.	Persons who are not RPs and who do not hold a 10 percent (or greater) interest in the ownership or operations and who intend to buy a site and have documented their intent to buy in writing and who have offered fair market value for the site in the contaminated state.

In short, the sites must be "low-priority contaminated properties." At least eighty sites have enrolled in the Industrial Property Remediation and Reuse Program established by the 1995 legislation. To be eligible for consideration in the program, a site must meet two broad criteria:[154]

some significant level of economic investment or redevelopment must be proposed for the site; and

environmental contamination at the site is a barrier to that economic investment.

Site investigation and remediations are conducted under the state's Site Remediation Regulations. These require RPs to notify DEM of accidental releases of hazardous substances to the environment.

The state of Rhode Island may issue a covenant not to sue at the completion of the site cleanup. The covenant will only be issued where:

(1) The site is in a condition suitable for the current and reasonably foreseeable land use; and

(2) The volunteer agrees to conduct cleanup activities in accordance with a RIDEM approved work plan.

---

[153]RIDEM Brownfields Fact Sheet, the Voluntary Clean-up Program (Sept. 1995), reprinted in KEY ISSUES IN U.S. EPA REGION I (Oct. 2, 1996, Boston, Massachusetts).
[154]*Id.*

Where issued, these covenants are transferable with the property title. The covenants can also be used to protect lenders and other secured creditors.

RIDEM is committed to public notification and participation. Thus, public notice and participation is available at each of the following stages of brownfields cleanup and redevelopment:

Form of Notice	Stage
Notice	Site investigation field activities
Notice	Site investigation is completed
Notice and comment	Cleanup is proposed (including encapsulation or treatment)
Notice and fourteen-day comment period	All settlement agreements
Public record review	Of all sites

Business Transfer Statute[155]

The Rhode Island Hazardous Waste Management Act (RIHWMA)[156] applies to any real property.[157] RIHWMA disclosure requirements are triggered by the transfer of real property.[158] "As soon as practicable" but not later than before a transfer agreement is signed, the seller must deliver a written disclosure to the buyer.[159] The disclosure must include information on hazardous waste on the property.[160]

Vermont

Voluntary Cleanup Program

In 1995, Vermont created the Redevelopment of Contaminated Properties Program (RCPP).[161] The RCPP is overseen by the Vermont Agency of Natural Resources (ANR), which began taking program applications on January 1, 1996. The benefits of the program include:

providing certainty as to liability;
easing the burdensome process of negotiating agreements (waivers of liability) on a case-by-case basis;
encouraging site investigation and cleanup that might not otherwise occur;

---

[155]R.I. GEN. LAWS § 23-19.1-6 (1993).

[156]Id. Compare R.I. GEN. LAWS § 23-19.1-6 Rules & Reg. for Haz. Waste Gen., Treatment, Storage & Disposal, § 9.16 (for disclosure requirements under RCRA).

[157]Id.

[158]Id.

[159]Id.

[160]Id.

[161]VT. STAT. ANN. tit. 10 § 6615a (f)(1)(B). Compare VT. STAT. ANN. tit. 10 § 6606(5) (for disclosure requirements under RCRA).

establishing a clear process for submittal and approval of site investigation
plans and corrective action plans;

bringing back unused property onto the tax rolls; and

encouraging the redevelopment of brownfields instead of undeveloped land.[162]

Eighty-seven percent of the contaminated sites in Vermont are the result of
leaking underground petroleum tanks. Accordingly, the RCPP is operated under
Vermont's UST program, rather than the state equivalent of CERCLA.

To be eligible for the program, the applicant may not be liable or potentially
liable for the contamination at the site, nor should the applicant have had previous
contact with the property.[163] In other words, "the applicant will be excluded from
the program if that person has in any way been previously involved with hazardous
waste activities at the property."[164] Thus, applicants "who acquire an ownership
interest in the property during the period between submittal of the application, but
prior to acceptance into the program assume the risk of denial."[165]

For the site to be eligible for RCPP, the contamination must be caused by a
release or threatened release of hazardous materials[166] and the property must be
"vacant, abandoned, or substantially underutilized" or "be acquired by a municipality."[167] The program excludes sites that are on CERCLA's NPL, are subject to
RCRA corrective action, or fall under Vermont's underground storage tank enforcement provisions.[168]

Eligible Sites	Ineligible Sites
Vacant; abandoned; substantially underutilized; to be acquired by a municipality.	A property at which the only release is a petroleum release relating to a UST that is subject for the Petroleum Cleanup Fund; any property listed on the CERCLA NPL; TSD facilities that have a certificate from the state of Vermont and are undergoing corrective action under RCRA.

Applicants must include an initial application fee of $500. In addition, the ap-
plicant must complete an application package that includes three basic elements:

a preliminary environmental assessment;

a certification from each person who would benefit from liability protection;
and

---

[162]Cleanup of Hazardous Waste Sites under the Vermont Redevelopment of Contaminated Properties Program, reprinted in KEY ISSUES IN U.S. EPA REGION I (Oct. 2, 1996, Boston, Massachusetts).

[163]VT. STAT. ANN. tit. 10 § 6615a (f)(1)(B).

[164]Cleanup of Hazardous Waste Sites under the Vermont Redevelopment of Contaminated Properties Program, reprinted in KEY ISSUES IN U.S. EPA REGION I (Oct. 2, 1996, Boston, Massachusetts).

[165] *Id.*

[166]VT. STAT. ANN. tit. 10 § 6615a (f)(1)(A).

[167]*Id.* § 6615a(f)(1)(c). Note that definitions of each of these terms can be found in the document
"Procedures for Determination of Site Eligibility."

[168]*Id.* § 6615a(f)(2)(A-C).

any other information requested by the secretary of natural resources regarding the site.[169]

The following table depicts the data that must be included in each of these documents:

Application	Description
Program fee	Five-hundred dollars due with application. Upon acceptance and submittal of the site investigation plan, an additional fee of $5,000 is required.
Preliminary environmental assessment	A legal description of the property;
	A description of the physical characteristics of the property (including size, location, topography, etc.);
	All information concerning the operational history of the property that is known to, or in the control of, the applicant;
	The nature and extent of releases and threatened releases;
	A description of the proposed redevelopment and use of the property.
Certification	Under oath and notarized, from each person who would benefit from liability protection, the certificate must state:
	That each person accurately disclosed to the secretary all information currently known to the person, or in the person's possession or control, which relates to releases or threatened releases of the hazardous materials at the property;
	That neither the person nor any of its principals, owners, directors, affiliates, or subsidiaries:
	Currently holds or ever held an ownership interest in the property or in any related fixtures or appurtenances, excluding a secured lender's holding indicia of ownership in the property primarily to assure the repayment of a financial obligation;
	Directly or indirectly caused or contributed to any releases of hazardous materials at the property;
	Currently operates or controls, or ever operated or controlled the operation, at the property, of a facility for the storage, treatment, or disposal of hazardous materials from which there was a release;
	Disposed of, or arranged for the disposal of, hazardous material at the property;
	Generated hazardous materials that were disposed of at the property.
Other information	Any information requested by the state.

Once the party is accepted into the Vermont program, then the participant must make a $5,000 deposit toward oversight costs. The state may charge additional oversight costs once this initial deposit is exhausted. In addition, the state retains the right to pursue cost recovery against other PRPs associated with the site.

Once both the party and the site are deemed eligible for the program, the application process is similar to programs in other states: the application consists of oversight fees and a work plan.[170] After considering the work plan, the applicant

---

[169]Cleanup of Hazardous Waste Sites under the Vermont Redevelopment of Contaminated Properties Program, reprinted in KEY ISSUES IN U.S. EPA REGION I (Oct. 2, 1996, Boston, Massachusetts).
    [170]*Id.* § 6615a(g)(1)-(3).

must provide a site investigation report and later provide a corrective action plan.[171] The site investigation must:[172]

> define the nature, source, degree, and extent of contamination;
> define all possible pathways for contamination migration;
> present data that quantifies the amounts of contaminants migrating along each pathway;
> define all relevant sensitive receptors, including but not limited to public or private water supplies, surface waters, wetlands, sensitive ecological areas, outdoor and indoor air, and enclosed spaces such as basements, sewers, and utility corridors;
> determine the risk of contamination to human health and the environment;
> gather sufficient information to identify appropriate abatement, removal, remediation, and monitoring activities;
> gather sufficient information to provide a preliminary recommendation, with justification, for abatement, removal, remediation, and monitoring activities.

Cleanup actions begin upon approval of the corrective action plan.[173] When cleanup is completed, ANR provides the party with a certificate of completion. This certifies that:

> (1) the cleanup was conducted and completed per the corrective action plan,
> (2) no further activity is required to remediate the site, and
> (3) the party is shielded from liability arising under the Hazardous Waste Management Act.[174]

The state retains the right to reopen the case in certain circumstances, including fraud.

Waivers of Liability/ Forbearance Provision	Limitations to the Waiver of Liability
Additional contamination discovered as a result of changes in recognized standard methods of technology; Changes in regulatory standard that occur after the date of CAP approval; A new release or worsened release caused during the cleanup but that is cleaned up prior to issuance of the certificate of completion.	New release that occurs after the applicant has submitted an application for the program, unless the new release is caused during the cleanup and is in fact cleaned up prior to issuance of the certificate of completion; any preexisting release that the participant did not bring to the secretary's attention prior to the issuance of the certificate of completion; third-party liability; liability under any other state or federal including, but not limited to, RCRA and CERCLA.

---

[171]*Id.* § 6615a(g)(4)-(6);(h).

[172]Cleanup of Hazardous Waste Sites under the Vermont Redevelopment of Contaminated Properties Program, reprinted in KEY ISSUES IN U.S. EPA REGION I (Oct. 2, 1996, Boston, Massachusetts).

[173]*Id.* § 6615a(i).

[174]*Id.* § 6615a(k).

Cleanup standards for the Vermont RCPP are the same as those used by the Department of Conservation's (DEC) Hazardous Materials Management Division for other DEC programs. Specifically, the cleanups must comply with the Vermont Groundwater Protection rules and EPA Region III's risk-based tables for contaminated soils.[175]

RCRA

Vermont, like other states, has a state statute requiring that some form of general notice be placed on the public record or in the chain of title to alert prospective purchasers and the public of a site's status as formerly managing hazardous waste.

---

[175] *See* Charles Bartsch & Elizabeth Collaton, COMING CLEAN FOR ECONOMIC DEVELOPMENT at 82.

CHAPTER 5

Region II:
New Jersey, New York, and Puerto Rico

The states in Region II, principally New Jersey and (to a lesser extent) New York, pioneered the concept of transaction-triggered cleanup statutes. Experience with these statutes has caused both states to develop voluntary cleanup programs. The following chart describes the more recent programs developed in Region II to encourage voluntary cleanup:[1]

States	Brownfields Financing Programs	VCP with Written Assurances	Incentives to Attract Private Development
New Jersey	X	X	X
New York	X	X	
Puerto Rico			

These programs enable those who anticipate a property transfer to undertake cleanup in a context that does not necessarily involve property transfer. The VCP programs generally enable certain parties to obtain the benefits of state oversight even if there is no current plan to transfer property.[2] The following table briefly compares the voluntary cleanup programs in New York and New Jersey:

VCP	New York	New Jersey
Oversight Authority	NYDEC General authority of the NYDEC	NJDEP ISRA; Brownfield & Contaminated Site Remediation Act
Established	1994	1993, amended in 1997
Eligibility	Sites subject to state Superfund and UST enforcement actions; PRP lenders that would like to improve their collateral, municipalities foreclosing on sites due to tax arrears, and municipal financier (e.g., industrial development agencies)	Low-priority sites that are not subject to UST laws and are not landfills

(continued)

---

[1]*See* Charles Bartsh, BROWNFIELDS STATE OF THE STATE REPORT: 50-STATE PROGRAM ROUNDUP (Northeast-Midwest Inst. January 28, 1998).

[2]*See generally* KEY ENVIRONMENTAL ISSUES IN U.S. EPA REGION II (American Bar Association Section of Natural Resources, Energy, and Environmental Law (June 28, 1996 New York, New York).

Sites not eligible	Sites listed on the NPL, persons subject to RCRA enforcement	A site subject to a NNJDEP permit may be eligible
Excluded persons	PRP or cleanup volunteers intending to convey or lease the property to a PRP	Nobody. Any party, including a PRP may become a volunteer
Initiating the process	NYDEC reviews the volunteer's application & determines eligibility. Approved volunteer entered into either an agreement or consent order with NYDEC, outlining remedial plan, cleanup levels and actions to be taken on site	Fully voluntary program Entering into a MOU is only necessary if the volunteer wants to apply for an NFA letter. Volunteer can approach NJDEP at any point during cleanup to enter into a NJVCP MOA
Oversight costs	Volunteer must promise to pay state oversight costs	Volunteer must promise to pay oversight costs
Scope of the agreement/ consent order	Must include: An outline of remedial activities; procedural guidelines for completing the remedial activities; a clause permitting NYDEC staff access and oversight of site; a clause holding the state harmless with respect to remediation activities on site; and a promise to indemnify the state for the costs of oversight.	MOA contains enforcement provisions; right of state or volunteer to terminate at any time; extent of oversight varies with contamination (remedial action report or remedial work plan); cleanup levels are defined by the state (soil, groundwater, technical, engineering and institutional controls); volunteer must provide any and all data developed during its efforts to NJDEP.
Effect on continuing liability	NFA letter issued where volunteer completes remediation; NFA releases volunteer from a future NYDEC enforcement action concerning known contamination	NJDEP may issue an NFA letter to volunteer that successfully remediated; NFA letter does not provide an explicit release from future liability.
Reopeners	State adopts new, more stringent cleanup standards; Intended use of site changes; State determines volunteer withheld information from NYDEC or committed fraud	NFA may easily be reopened
Grant		Municipalities may be eligible for loans or grants up to $2 million per year

New Jersey

Voluntary Cleanup Program

Although the state of New Jersey historically approached the cleanup of contaminated property through the use of a mandatory program under the New Jersey

Environmental Cleanup Responsibility Act (ECRA),[3] in 1991 the state passed legislation to encourage voluntary cleanup as well. In 1997, New Jersey again supplemented its voluntary cleanup laws with the passage of the Brownfield and Contaminated Site Remediation Act.[4] The 1991 laws provided certain tax incentives for volunteers. It was designed to complement the earlier business transfer statutes, which by themselves were insufficient to deal with New Jersey's brownfield redevelopment problems. The 1997 law focused on the remediation standards required to ensure protection of public health and yet also ensure finality in the brownfields cleanup process.[5]

In addition, the 1997 legislation created a Brownfields Redevelopment Task Force. The purpose of this task force is to create and update an inventory of brownfield sites found in the state of New Jersey. After locating those properties which qualify as brownfields, the task force must then:

(1) coordinate state brownfields policy;
(2) prioritize the sites on the inventory site based on economic potential and then help market these sites to prospective purchasers;
(3) prioritize the sites based on ecological risk; and
(4) make further recommendations on brownfields redevelopment to the governor and the state legislature.[6]

New Jersey's Department of Environmental Protection (NJDEP), now administered under the title Department of Environmental Protection and Energy (DEP), developed the voluntary cleanup program.[7] This program was, in turn, codified by the state legislature in 1997.[8]

Parties that wish to participate in the program apply to NJDEP. Once an application is accepted, the New Jersey Voluntary Cleanup Program requires parties to enter into a Memorandum of Agreement (MOA) with the NJDEP. The MOA does not protect against liability, although it does expressly provide that performance under the MOA does not constitute any admission of fact or liability.[9] The MOA is, further, considered a voluntary agreement that can be terminated, without penalty, by either party.

To be eligible sites must be considered a low DEP priority. LUST sites and landfills may not participate in the program. Any person, including current PRPs, may apply to the program. Applicants are charged at the rate of approximately $75 per hour for oversight costs.

Where the scope of cleanup work is small, DEP oversight is limited. For sites in which there is only soil contamination and the cleanup will take less than five

---

[3]N.J. STAT. ANN. § 13:1K-6 to 11; N.J. ADMIN. CODE, tit. 7, § 26C-E.
[4]1997 N.J. ALS 278; 1997 NJ Laws 278; 1996 NJ S.N. 39.
[5]*Id.*
[6]*Id.*
[7]Procedures for Department Oversight of Contaminated Sites, N.J. ADMIN. CODE 7:26C *et seq.*
[8]1997 N.J. ALS 278; 1997 NJ Laws 278; 1996 NJ S.N. 39.
[9]I. Leo Motiuk & Sean T. Monaghan, New Jersey, at 4 (unpublished manuscript, on file with author).

years, an applicant need only submit a remedial action report to DEP. If there is groundwater contamination or a cleanup is expected to take longer than five years, then the applicant must also submit a remedial work plan to DEP. Applicants can request additional DEP oversight if they so desire. This request is generally codified in the MOA between DEP and the applicant. The following table describes points of oversight that have been included in MOAs negotiated by DEP to date:

Points of Oversight	Description
Preliminary Assessment	Involves review of site history, data on past discharges, past permits, and enforcement actions.
Site investigation work plan and report	Determines what the existing contaminants are and which are presentat levels above the acceptable thresholds.
Remedial investigation work plan and report	Extracts a more detailed picture of the contamination on-site.
Remedial action work plan	Includes the cleanup levels to be achieved and outlines the sort of remedies to remove the extant contamination. Where applicable, it is to include a feasibility study and risk assessment.
Remedial action report	Shows the work done and the results achieved.

Upon completion, the applicant may received a No Further Action (NFA) letter from DEP. These letters are not an absolute bar to liability. Certain standard reopeners include:

the discovery of new information about a site;
failure of the proposed remedy;
substantial raising of cleanup standards by DEP above the negotiated standards (for example, by a level of one order of magnitude or more).

Financing Voluntary Cleanups

Various funds are available in New Jersey to help finance the voluntary cleanup programs. In 1993, the New Jersey Hazardous Discharge Site Grant Program was appropriated $55 million to accomplish voluntary cleanups. Private parties are eligible for loans of up to $1 million at 5 percent interest, payable in up to ten years. Municipalities are also eligible for both grants and loans of up to $2 million per year. The municipalities must, however, be working to clean up abandoned, orphaned sites or sites obtained through tax sale certificates.[10]

In 1995, New Jersey enacted the Landfill Reclamation Act. This law allowed municipalities in New Jersey to finance landfill closure and pollution abatement costs. It also allowed the municipality to encourage and finance, in part, the rede-

---

[10]See Charles Bartsch & Elizabeth Collaton, COMING CLEAN FOR ECONOMIC DEVELOPMENT at 44. See generally Todd S. Davis & Kevin D. Margolis, BROWNFIELDS: A COMPREHENSIVE GUIDE TO REDEVELOPING CONTAMINATED PROPERTY at 518-527 (1997).

velopment of old landfills that had not been properly closed. The legislation was so successful that the program was expanded in 1996 by the enactment of the Large Site Landfill Reclamation and Improvements Law.

In 1996, the legislature also enacted the New Jersey Environmental Opportunity Zone Act. This law allows municipalities to establish "environmental opportunity zones" that are eligible for assistance under the New Jersey Special Municipal Aid Act. To take advantage of this law, a developer would have to entice a municipality to declare the target site within an urban enterprise zone defined as an "environmental opportunity zone" under New Jersey law. The municipality can then provide for a tax abatement for the developer of the contaminated property for a term of up to ten years. Payments would then be made to the municipality in lieu of property taxes based on a sliding scale computation. The scale would start at zero percent and then increase by 10 percent per year until the tenth year when the equivalent of full property taxes would be paid to the municipality.

A companion to this bill is the New Jersey Urban Redevelopment Act, which was also passed in 1996. This later law attempts to entice prospective purchasers to redevelop properties in areas targeted by municipalities as environmental opportunity zones. the statue also classifies a new development target termed a "special needs district."

Business Transfer Statutes

Enacted in 1983, the New Jersey Environmental Cleanup Responsibility Act (ECRA) was one of the first of the state transaction-triggered statutes. The ECRA imposed preconditions upon the consummation of a transaction to close, terminate, or transfer the operations of an industrial establishment in New Jersey. The preconditions included submitting to the New Jersey Department of Environmental Protection a notification form, a site evaluation form, and a declaration essentially stating that the property was not contaminated at the time of the transfer or, alternatively, a cleanup plan backed by financial security, which the department must approve. The sanctions for noncompliance with the Act were severe—avoidance of the transaction and imposition of damages and stringent, escalating penalties.[11]

From the beginning, New Jersey's Department of Environmental Protection and Energy was criticized for its handling of the program. These attacks gained in severity as New Jersey's economy lagged behind the nation in the most recent and continuing recession. After many months of effort in New Jersey's legislature, senate S.1070 passed the senate and assembly on June 10, 1993, and was

---

[11]Under the Industrial Site Recovery Act (ISRA), only private parties have the power to void a transaction. Although it did under ECRA, ISRA took away the state's authority to void a transaction. The legislative history indicated that since private parties had attempted to use this remedy and the state had not the legislature decided to continue the remedy for private parties but not for the state.

promptly signed by Governor Florio on June 16, 1993. Thus, New Jersey enacted ISRA to replace the ECRA program.[12]

The changes to ISRA were intended to improve the efficiency of the program and better balance legitimate business concerns with protection of the environment. Many of the studies required by ISRA are certain to continue the debate on environmental policy both in and out of New Jersey. As these studies are delivered, the legislature will revisit ISRA itself, conceivably as early as 1994. DEP's actual implementation of ISRA will then be clear, and the legislature should be able to assess its actual impact on, and reception by, the regulated community.

The impact of ISRA's changes may not be limited to New Jersey. Because of ISRA's continued extraterritorial effect, out-of-state counsellors involved in New Jersey transactions must be educated as to the implications of ISRA. In addition, other state legislatures may consider New Jersey's experience under ISRA when evaluating their own environmental laws and policies.

Enacted in 1983,[13] ECRA represented the New Jersey Legislature's response to the problem of abandoned hazardous waste sites. *New Jersey Department of Environmental Protection v. Ventron*,[14] a suit brought by NJDEP against the current owners of a decaying chemical plant and their predecessors in interest, illustrated

---

[12]N.J. STAT. ANN. § 13:1K-7, amended by ISRA Section 2. The legislature clearly announced their probusiness purposes in enacting the changes:

> . . . [I]t is in interest of the environment and the State's economic health to promote certainty in the regulatory process . . . to create a more efficient regulatory structure and to allow greater privatization. . . . it is the policy of this State to protect the public health, safety, and the environment, to promote efficient and timely cleanups and to reduce any unnecessary financial burden of remediating contaminated sites; . . . by streamlining the regulatory process, by establishing summary administrative procedures for industrial establishments that have previously undergone environmental review, and reducing oversight . . . where less . . . review will assure the same degree . . . of protection to public health, safety and the environment; and that new procedures . . . shall be designed to guard against redundancy . . . and minimize governmental involvement in certain business transactions.

The legislative history contained several other statements of note (unless otherwise noted, all citations will be to the Assembly Policy and Rules Committee Statement to S.1070 of June 3, 1993, and will be cited as "*APRC Statement*"):

    (1) "[U]ntil regulations are adopted . . . [NJDEPE shall] act reasonably in the interim period when reviewing applications and petitions and in all other interactions with the public." *APRC Statement* p. 17, 7th bullet.

    (2) NJDEPE should: "[a]llow cleanups that do not remediate property to pristine levels, provided that appropriate and DEPE approved engineering or institutional controls are implemented." *APRC Statement* p. 17, 8th bullet.

    (3) NJDEPE should: "[p]rovide, to the greatest extent possible, finality to compliance with ISRA." *APRC Statement* p. 17, last bullet.

[13]The statute became effective on December 31, 1983. Compare N.J. STAT. ANN. § 13:1E-56 N.J. ADMIN. CODE tit. 7, § 26-9.9(n) (describing the RCRA disclosure requirements).

[14]468 A.2d 150 (N.J. 1983).

the difficulties that governmental agencies experienced in cleaning up abandoned facilities. NJDEP was compelled to engage in expensive and time-consuming litigation to remedy the environmental damage at the plant. In similar situations, state and federal agencies were forced to initiate comprehensive cleanup efforts and legal actions against previous owners. ECRA's sponsor, state Senator Raymond Lesniak, explained that contaminated property "leav[es] a blight on our State and health hazards and economic burdens to our residents."[15] ECRA was passed in response to Ventron-type problems at hazardous industrial facilities.

ECRA's primary purpose was to impose liability for cleanup costs on property owners without protracted litigation. As courts have recognized, "[t]he essential goal of ECRA is to secure the cleanup of contaminated industrial sites at the earliest possible date."[16] This motive was expressed in the legislature's finding that the generation, handling, storage, and disposal of hazardous substances and wastes pose an inherent danger of exposing citizens, property, and natural resources of this state to substantial risk of harm or degradation; that the closing of operations and the transfer of real property utilized for the . . . hazardous substances and wastes should be conducted in a rational and orderly way, so as to minimize potential risks; and that it is necessary to impose a precondition on any closure or transfer of these operations by requiring the adequate preparation and implementation of acceptable cleanup procedures.[17] As the court observed in *Superior Air Products v. NL Indus.*,[18] ECRA imposes "a self-executing duty to remediate without the necessity and delay" of a determination of liability through litigation.[19]

ECRA required all "industrial establishments"[20] to obtain prior NJDEP approval of either a negative declaration that the site is free from possible contamination or a cleanup plan as a precondition to the sale, closure, or transfer of business operations.[21] Any "owner or operator" of an industrial establishment planning to sell or close operations must notify NJDEP within five days after public release of its decision to close or within five days of the execution of any agreement of

---

[15]*See* Lesniak, *ECRA is Coming*, 104 N.J. Law 41 (1983).

[16]Dixon Venture v. J. Dixon Crucible, 561 A.2d 663 (N.J. Super. Ct. App. Div. 1989), *aff'd as modified*, 584 A.2d 797 (N.J. 1991).

[17]N.J. Stat. Ann. § 13:1K-7.

[18]522 A.2d 1025 (N.J. Super. App. Div. 1987).

[19]*Id.*

[20]An *industrial establishment* is defined as: "any place of business engaged in operations which involve the generation, manufacture, refining, transportation, treatment, storage, handling, or disposal of hazardous substances or wastes on-site, above or below ground, having a [specified] Standard Industrial Classification number." N.J. Stat. Ann. § 13:1K-9. A Standard Industrial Classification (SIC) number is assigned by the federal Office of Management and Budget and describes the type of operation. For example, number 22 refers to textile mill products and number 23 refers to apparel and similar materials. ECRA initially included SIC designation codes covering a wide range of businesses. In 1986, pursuant to its authority under the statute, NJDEP adopted exemptions for several SIC codes originally included in ECRA.

[21]N.J. Stat. Ann. § 13:1K-9.

sale or option to buy, as applicable.[22] The NJDEP regulations describe the type of information to be submitted, including inventories of hazardous substances, description of current operations, copies of all soil, groundwater, and surface water sampling results, and details of the impending transaction.[23]

After conducting a preliminary site inspection of all industrial establishments that have notified the department, NJDEP issues a preliminary report describing site conditions and providing guidance for ECRA compliance. The department supplements all initial notice submissions with its own review of available records of relevant federal, state, county, and municipal agencies. Following approval and implementation of an appropriate sampling plan, the owner or operator of the establishment must submit either a cleanup plan or a negative declaration. All cleanup plans must be accompanied by financial security, guaranteeing the performance of the cleanup plan, in an amount equal to the cost estimate approved by NJDEP for the plan's implementation.[24]

ECRA provided stringent penalties for failure to comply with the Act or regulations. Failure to comply allows NJDEP or the transferee to void the sale or transfer of the industrial establishment or real property.[25] The transferee is also entitled to recover damages due to the voiding of the transaction, and the transferor is held to be strictly liable for all cleanup and removal costs.[26] In addition, NJDEP may seek penalties of a maximum of $25,000 per day from any person giving false information or otherwise failing to comply with ECRA.[27] Any corporate officer who knowingly directs or authorizes violations of the Act will be personally liable for ECRA penalties.[28]

ECRA served as a major incentive for businesses to plan and conduct environmentally safe industrial operations. In its six years of existence, ECRA was responsible for more than two hundred million dollars in privately funded cleanups.[29]

ECRA was the first state law that imposed wide-ranging liability on owners and operators of commercial and industrial real estate prior to transfer or closing of business operations. As the New Jersey Supreme Court has noted, ECRA is quite unlike other environmental regimes in that it uses market forces to bring about the reversal of environmental pollution. It recognized that some environmental conditions not posing an imminent hazard to air or water resources of the state may

---

[22]*Id.*

[23]N.J. ADMIN. CODE § 7:1-3.7(d).

[24]N.J. STAT. ANN. § 13:1K-9(a)(2) and (b)(3).

[25]*Id.* § 13:1K-13(a) and (b).

[26]*Id.* § 13:1K-13(a).

[27]*Id.* § 13:1K-13(c).

[28]*Id.*

[29]*See* Motiuk & Sheridan, *New Jersey's ECRA: Problems, Policies, Future Trends,* 21 Env't. Rep. (BNA) No. 13, at 549 (July 27, 1990). "Already some 370 sites have been cleaned up or investigated under the program, and companies have agreed to provide $475 million to clean up another 480 sites in coming years." Barrett, *New Jersey, After Federal Leadership Waned, Has Become Environmental Protection Pioneer,* WALL ST. J., July 5, 1988, p. 44, col. 4.

safely attend economic activities.[30] Inspired by ECRA, Connecticut,[31] Iowa,[32] Missouri,[33] and Illinois[34] passed similar statutes.[35]

Dissatisfied with the pace of cleanup actions, New Jersey elected to regulate directly all business transactions involving the "closing, terminating, or transferring" of operations of industrial establishments that may pose an environmental risk by imposing *preconditions* upon the consummation of such transactions. New Jersey enacted the Environmental Cleanup Responsibility Act, effective in 1983, and rules implementing ECRA.[36] The preconditions upon the consummation of a closure, transfer, or termination of operations include the execution of a cleanup plan approved by the NJDEP and the approval by the NJDEP of a declaration that there has not been a discharge of hazardous substances or waste or, if there has been, that the property has been cleaned up pursuant to NJDEP procedures.[37]

ISRA continues to apply to industrial establishments, essentially as previously defined under ECRA (i.e., a place of business with a subject Standard Industrial Code (SIC) number having hazardous substances or wastes). ISRA is triggered when the owner or operator of an industrial establishment is involved in either "closing operations" or "transferring ownership or operations." These triggers are not dramatically different from ECRA's recent regulations.

Under ECRA and the rules governing the implementation of ECRA, sanctions for noncompliance could be severe—avoidance of the transaction and imposition of damages and stringent, escalating penalties.[38] This remedy existed in

---

[30]Dixon Venture v. Joseph Dixon Crucible Co., 584 A.2d 797 (N.J. 1991) (per curiam).

[31]CONN. GEN. ANN. §22a-134.

[32]IOWA CODE ANN. § 445B (1), (2).

[33]MO. ANN. STAT. § 260.465B (1), (2).

[34]ILL. REV. STAT. ch. 30, para. 901.

[35]*See* Motiuk & Sheridan, *New Jersey's ECRA: Problems, Policies, Future Trends,* 21 Env't. Rep. (BNA) No. 13, at 549 (July 27, 1990). State legislatures in New York, Pennsylvania, and Massachusetts considered ECRA-type legislation.

[36]N.J. STAT. ANN. § 13:1K-6; N.J. ADMIN. CODE § 7:26B-1.1 (Supp. 8-21-89).

[37]N.J. STAT. ANN. § 13:K-9.

[38]ECRA provided that:

[f]ailure of the transferor to comply with . . . this act is grounds for voiding the sale or transfer of an industrial establishment or any real property utilized in connection therewith by the transferee, entitles the transferee to recover damages from the transferor, and renders the owner or operator of the industrial establishment strictly liable, without regard to fault, for all cleanup and removal costs and for all direct and indirect damages resulting from the failure to implement the cleanup plan.

Thus, if the owner or operator of an industrial establishment failed to comply with ECRA, the NJDEP or the purchaser could have voided the transaction. The purchaser is entitled to recover damages resulting from avoidance of the transaction. Additionally, the owner or operator is strictly liable for cleanup and removal costs and damages caused by not implementing, or by implementing improperly, the cleanup plan.

IRSA ended this state right and determined that only private parties may sue to void the transaction.

no other state statute requiring environmental compliance on the happening of a business deal. It provided for the broad power of a transferee to void a transfer for any violation of ECRA.

This provision has been changed in ISRA to permit a right to void only (i) if the transferor fails to perform a remediation and obtain DEP approval and (ii) after the transferee gives notice to the transferor of that failure, affording the transferor a reasonable period of time to comply (i.e., a cure period). Although arguments had been made that this remedy should be eliminated, the legislature recognized that some actions have been filed by transferees seeking to void transactions and that transferees might need such protection in some circumstances.

DEP has lost its separate right to void a transfer for a failure to file a negative declaration or cleanup plan. The legislature was convinced that the threat of voiding was not a weapon needed by DEP because it had never sought to use it. It is not clear whether this loss applies retroactively to violations occurring prior to ISRA's effective date. DEP retains its power to fine up to $25,000 per day. With the loss of DEP's right to void, the regulated community may be more willing to engage in transactions that strain past interpretations of what events necessitate compliance with ISRA. DEP has a Remediation Guarantee Fund, which it can use to remediate a site when someone required to post a remediation funding source fails to do the required work.[39]

DEP can assert a lien against all property owned by the person required to do the work, and the discharger, to the same extent as the Spill Act allows under NJSA.[40] It can also seek trebles damages.[41] Under ISRA, section 48, there will be a study by DEP and the attorney general of key concepts of environmental liability. A report is due to the legislature within nine months.[42]

ISRA contains two separate amnesty provisions designed to allow past violations to be corrected without fines and penalties.[43] Those guilty of ECRA violations occurring before the effective date of ISRA are given one year to seek amnesty.[44] To be eligible, violators must enter into an Administrative Consent Order (ACO) or Memorandum of Agreement (MOA) which will govern their belated

---

[39]ISRA Section 45.

[40]N.J. ADMIN. CODE 58:10-23.11f.

[41]ISRA Sections 45.e. & f.

[42]N.J. ADMIN. CODE § 7:26B-9.2(b).

[43]ISRA Section 21.

[44]ECRA provides stringent civil penalties for submitting false information or failing to comply with ECRA:

[a]ny person who knowingly gives or causes to be given any false information or who fails to comply with the provisions of this act is liable for a penalty of not more than $25,000.00 for each offense. If the violation is of a continuing nature, each day during which it continues shall constitute an additional and separate offense. Penalties shall be collected in a civil action by a summary proceeding. . . . Any officer or management official of an industrial establishment who knowingly directs or authorizes the violation of any provisions of this act shall be personally liable for the penalties. . . .

compliance with ISRA. If the party thereafter fails to honor the ACO or MOA, the DEP can enforce penalties from the date of the original violation.

Amnesty does not appear to resolve other enforcement issues resulting from past violations (e.g., rights of transferees). Also, in order to get amnesty a person has to remediate the site. This remediation may be more costly and difficult than it would have been at the time of the violation and may require dealing with adverse interests of the current owner and operator in order to perform as required. Conversely, it may be easier if the site has been through ECRA since the violation may be discovered by reason of some other trigger. Information tendered in the amnesty process cannot be used in a criminal matter. DEP has yet to adopt regulations on amnesty.[45] ECRA may void the transaction if the owner or operator does not submit a negative declaration or cleanup plan.[46]

When ISRA is triggered, the owner or operator must follow a specific process. Five days after (i) the actual closing of operations or release of its decision to close or (ii) execution of an agreement to transfer, the owner or operator must provide written notice to DEP. This notice is equivalent to the present ECRA general information submission.[47] Subsequent to this initial notice, the owner or operator must proceed to remediate the site (which term includes investigate).[48] No specific period is set, but regulations will likely provide DEP's target schedule. Remediation includes preliminary assessment (equivalent to ECRA's site evaluation submission), site investigation and remedial investigation (collectively equivalent to ECRA's use of successive sampling plans to characterize a site's areas of concerns), and, when needed, a remedial action work plan (like ECRA's cleanup plan) and remedial action (the cleanup itself).

When a site is clean, an owner or operator can submit a negative declaration.[49] Upon approval, a deal that has not yet closed (e.g., under a remediation agreement) can proceed.[50] When a site is not clean, an owner or operator can submit a remedial action work plan, and once it is approved and the remediation funding source is posted the deal can occur. When all of the remediation at a contaminated site has been completed, DEP may issue a No Further Action letter.

N.J. Stat. Ann. 13:1K-13(c); *see also* N.J. Admin. Code § 7:26B-9.3(a)-(c). Thus, failure to comply with ECRA may result in stringent, escalating civil penalties. Moreover, any officer or manager who knowingly participates in violating ECRA is liable for the penalties. N.J. Admin. Code § 7:26B-9.3(c).

[45]Spill Act, ISRA Section 39. There is a separate amnesty procedure for certain violations of the Spill Act (unreported discharges) if a directive has not yet been issued, and if that person enters into an MOA or ACO to remedy the discharge. Information tendered in this process also cannot be used in a criminal matter. However, this immunity may be of no value if the violation of the Spill Act was also a violation of federal law. Given NJDEPE's recent interpretations concerning discharge-reporting obligations, this provision may be very useful in eliminating this liability. Once the amnesty period passes, NJDEPE enforcement will be more strenuous.

[46]N.J. Stat. Ann. 13:1K-13(b); N.J. Admin. Code § 7:26B-9.1(b).

[47]ISRA Section 4.a.

[48]ISRA Section 4.b.(1).

[49]ISRA Section 4.d.

[50]ISRA Section 4.c.

Concomitantly with the heavy burden put upon businesses and individuals, the NJDEP has had to cope with an enormous work load in administering ECRA. Recognizing that many operations fall within the broad industry groupings specified in ECRA, the NJDEP has exempted various subgroups within the interdicted primary SIC groupings that do not pose a significant threat to the environment. Some of the operations and transactions not subject to ECRA are the following: any business engaged principally in agricultural production or distribution; "[a] sale or transfer of assets of an industrial establishment . . . in the ordinary course of business"; and forty-three subgroups within SIC numbers 22-39, 46-49, 51, or 76, such as a water supply and an irrigation system.[51]

The NJDEP has adopted the practice of issuing applicability determinations, or letters of nonapplicability, which have become a safe harbor for many business transactions. Out of 5,000 of these requests processed in 1987, 93 percent resulted in exemptions or findings of nonapplicability. The NJDEP now routinely processes applications within a few days, and generally in no more than three weeks in the most complicated situations. A person seeking an applicability determination must submit a required form and fee to the NJDEP, must permit to enter and to inspect the property, and must show the NJDEP that the ECRA does not apply.[52]

Although ISRA is silent on the process, presumably DEP will continue its helpful process of issuing letters of nonapplicability. Letters of nonapplicability explain DEP's position as to whether a particular transaction or event triggers the need to comply with ECRA.

Even after granting exemptions, the NJDEP faces numerous case filings a year, which require NJDEP notification and individual compliance. An owner or operator should consider the following: (1) seeking an exemption for a low-volume generator; (2) filing a negative declaration, if possible; and (3) seeking an administrative consent order, which will permit a transaction to proceed, pending ECRA compliance, but places the applicant at the mercy of the NJDEP, which can require a substantial bond or other form of financial security.

Environmental Lien

New Jersey[53] is one of a minority of states that has enacted a superlien.

---

[51]N.J. ADMIN. CODE. § 7:26B-1.8(a)(3), (6), (b).

[52]*Id.* § 7:26B-1.9 (Supp. 8-21-89).

[53]*See* Kessler v. Tarrats, 476 A.2d at 326 (the lien was a proper exercise of police power and served a legitimate public purpose; it also withstood the contracts clause of the U.S. Constitution, which provides that "No state shall . . . pass any . . . Law impairing the Obligation of Contracts," and did not violate the secured creditor's private property rights).

New York

Voluntary Cleanup Program

The New York legislature first promulgated legislation to provide for the financing of brownfield redevelopment studies with the state in 1994.[54] This legislation has been carried over to today.[55] The bill provides for financial assistance to municipalities within the state for the undertaking of Brownfields Economic Redevelopment Studies, in order to assess and develop strategies for the promotion of voluntary cleanups of properties contaminated with hazardous substances.[56]

New York's Department of Environmental Conservation (NYDEC or DEC) has initiated its Voluntary Cleanup Program under its enforcement discretion.[57] To date, at least eight sites have been cleaned up under the program.[58] Under this informal program, parties may enter into agreements with the NYDEC to clean up limited sites while paying NYDEC's oversight fees.

These agreements may take the form of a consent order. In either form, the agreement must contain the following elements:

a description of activities to be performed on the site sufficient that DEC can determine whether it will be properly remediated;
the procedural guidelines to be followed in completing the remediation activities;
a requirement allowing DEC staff access to and oversight of the site;
a "hold harmless" clause from the applicant to the state; and
a provision indemnifying the state for its oversight costs.

Cleanup standards established in the agreement are determined on a case-by-case basis. DEC's Soil Standards Guidelines and the state's Groundwater Standards Regulations will almost always be consulted in calculating the appropriate measure. Land use will also be considered as a factor. Institutional controls, such as deed restrictions, may be required in combination with certain cleanup standards.

It is the policy of the DEC that neither an agreement or a consent order can address penalties assessed against a party for other environmental obligations beyond the scope of the agreement. Parties participating in the program are expected to pay all DEC oversight costs.

---

[54]*See* 1994 S.B. 7787, 215 Gen. Ass., 2d Sess. N.Y. Laws. *See also* Robert S. Berger, *Brownfields: The New York Approach* 6 ENVTL. LAW IN NEW YORK 1 (January 1995) (a discussion of the proposal to study brownfields legislation within the state of New York).

[55]1995 S.B. 5607, 219 Gen. Ass., 2d Sess., 1996.

[56]*Id.* § 24, tit. 17.

[57]Margaret Murphy, New York, at 1 (unpublished manuscript, on file with author).

[58]*See* Charles Bartsch & Elizabeth Collaton, COMING CLEAN FOR ECONOMIC DEVELOPMENT at 84.

Once cleanup is complete, the NYDEC provides a No Further Action (NFA) letter.[59] This NFA letter contains numerous reopeners. These include the following factors:

> contamination not found in the site assessment is subsequently discovered on the site;
> the state adopts new, more stringent cleanup standards;
> the site remediation is not deemed adequate for the site's intended use;
> a future owner changes the site use to one requiring higher levels of remediation;
> information was withheld from DEC; or
> the party enters into the agreement or consent order in a fraudulent manner.

Despite efforts, the New York agency-driven program has yet to be codified.[60] As such, no financial assistance is currently available to fund the New York projects.

Business Transfer Statute

Before developing the VCP, New York[61] enacted a statute affecting the transfer of hazardous waste sites and requiring notice and disclosure.

The statute requires approval of state environmental officials of transfers in New York.[62] Under New York law, a person may not "substantially change" the way "in which an inactive hazardous waste disposal site for which the commissioner of health [has issued] a declaration" without approval by the commissioner.[63] The commissioner will disapprove a change of use if the new use will hamper substantially a proposed, pending, or completed remedial action at the site or will harm the environment or health.[64]

In New York, a person may not "substantially change" the way "in which an inactive hazardous waste disposal site . . . is used" without notifying the New York Department of Environmental Conservation at least sixty days before the change.[65] A "substantial change" includes constructing a facility or paving a parking lot upon the property.[66] A "substantial change" also includes transferring title to or leasing the property.

In addition, the New York Inactive Hazardous Waste Disposal Sites Rules (N.Y. Rules)[67] apply to inactive hazardous waste disposal sites listed on the state's

---

[59]*Id.* at 2.
[60]*Id.* at 1.
[61]N.Y. ENVTL. CONSERV. LAW § 27-1317.
[62]*Id.* Compare N.Y. ENVTL. CONSERV. LAW. § 27-09-18 N.Y. COMP. CODES, RULES & REGS. tit.6, § 373-2.7(i) (2)(RCRA disclosure requirements).
[63]N.Y. ENVTL. CONSERV. LAW § 27-1317.
[64]*Id.*
[65]*Id.*
[66]*Id.*
[67]*Id.*

registry of those sites.[68] The N.Y. Rules are triggered by the transfer of title or lease of the property or before any "substantial change" in the use of the property.[69]

Prior to transferring, leasing, or substantially changing the use of a listed site, the owner must notify the NYDEC and the commissioner of health (the commissioner) in writing by certified mail.[70] If the commissioner issues a declaration that there is a condition on the property dangerous to life or health, the owner may not transfer, lease or substantially change its use without prior written approval of both the commissioner and the NYDEC.[71] The commissioner and NYDEC are required to approve or disapprove within sixty days following notification by the owner.[72]

Environmental Superlien

Efforts to create environmental superliens in New York were defeated.[73]

Nuisance Lien

New York state has been among the few states that has codified its nuisance law into a cleanup statute. This allows the state of New York to undertake cleanup actions in sites, such as those contaminated by petroleum, that are not authorized under other, more specific, environmental laws.

Under New York law, the term *nuisance* embraces whatever is dangerous to human life or detrimental to health. The term expressly includes, but is not be limited to:

(1) a public nuisance as known at common law, at statutory law, and in equity jurisprudence, and

(2) a dwelling that, in violation of any state or local law, ordinance, or regulation, does not have adequate egress, safeguards against fire, electrical service, installation and wiring, structural support, ventilation, plumbing, sewerage, or drainage facilities, is overcrowded or inadequately cleaned or lighted, and the condition constituting such violation is dangerous to human life or detrimental to health.

All such nuisances are considered unlawful.[74] Whenever the department declares that a dwelling is a nuisance, it must serve a notice or order reciting the facts

---

[68]*Id.* at § 375.9(a).

[69]*Id.*

[70]*Id.*

[71]*Id.* at § (b).

[72]*Id.*

[73]Robert S. Bozarth, *Environmental Liens and Title Insurance,* 23 U. RICH. L. REV. 305, 324 (Spring 1989). As were similar efforts in Pennsylvania and Kansas.

[74]N.Y. MULT. RESID. LAW § 305(1) (Consol. 1994) (Nuisances).

constituting such nuisance, specifying in what respect the dwelling is dangerous to human life or detrimental to health, and requiring the owner to remove such nuisance within thirty days after service of such notice or order, or such lesser period of time where an emergency exists as may be determined by the department head.[75] Such notice or order must provide that, if the owner fails to remove the specified nuisance within the proscribed compliance period, the department may remove or cause the removal of such nuisance by cleansing, repairing, vacating, demolishing, or taking other corrective action deemed necessary. In the event the state decides to step in and take corrective action, the state must notify the owner of his right to a hearing.[76]

Wherever notice of the presence of a nuisance is given, the owner may request a hearing before the head of the department charged with enforcement. The hearing must be given to the owner prior to the expiration of the prescribed period of compliance.

If the specified nuisance is not removed by the owner within the time prescribed after service of such notice or order, the department may proceed with the removal of such nuisance as provided in the notice or order.[77] If the owner refuses to permit the department to remove or cause the removal of such nuisance by cleansing, repairing, vacating, demolishing, or taking such other corrective action as may be necessary, or interferes in any way with the department or causes delay to the taking of corrective action, the department may cause such dwelling, in whole or in part, to be vacated and sealed up or vacated and demolished.[78] In such a case, the department must commence a special proceeding in the supreme court for the relief. During the pendency of the special proceeding, the department may obtain a temporary order for the immediate vacating of the dwelling, upon proof of a present danger to human life or detriment to health. In addition to the owner, all tenants, mortgagees, and lienholders of record must be necessary parties to such special proceedings.[79]

If the department proceeds to execute a notice or order issued by it or by the court for the removal of a nuisance, the department may let contracts therefor, in accordance with the provisions of any local laws, ordinances, rules, and regulations of the municipality applicable to the letting of contracts for municipal improvements.[80] The cost of executing such notice or order(s) whether or not carried out pursuant to court order must be met from any appropriation made therefor, or if such appropriation has not been made or is insufficient, from the proceeds of the sale of obligations pursuant to the local finance law. The department must keep a record of such notices and orders together with the acts done and the items of cost incurred in their execution.

---

[75]This notice must be prescribed in the manner mandated by law. *See* § 306 of this chapter.

[76]§ 305(2).

[77]§ 305(3a).

[78]§ 305(b).

[79]§ 305(3b).

[80]§ 305(4a).

The municipality will have a lien upon the premises of the dwelling for the cost of executing such notice or order or orders for the removal of a nuisance and must file a notice of such lien in the office of the clerk where notices of mechanics' liens are filed.[81] All proceedings with respect to such lien, its enforcement, and discharge must be carried on in the same manner as proceedings with respect to mechanics' liens under the lien law.

Notwithstanding the foregoing and in addition to any other remedy available, the department may maintain an action against the owner to recover the cost of executing such notice or order or orders.[82]

Puerto Rico

Puerto Rico has no brownfields redevelopment program.

---

[81]§ 305(4b).

[82]§ 305(4c). *See generally* HOUSING CODE ENFORCEMENT: SANCTIONS AND REMEDIES, 66 COLUM. L. REV. 1254. *See also* Motyka v. Amsterdam, 204 N.E.2d 635 (N.Y. 1965) (no liability is cast upon any public authority or municipal corporation by a failure to enforce the above statute); Stranger v. New York State Elec. & Gas Corp., 268 N.Y.S.2d 214 (N.Y. App. Div. 1966) (no liability is cast upon public authority by the Multiple Residence Law for private damage resulting from failure to enforce its directory provisions); Poughkeepsie v. Clifford, 502 N.Y.S.2d 768 (N.Y. App. Div. 1986) (city's application for authorization to demolish building as nuisance and to impose lien upon premises to cover costs of demolition, based upon delay by owner in renovation of building, must be denied where delay in renovation results from city's improper refusal to grant permit for renovation); Poughkeepsie v. Bastille Dev. Corp., 537 N.Y.S.2d 982 (N.Y. Sup. Ct. 1989) (owner of rental property was entitled to dismissal of proceeding brought by city pursuant to city code and N.Y. Mult. Resid. Law § 305 to have property vacated for failure to correct unfit and hazardous violations since the city code mandated a special proceeding pursuant to the Multiple Residence Law and the city failed to comply with procedural requirements of § 305, which did not provide for fulfillment of its procedural requirements by fulfillment of the city's procedural requirements); City of Poughkeepsie v. Bastille Dev. Corp., 537 N.Y.S.2d 982 (N.Y. Sup. Ct. 1989) (a proceeding brought by petitioner city of Poughkeepsie to have respondents' apartment building vacated because of the failure to correct hazardous violations, must be dismissed since the city ordinance relied upon by petitioner mandates, a "special proceeding pursuant to the Multiple Residence Law . . . to cause such building to be vacated" and petitioner's failure to comply with the procedural requirements of section 305 of the Multiple Residence Law as a condition precedent to the commencement of a special proceeding to declare the building a nuisance renders the proceeding defective; by requiring the commencement of a special proceeding pursuant to Multiple Residence Law § 305, petitioner is thereby required to follow the procedural dictates of that statute; a village may adopt a local law to remove unsafe or dangerous buildings constituting a nuisance, provided that the local law is consistent with constitutional due process requirements, as, for example, shown by the Multiple Dwelling Law, § 309[1][a] through [f], Multiple Residence Law, § 305[2], or General City Law, § 20[35]).

CHAPTER 6

Region III:
Delaware, District of Columbia, Maryland, Pennsylvania, Virginia, and West Virginia

The following table describes the more recent programs developed in Region III to encourage voluntary cleanup:[1]

States	Brownfields Financing Programs	VCP with Written Assurances	Incentives to Attract Private Development
Delaware	X	X	X
District of Columbia			
Maryland	X	X	X
Pennsylvania	X	X	X
Virginia		X	X
West Virginia	X	X	

Delaware

Voluntary Cleanup Program (VCP)

In 1993, Delaware began to operate a voluntary cleanup program under the auspices of its Hazardous Substances Cleanup Act (HSCA).[2] HSCA is the Delaware state equivalent of CERCLA. HSCA is administered by the Delaware Department of Natural Resources and Environmental Control (DNREC).

"The VCP is designed to provide owners of contaminated properties and persons interested in acquiring contaminated properties with an efficient, flexible process to complete an approved cleanup."[3] The Delaware VCP is, however, limited to properties that "pose no immediate threat to human health and the environment." Thus, certain threshold eligibility must be demonstrated to the agency by an applicant before a site can participate in the VCP program. This demonstration is generally conducted through a formal application to DNREC. The application process

---

[1]See Charles Bartsh, BROWNFIELDS STATE OF THE STATE REPORT: 50-STATE PROGRAM ROUNDUP (Northeast-Midwest Inst., January 28, 1998).

[2]DEL. CODE ANN. tit. 7, ch. 91 (1991).

[3]Todd S. Davis & Kevin Margolis, BROWNFIELDS: A COMPREHENSIVE GUIDE TO REDEVELOPING CONTAMINATED PROPERTY at 377 (1997).

is designed to aid DNREC and the applicant to determine appropriate remedial actions as well as site eligibility.

Certain sites are ineligible for participation. These include sites:

subject to RCRA or LUST;

affected by soil or groundwater contamination levels with a cancer risk of greater than 10^{-4} of a hazard index equal to 1;

with a well contaminated at or above the MCLs;

with soil or groundwater contamination within 300 feet of a drinking well;

where contamination affects a surface drinking water source and contamination levels are at or above MCLs;

where contamination affects surface water quality in excess of one order of magnitude over the ambient water quality;

where there may be any other reasons that the DNREC deems sufficient to reject the party's involvement with the VCP.

Sites that are rejects from the program or do not meet the jurisdictional requirements of the program because they contain one or more of the above site characteristics, are referred to the state CERCLA equivalent, the Hazardous Sites Control Act.

Applicants with mildly contaminated sites that are accepted into the program have two options. They can enter into a VCP agreement with DNREC. This agreement must include:

the name of the environmental consultant and laboratory to be used in site remediation; and

a detailed work plan for site investigation and remediation.

The work must then be completed by the applicant in accordance with the VCP agreement.

In the alternative, a party may enter into a letter agreement with DNREC whereby the party agrees to pay for oversight costs incurred by the agency. The party then proceeds to seek approval of an environmental consultant and laboratory from DNREC. Once approved, the party then can obtain approval of the work plan before entering into the ultimate VCP agreement.

All applicants must pay an initial deposit of $5,000 for oversight costs. Additional oversight costs over this initial deposit must be paid quarterly. The rate charged for oversight costs ranges from approximately $45 to $65 per hour.

Studies done by the applicant must conform to DNREC guidelines, which are described in the following table:

For	The Applicant Must Follow
Phase I studies	ASTM standards
Phase II studies	Delaware Rules and Regulations on facility studies

Parties participating in the Delaware program can select between two different cleanup standards: (1) trigger levels derived from the EPA Region III Risk-Based Concentration Tables or (2) site-specific cleanup standards. In general, use of the latter requires that the applicant also adopt certain institutional and/or engineering controls.

Once the remediation is complete, the applicant must submit information sufficient to demonstrate that the cleanup was conducted according to the approved plan and DNREC guidelines. The DNREC will then issue a NFA letter.

Tax Incentives

In addition to the VCP, Delaware enacted two major corporate tax incentives in July 1995 to encourage brownfields redevelopment. These laws are the Blue Collar Jobs Credit[4] and the Targeted Areas Credit.[5]

These laws allow for investment and employment credit against corporation income tax.[6] Thus, if a site is a qualified facility located on a brownfield, then the taxpayer can treat "as additional qualified investment all amounts expended by the taxpayer for environmental investigation and remediation of the brownfield." The total incremental credits allowable to the taxpayer under this law may not, however, exceed the aggregate amount expended by the taxpayer for environmental investigation and remediation of the brownfield.[7]

For purposes of this law, a *brownfield* is defined as:

> a vacant or unoccupied site with respect to any portion of which the taxpayer has reasonable cause to believe may, as a result of any prior commercial or industrial activity by any person, have been environmentally contaminated by the release or threatened release of a hazardous substance . . . in a manner that would interfere with the taxpayer's intended use of such site; provided, however, that such term shall not include any site or facilities with respect to any portion of which enforcement action has been initiated against any person pursuant to . . . [Delaware environmental law].[8]

The term *qualified facility* is defined as "any qualified property located within this State that constitutes a new facility or an expanded facility and that is used by the taxpayer in or in connection with a qualified activity."[9]

---

[4]DEL. CODE ANN. tit. 30 §§ 2010-14 (Supp. 1996).

[5]DEL. CODE ANN. tit. 30 §§ 2020-24 (Supp. 1996).

[6]30 Del. C. § 2011. I (1996).

[7]*Id. See also* 30 Del. C. § 2021(d) (1996).

[8]30 Del. C. § 2010(16) (1996). For purposes of establishing what is or is not a brownfield, the term *hazardous substance* is defined under 7 Del. C. c. 91. Enforcement actions that preclude participation in the program are those undertaken pursuant to Chapter 63, Chapter 74, or Chapter 91 of Title 7; 42 U.S.C. § 6901 et seq.; or 42 U.S.C. § 9606 or § 9607.

[9]30 Del. C. § 2010(1)(1996).

For practical purposes, the tax credit for sites demonstrating the necessary "development potential" amounts to $500 per year for each job created through the redevelopment of contaminated property. The tax credit runs for as long as necessary for the cost of cleanup to be recouped by the taxpayer.

Specialized Stakeholder Laws

Delaware was one of the states that amended its state equivalent of CERCLA to include some specialized lender liability protection.

Delaware law exempts, inter alia, from liability under its mini-CERCLA statute, any "commercial lending institution which acquires ownership or control of a property to realize on a security interest held by the person in that property or a fiduciary which has a legal title to or manages any property for purposes of administering an estate or trust of which such property is part."

A person who expends money performing a remedy or any remedial action may bring an action against any responsible party who has not entered into a settlement agreement with the secretary. The person bringing the action must be entitled to reimbursement for the costs incurred that are consistent with state law and contribution for money expended to reimburse the state for its expenses.

In addition, Delaware law contains a provision protecting past owners or operators "who can establish that at the time the facility was acquired or operated by the person, the person had no knowledge or reason to know of any release or imminent threat of release." To establish that a person had no knowledge or reason to know, the person must have undertaken, at the time of acquisition, "all appropriate inquiry into the previous uses of the property, consistent with good commercial or customary practice, in an effort to minimize liability." In determining what constitutes an appropriate inquiry, the secretary must take into account any specialized knowledge or experience on the part of the person, the relationship of the purchase price to the value of the property if uncontaminated, commonly known or reasonably ascertainable information about the property, the obviousness of the presence or likely presence of contamination at the property, and the ability to detect such contamination by appropriate inspection. The standard for determining knowledge or reason for knowledge under Delaware state law is expressly based on standard engineering practices at the time of operation or ownership of the facility. This defense is "not available to any person who had actual knowledge of the release or imminent threat of release when the person owned the real property and subsequently transferred ownership of the facility without first disclosing such knowledge to the buyer."[10]

---

[10]Delaware law also expressly allows a private cause of action under its mini-CERCLA statute:

A person who expends moneys performing a remedy or any remedial action under this chapter or reimbursing the State for any remedial action may bring an action against any responsible party as defined in subsection (a) of this section who has not entered into a settlement agreement with the Secretary. In an action authorized by this section, the person bringing the action shall be entitled to

Business Transfer Statute

The Delaware Hazardous Substance Cleanup Act (DHSCA)[11] covers facilities and property on which a release of a hazardous substance has occurred.[12] If the secretary determines the release is a threat to public health or the environment, DHSCA's requirements are triggered.[13]

DHSCA requires that the owner of the property place a notice in the real property records in the county where the property is located.[14] The notice must identify the facility[15] and the owner and recorder of the notice,[16] state that a release has occurred[17] and the date of the release,[18] and indicate that further inquiries should be directed to the secretary.[19]

District of Columbia

The District of Columbia has no brownfields redevelopment program. The District's Pennsylvania Avenue Development Corporation has, however, been working since the inauguration of President Kennedy to restore the Pennsylvania Avenue corridor located between the nation's Capitol Building and the White House to its former vigor.

Maryland

Voluntary Cleanup Program

In 1996, each house of the Maryland legislature enacted voluntary cleanup legislation.[20] The two bills could not, however, be reconciled before adjournment of the legislative session.

The legislation was reintroduced in 1997. The General Assembly of Maryland unanimously passed emergency legislation establishing both a statutory voluntary cleanup program and a financial incentives program.[21]

---

reimbursement for the costs incurred which are consistent with this chapter and contribution for moneys expended to reimburse the state for its expenses.

Id. (d).
[11]*Id.*
[12]DEL. CODE. ANN. tit. 7, § 9115(a).
[13]*Id.*
[14]*Id.*
[15]*Id.* at § (a)(1).
[16]*Id.* at § (2).
[17]*Id.* at § (3).
[18]*Id.* at § (4).
[19]*Id.* at § (5).
[20]Senate Bill 205; House Bill 5.
[21]Senate Bill 340; House Bill 409.

The purpose of the Maryland brownfields revitalization incentive program is to:

(1) provide financial incentives for redevelopment of properties previously used for commercial or industrial purposes;
(2) provide financial incentives for redevelopment of properties within locally designated growth areas;
(3) prevent urban sprawl;
(4) encourage economic revitalization;
(5) expand employment opportunities; and
(6) provide financial incentives for qualified brownfields sites.[22]

Maryland's Voluntary Cleanup Program will be administered by the Maryland Department of the Environment (MDE). The enabling legislation established a Voluntary Cleanup Fund to be administered by MDE.

The legislation requires MDE to adopt certain regulations to establish the VCP. The items to be addressed in implementing the regulation include:

establishing certain application requirements, including the payment of a certain fee;
establishing certain requirements for MDE's determination of whether to approve an application;
authorizing MDE to notify an applicant of a certain determination of no further requirements at the applicant's property, subject to certain conditions;
establishing procedures for the payment of certain additional costs or the refund of certain application fees under certain circumstances;
establishing certain requirements upon approval of an application;
establishing certain requirements and procedures for the proposal and approval of response action plans;
requiring a participant to file a certain bond or other security for certain purposes;
requiring MDE to review certain standards in a certain time period and authorizing MDE to revise the standards;
establishing certain procedures for public participation in MDE's process of approving response action plans;
establishing certain requirements for MDE's decision on a proposed response action plan, including the issuance of a certain letter that provides the participant with certain assurances;
establishing certain procedures and requirements for the withdrawal of an application or response action plan;
requiring a participant to notify MDE that a response action plan has been completed;

---

[22]MD. ANN. CODE art. 83A § 3-901 et seq. (1997).

requiring MDE to issue a certificate of completion, that includes certain information and certain assurances, under certain circumstances;

requiring MDE to send a copy of a certificate of completion to the director of
the department of assessments and taxation within a certain time period;

establishing certain conditions under which MDE retains enforcement authority against certain persons;

authorizing the transfer of certain documents under certain circumstances;

requiring certain documents to be recorded in certain land records under certain circumstances;

providing that the brownfields law does not affect the planning and zoning
authority of a county or municipal corporation and certain tort actions.

The law also establishes a Brownfields Revitalization Incentive Program in
the Department of Business and Economic Development (DBED). DBED selects
the brownfields sites that will participate in the VCP based on certain criteria established under regulations to be promulgated by DBED. DBED will also administer the Brownfields Revitalization Incentive Fund.[23]

The law further encourages brownfields redevelopment be requiring taxing
jurisdictions participating in the program to establish a certain property tax credit
for a certain period of time and for a certain amount. In addition, the local taxing
authority can grant an additional property tax credit up to a certain set amount.
County or municipal corporations can also grant a tax abatement for certain taxes
for certain properties.

For purposes of Maryland law, a *brownfields site* is defined as a site owned
or operated by an inculpable person[24] and located in a taxing jurisdiction that has
elected to participate in the brownfields revitalization incentive program.[25] The
term may also include a property where there is a release, discharge, or threatened
release of oil;[26] the site is subject to a corrective action plan approved by MDE,
and the property is located in a taxing jurisdiction that has elected to participate in
the brownfields revitalization incentive program. The term *brownfields site* does
not include property that is owned or operated by a responsible person or a person
responsible for the discharge.

A *qualified brownfields site* is defined as a brownfields site that has been determined by the DEBE to be eligible for financial incentives.[27]

After a person applies to participate in the voluntary cleanup program or
receives approval from MDE of the environment for the implementation of a
corrective action plan, the person may submit a request to the DBED to determine whether the person qualifies for financial incentives for the potential

---

[23]MD. ANN. CODE art. 83A § 3-903 (1997).

[24]*See* § 7-501.

[25]§ 3-901.

[26]*See* § 4-401.

[27]*See* § 3-903.

redevelopment of a brownfields site. Within thirty days after receipt of a such a request, DBED must notify an applicant whether, if approved to participate in the voluntary cleanup program or a corrective action plan, the applicant qualifies for financial incentives for the redevelopment of a brownfields site.

DBED must determine the eligibility of a site as a qualified brownfields site based on whether:

(1) the property is located in a densely populated urban center and is substantially underutilized; or

(2) the property is an existing or former industrial or commercial site that poses a threat to public health or the environment.

DBED may consider the following criteria when selecting a qualified brownfields site:

(1) the feasibility of redevelopment;

(2) the public benefit provided to the community and the state through the redevelopment of the property;

(3) the extent of releases or threatened releases at the site and the degree to which the cleanup and redevelopment of the site will protect public health or the environment;

(4) the potential to attract or retain manufacturing or other economic base-type employers;

(5) the absence of identifiable and financially solvent responsible persons; or

(6) any other factor relevant and appropriate to economic development.

During the course of evaluating potential qualified brownfields sites, DBED must consult with:

(1) MDE, the office of planning, and relevant local officials;

(2) the neighboring community and any citizen groups located in the community;

(3) representatives of state and local environmental organizations;

(4) public health experts; and

(5) any other person the department considers appropriate.

In DBED's notice of an applicant's qualification for financial incentives, DBED must specify which of the criteria the applicant met.

MDE and DBED must each report to the governor and certain committees of the General Assembly on the progress of (1) Voluntary Cleanup Program and Fund and (2) the Brownfields Revitalization Incentive Program, respectively, by certain dates.

Environmental Lien

Maryland has enacted a nonsuperlien environmental statute.[28]

Specialized Stakeholders Protection

Maryland was the very first state to amend its state equivalent of CERCLA to include some specialized lender liability protection.

Legislation protecting lenders began to be enacted at the state level in 1986, when the Maryland legislature amended its CERCLA equivalent[29] to legislatively veto the Federal District Court for Maryland's finding of potential lender liability on motion for summary judgment.[30] The trend accelerated following the Eleventh Circuit's decision in *Fleet Factors*, at which time the lending community's concern over CERCLA was heightened due to dicta in that case that declared that the unexercised ability to control the affairs of the debtor was sufficient to void the secured creditor exemption.[31] Of course, such legislation is ineffective against any federal CERCLA actions, public or private.

Pennsylvania

Voluntary Cleanup Program

Pennsylvania's progressive voluntary cleanup program, the Land Recycling Program (LRP), is comprised of three different statutes, which were signed into law as a package on May 19, 1995. These laws are:

(1) Land Recycling and Environmental Remediation Standards Act (Act 2);
(2) Economic Development Agency, Fiduciary and Lender Environmental Liability Protection Act (Act 3); and
(3) Industrial Sites Environmental Assessment Act (Act 4).

The LRP is administered by the Department of Natural Resources (DNR) in consultation with the Department of Commerce (DOC). The former handles environmental cleanup and oversight matters. The latter addresses financial support for LRP program participants. The laws have three purposes:

---

[28]Other states with nonsuperlien statutes include: Alaska, Arkansas, Illinois, Indiana, Iowa, Kentucky, Maryland, Michigan, Minnesota, Montana, Ohio, and Oregon.

[29]MD. CODE ANN., HEALTH ENVTL. § 7-201.

[30]United States v. Maryland Bank & Trust Co., 632 F. Supp. 573 (D. Md. 1986).

[31]*See, e.g.,* CAL. CIV. PROC. CODE § 726.5; COLO. REV. STAT. § 13-20-703; DEL. CODE ANN. tit. 7, § 9105(c)(3); HAW. REV. STAT. ch. 415, § 5/22.2; KEN. REV. STAT. ANN. § 224.01-400(26)-(27); ME. REV. STAT. ANN. tit. 38, §§ 1362 & 1367-A; MICH. COMP. LAWS ANN. § 299.612a; MINN. STAT. ANN. § 115B.03; MO. ANN. STAT. § 427.031; OR. REV. STAT. § 465.255; S.D. CODIFIED LAWS ANN. § 34A-15-3 through 5.

to make contaminated sites safe based on sound science;
to return these sites to productive use; and
to preserve farmland and green space.[32]

The Pennsylvania LRP is widely regarded as having one of the best brown-fields programs in the country.[33] In the short time since Governor Ridge signed the Land Recycling Act into law in May 1995, over 195 sites have begun the formal process toward redevelopment and a total of 64 have been completely re-mediated. "Compare that to the Federal scorecard for cleaning up contaminated sites in Pennsylvania under [CERCLA]. In 16 years, only 8 of Pennsylvania's 103 Superfund sites have been cleaned up and removed from the National Priority List."[34] James Seif, secretary of the Pennsylvania Department of Environmental Protection said:

Land recycling is the most significant environmental innovation developed in the last decade—an innovation pioneered by states in response to unreal-istic federal policies that actually encourage the abandonment of contami-nated properties. Returning properties to productive reuse free from envi-ronmental liabilities has not only obvious environmental benefits but economic benefits as well. And by encouraging businesses to locate on old industrial sites in towns and cities, land recycling may also turn out to be a major factor in reducing sprawl development and preserving open space and farmland.

The first statute, entitled the Land Recycling and Environmental Remediation Standards Act (Act 2), creates incentives to encourage responsible persons to volun-tarily develop and implement cleanup plans without the use of taxpayer funds or the need for adversarial enforcement actions.[35] Arguably as comprehensive as the Ohio Voluntary Action Program, the Act provides for the recycling of existing industrial and commercial sites, further defines cleanup liability of new industries and tenants, and establishes a framework for setting environmental remediation standards.[36] The Pennsylvania program also created an Industrial Land Recycling Fund[37] and

---

[32]*Pennsylvania's Land Recycling Program: Six Month Progress Report* at 1 (Feb. 1996), reprinted in KEY ENVIRONMENTAL ISSUES IN EPA REGION III (Philadelphia, Pennsylvania, April 9-10, 1996).

[33]Prepared Testimony of James M. Seif, Secretary, Pennsylvania Department of Environmental Pro-tection before the Senate Environment and Public Works Committee Superfund, Waste Control, and Risk Assessment Subcommittee (March 1997) ("Pennsylvania's Land Recycling Program is a major environmental success story for the Ridge Administration and has been selected as a national model by the American Legislative Exchange Council").

[34]*Id.*

[35]1995 PA S.B. 1.

[36]*See* Preamble, 1995 PA S.B. 1.

[37]1995 PA S.B. 1, § 701.

Voluntary Cleanup Loan Fund[38] to aid industrial site remediation. There are four cornerstones to Act 2:[39]

Priority	Description
Uniform cleanup standards	Establishes environmental remediation standards to provide a uniform framework for both voluntary and mandatory cleanups. Allows choice between three cleanup standards: background, statewide health, or site-specific.
Standardized review procedures	Describes the submission and review procedures to be used at sites using each of the three types of cleanup standard.
Releases from liability	Provided releases from liability for owners or developers of a site that has been remediated according to the standards and procedures in the law.
Financial assistance	Establishes the Industrial Sites Cleanup Fund, which is designed to help innocent persons conduct voluntary cleanup. Grants or low-interest loans are provided to cover up to 75 percent of the cost completing an environmental study and implementing a cleanup plan. The Department of Commerce administers the program.

Under the Land Recycling and Remediation Standards Act, volunteers must select and attain compliance with one or more specified cleanup standards.[40] These standards include background standards, statewide health-based standards, site-specific standards, or a combination thereof.[41]

---

[38]*Id.* at § 702. The purpose of the Voluntary Loan Fund is to assist volunteers through providing funding to persons undertaking voluntary remediation of a property. *Id.* at § 702(b). The following parties are eligible for low interest loans and grants for up to 75 percent of the costs incurred for conducting an environmental study and for implementing a cleanup plan:

local economic agencies;
municipalities;
public agencies; and
persons not responsible for site contamination.

[39]*Land Recycling Program Fact Sheet 1: Overview of the Land Recycling Program* at 1 (2530-FS-DEP1843 9/95), reprinted in KEY ENVIRONMENTAL ISSUES IN EPA REGION III (Philadelphia, Pennsylvania, April 9-10, 1996).

[40] *Id.* at § 302. The remediation standards established under this Act will be deemed to applicable, relevant, and appropriate requirements (ARARs) for the Commonwealth of Pennsylvania under federal CERCLA, as amended, 42 U.S.C. § 9601 *et seq.* and the Pennsylvania Hazardous Sites Cleanup Act (P.L. 169, No. 32). *Id.* at § 106. *See also* PENNSYLVANIA'S LAND RECYCLING PROGRAM TECHNICAL GUIDANCE MANUAL (2500-BK-DEP1853 7/18/95).

[41]*Id.* at § 301. Attainment of background standards must be demonstrated by the collection and analysis of representative samples of contaminated media located on site, including soils and groundwater in aquifers, through the use of statistical tests or other methods of generally recognized attainment procedures. *Id.* at § 302(a)(1). Statewide health standards, adopted by the Pennsylvania Environmental Quality Board, are designed to remove the threat of any substantial present or probable future risk to human health and the environment. *Id.* at § 301(a)(2). Site-specific standards achieve remediation levels based upon risk assessment procedures designed to eliminate or reduce any actual or potential threat to human health and the environment. *Id.* at § 301(a)(3).

Standard	Deed Restriction	Description
Background	None required provided background standard is attained for all regulated substances.	Most stringent standards; requires restoration of the site to its condition before contamination occurred.
Statewide health	Required only if nonresidential exposure factors are used.	Risk assessment goals must fall within certain ranges set for residential and nonresidential uses.
Site-specific	Required.	Party may chose to undertake a detailed risk assessment of the site to achieve a site-specific remedy for contamination. Standards may be no lower than federal standards. Limits are put on use of institutional controls.

For contaminated sites remediated to a level consistent with statewide cleanup standards, volunteers may meet either residential or nonresidential cleanup standards.[42] Compliance with background standards and state health-based standards are not subject to deed notice requirements, provided certain conditions are met.[43] Eliminating deed restriction requirements rewards the volunteers. Subsequent transfer of remediated property is not subjected to the stigma of being a formerly contaminated site. Therefore, a formerly impaired site can be sold and redeveloped free and clear of any indicia of past or present contamination.

---

[42]*Id.* at § 303(b). For statewide residential cleanup standards, the concentration of regulated substances in the soil cannot exceed either the direct contact medium-specific concentration based on residential exposure factors with a depth of fifteen feet from the existing ground surface or the soil-to-groundwater pathway numeric value throughout the soil column. *Id.* at § 303(b)(4). Residential soil-to-groundwater pathway numeric values may be determined by one of the following methods:

a value 100 times the medium specific concentration;
soil concentrations not producing leachate in excess of medium specific concentrations for groundwater in an aquifer under the Synthetic Precipitation Leaching Procedures developed by U.S. EPA; or
a generic value determined not to produce a concentration in groundwater in an aquifer in excess of medium specific concentrations based upon valid, peer-reviewed scientific methods.

Id. at § 303(b)(4)(i)-(iii).
[43]*See id.* at §§ 302(d) & 303(g). Persons attaining or demonstrating compliance with background or residential statewide standards for all regulated substances subject to remediation are not subject to deed acknowledgment requirements set forth in the Pennsylvania Solid Waste Management Act (P.L. 380, No. 97) and the Hazardous Sites Cleanup Act (P.L. 756, No. 108). For parties meeting residential statewide cleanup standards, an existing acknowledgment in a deed prior to completion of voluntary remedial activities may be removed from the chain of title. *Id.* Parties remediating sites to nonresidential statewide cleanup standards are required to meet the deed notice requirements in both the Solid Waste Management Act and Hazardous Sites Cleanup Act. *Id.*

Site-specific cleanup standards will require the development and implementation of specified procedures, including remedial investigation,[44] (if required) risk assessment procedures,[45] cleanup plans,[46] and a final report demonstrating completion of the approved remedial activities.[47] For contaminants known or suspected to be carcinogens, soil and groundwater cleanup standards must meet specific exposure factors.[48] Specifically, these standards should represent an excess upper-bound lifetime risk of between 1×10^{-4} and 1×10^{-6}.[49] The cumulative excess risk to populations exposed to brownfields, including sensitive subgroups, cannot exceed 1×10^{-4}.[50] For systemic toxicants, soil and groundwater cleanup standards must represent levels to which daily human exposure could occur without appreciable risk of deleterious effects.[51]

Volunteers satisfying program requirements are shielded from state environmental enforcement actions for the unanticipated future cleanup of contaminants.[52] Furthermore, volunteers should also receive protection from lawsuits brought by citizen groups and contribution actions brought by third parties responsible for site contamination.[53] The Act's liability protection provision extends to the current and future owners of contaminated industrial property, developers, successors, and assigns of any person to whom liability may apply, including public utilities performing activities on contaminated sites.[54] Nevertheless, the state retains certain reopeners:

---

[44]*Id.* at § 304(l)(1). Specifically, a remedial investigation report should include:

a description of the procedures and conclusions used to characterize the nature, extent, direction, rate of movement, volume, and composition of regulated substances;

concentrations of regulated substances in contaminated media, including summaries of sampling methodology, analytical results, and any information obtained from attempts to comply with either background or statewide standards;

a description of existing or potential public benefits of use or reuse of the property;

a fate and transport analysis demonstrating no present or future exposure pathways exist; and

if no exposure pathways exist, a risk assessment report and cleanup plan are required and no remedy is required. *Id.* at § 304(l)(1)(i)-(v).

[45]*Id.* at § 204(l)(2). If required, a risk assessment report should demonstrate the potential adverse effects under both current and planned future conditions caused by the presence of a regulated substance in the absence of any further control, remediation, or mitigation measures. *Id.*

[46]*Id.* at § 304(l)(3). The cleanup plan should select a remedy that achieves the requirements necessary to meet the applicable cleanup standards consistent with the intended future use of the facility.

[47]*Id.* at § 304(l)(4).

[48]*Id.* at § 304(b).

[49]*Id.*

[50]*Id.*

[51]*Id.* at § 304(c).

[52]*Id.* at § 501(a).

[53]*Id.*

[54]1995 PA S.B. 1, at § 501(a)(1)-(4).

DNR demonstrates that the executed remedy does not work;
contamination not previously found on the site is discovered in levels that increase health risks beyond the applicable risk assessment standard;
stricter standards are adopted by DNR in the future (unless the site was remediated during the "transition period");
land use of the site changes;
a discharge occurs on a nonindustrial site after the effective date of the legislation.

Financial Aid

The state of Pennsylvania has designated $17 million in financial aid for LRP pro-
gram participants. Five million dollars of this is specified for grants to municipalities. Private parties may obtain loans from the state, provided no more than
$200,000 can be used on any one assessment project and no more than $1 million
may be used in any given remediation project. The loans are payable at a rate of 2
percent over fifteen years. A caveat to all financial aid recipients is that they must
obtain at least 25 percent matching funds for the cleanup project. "All loans must
be sufficiently secured as determined by the Department of Commerce."[55]

In addition, the Infrastructure Development Program may also be used to finance site remediation. There are an additional $25.2 million in total funds avail-
able from this state initiative. Loans may not exceed $1.25 million per project. The
interest rate is 3 percent, again payable over fifteen years.

Funding Priority[56]	Other Criteria
Where contamination is reasonably suspected or known to exist;	Permanence of the remedy;
At sites for which there is a bona fide prospective purchaser, and/or at sites that present the greatest potential for redevelopment;	Financial need of the applicant; Ability of the applicant to repay the loan; Cost effectiveness of the project;
Which are local or regional development priorities;	The financial or economic distress of the area in which the project is being conducted;
Which will result in the cleanup of contamination that is significantly affecting the environment;	Project readiness.
Which have secured a high level of matching investment from other private and public sources.	

Funds may be used by eligible applicants for Phases I, II, and III environmental assessments and remediation of hazardous substances as described in the
following table:[57]

---

[55]INDUSTRIAL SITES REUSE PROGRAM GUIDELINES, "Investing in Pennsylvania's Future" at 4 (July 1995).
[56]Id.
[57]Id.

Term	Definition	Characteristics
Phase I environmental assessment	A qualitative review of the site, using readily available information, field observations, and sometimes limited screening of samples of soil and water.	Include preaudit, on-site and post-audit activities: review of regulations; records and document review;[58] on-site inspection; interviews with former employees, government agencies, and others familiar with the site; preparation of final report (to include recommendations for phase II audits).
Phase II and III environmental assessments	In-depth field investigation, site characterization, sampling, testing, and analysis to determine the source, nature, and extent of the problem, the risks involved, and the identification of the cost of remediation and possible cleanup alternatives.	Soil borings; UST testing; Installation of monitoring wells; Analysis of air, groundwater, surface water, and soil samples; Asbestos analysis; Transformer sampling; Analysis of dispersal pathways including in-depth studies of the site's geology, hydrogeology, and surface water resources; Development of remediation plans.
Remediation of hazardous substances	Removal and remediation of hazardous substances and contaminants.	Removal of containers, regulated substances, or contaminated media; on-site and off-site treatment, incineration, or destruction or segregation of wastes; Groundwater treatment, provision of alternative water supplies; Storage and containment; Covering; Neutralization; Recycling and reuse; Repair and replacement of containers or collection systems; Fencing and other security measures; Monitoring and maintenance.

Business Transfer Statute

Pennsylvania[59] has a statute affecting the transfer of hazardous waste sites and requiring notice and disclosure. The Pennsylvania Solid Waste Management Act[60]

---

[58]These include review of: abstracts and property maps; deed notices; liens and encumbrances; rights of way and easements; restrictive covenants; zoning records; subsurface rights; judgments; tax records, notices, and unpaid assessments; pending lawsuits; history of ownership and use; correspondence; building, sewage, water, and other permits; contracts for waste removal; location of sewage, drainage, electrical lines, waste processing, and storage areas; other modifications or improvements; and agreements or orders defining liability.

[59]PA. CONS. STAT. ANN. § 6018.405.

[60]*Id.*

and the Pennsylvania Hazardous Sites Act[61] require notice when property that has been exposed to hazardous substances is transferred. A grantor in a deed or other transfer instrument must disclose in the document the fact that real property has been used as a hazardous waste disposal site.[62] The law only applies if hazardous waste presently is being disposed of on the property or if the grantor has ever disposed of, or has actual knowledge of, any previous disposal.[63]

In addition, the Pennsylvania Hazardous Site Act (PHSA)[64] covers hazardous waste disposal sites.[65] PHSA is triggered if the grantor has actual knowledge that hazardous waste is being disposed of on the property or has been disposed of on the property in the past.[66] Once triggered, the grantor is required to place an acknowledgment of the disposal in the property description of the deed.[67] The acknowledgment should include, but is not limited to information on the size and location of the disposal site and a description of the hazardous wastes disposed.[68] The acknowledgment must be included in the deed for all conveyances and transfers once the notation is in place.[69]

Environmental Lien Law

Efforts to create environmental superliens in Pennsylvania were defeated.[70] The state lien statute is of the nonsuperlien variety. Pennsylvania has, however, a unique approach to providing for administrative or judicial review of the filing of an environmental lien. Pennsylvania has one of the most severe restrictions on postattachment review. In Pennsylvania, there is an absolute prohibition on judicial review of agency action prior to the state initiating an enforcement or cost recovery lawsuit.[71]

---

[61]*Id.* § 6020.512.

[62]*Id.*

[63]Pennsylvania is also currently considering several bills that impose additional ISRA-like obligations.

[64]PA. CONS. STAT. ANN. § 6018.405.

[65]*Id.*

[66]*Id.*

[67]*Id.*

[68]*Id.*

[69]*Id.*

[70]Robert S. Bozarth, *Environmental Liens and Title Insurance,* 23 U. RICH. L. REV. 305, 324 (Spring 1989).

[71]*See* MICH. STAT. ANN. § 13.32(16a); 35 PA. CONS. STAT. § 6020.509.

Virginia

Voluntary Cleanup Program

In 1995, The Commonwealth of Virginia first passed a bill to amending the Code of Virginia relating to the voluntary remediation of hazardous substances.[72] This law was amended in 1997 to give the state greater authority to regulate brownfields cleanup.[73] The Virginia Voluntary Cleanup Program allows persons who own, operate, have a security interest in, or enter into a contract for the purchase of contaminated property to voluntarily remediate releases of hazardous substances, hazardous wastes, solid wastes, or petroleum.[74] This program encourages the use of site-specific, risk-based standards "no more stringent than applicable federal standards" for the remediation of soil, groundwater, and sediments.[75] Virginia's program provides volunteers with flexibility regarding the selection of cleanup standards.[76] Each standard is based upon the proposed future use of the property and requires that the public health and environment can be assured.[77]

The Voluntary Cleanup Program permits the use of reasonably available and effective technology, as well as analytical quantification technology to achieve applicable cleanup standards.[78] Furthermore, institutional and engineering controls are also permitted remedial measures.[79] Procedures have been established for expedited administrative processes,[80] the issuance of Certificates of Completion,[81]

---

[72]1995 VA. H.B. 1847, VA. CODE ANN. § 10.1-1429.1-10.1-1429.3 (Michie 1995) (a bill to amend the Code of Virginia by adding Chapter 14 to Title 10.1 and Article 4.1, consisting of §§ 10.1-1429.1-10.1-1429.3, relating to voluntary remediation of hazardous substances).

[73]VA. CODE ANN. § 10.1-1429.1 (1997) (the state was allowed to administer its program on a case-by-case basis until it promulgated regulations).

[74]*Id.* § 10.1-1428.1(A).

[75]*Id.* § 10.1-1429.1(A)(1). The establishment of site-specific risk-based remediation standards must consider the following relevant factors:

protection of the public health and the environment;
future industrial, commercial, residential, or other uses of the property remediated and surrounding property;
reasonably available and effective remediation technology and analytical quantitation technology;
availability of institutional or engineering controls protective of human health and the environment; and
natural background levels.

Id.

[76]*Id.* § 10.1-1429.1(A)(1)(I)-(II).

[77]*Id.*

[78]*Id.* § 10.1-1429.1(A)(1)(III).

[79]*Id.* § 10.1-1429.1(A)(1)(IV).

[80]*Id.* § 10.1-1429.1(A)(2).

[81]*Id.* § 10.1-1429.1(A)(3). Certificates of Completion should be granted at the completion of remedial activities. This decision should be based on then present site conditions and available information satisfying applicable cleanup standards or where the state has determine that no further action is warranted. *Id.*

and the waiver or expedited issuance of any permits necessary to complete voluntary cleanups.[82] Finally, the Virginia program requires the remittance of a $5,000 application fee or 1 percent of the total cost of the remediation.[83] Satisfactory completion of voluntary cleanups and issuance of a Certificate of Completion will trigger statutory immunity from state environmental enforcement actions.[84]

Specialized Stakeholder Laws

Virginia was one of the states that amended its state equivalent of CERCLA to include some specialized lender liability protection.

Virginia law grants the fiduciary[85] additional powers and incorporates into any will or trust instrument the power to comply with environmental law:

(1) to inspect property held by the fiduciary, including interests in sole proprietorships, partnerships, or corporations, and any assets owned by any such business enterprise, for the purpose of determining compliance with environmental law[86] affecting such property and to respond to a change in, or any actual or threatened violation of, any environmental law affecting property held by the fiduciary;

(2) to take, on behalf of the estate or trust, any action necessary to respond to a change in, or prevent, abate, or otherwise remedy any actual or threatened violation of, any environmental law affecting property held by the fiduciary, either before or after the initiation of an enforcement action by any governmental body;

(3) to refuse to accept property in trust if the fiduciary determines that any property to be transferred to the trust either is contaminated by any hazardous substance or is being used or has been used for any activity directly or indirectly involving any hazardous substance which could result in liability to the trust or otherwise impair the value of the assets held therein;

(4) to disclaim any power granted by any document, statute, or rule of law that, in the sole discretion of the fiduciary, may cause the fiduciary to incur personal liability under any environmental law.

The fiduciary is entitled to charge the cost of any inspection, review, abatement, response, cleanup, or remedial action against the income or principal of the trust

---

[82]*Id.* at § 10.1-1429.1(A)(4).

[83]*Id.* § 10.1-1429.1(A)(5).

[84]*Id.* § 10.1-1429.2.

[85]The law defines a *fiduciary* as "one or more individuals or corporations having trust powers." A successor fiduciary will have all of the powers provided for the fiduciary named in the will or trust instrument.

[86]*Environmental law* is defined as "any federal, state, or local law, rule, regulation, or ordinance relating to protection of the environment or human health," and *hazardous substances* means "any substances defined as hazardous or toxic or otherwise regulated by any environmental law."

or estate. The fiduciary is not personally liable to any beneficiary or other party for any decrease in value of assets in trust or in an estate by reason of the fiduciary's compliance with any environmental law, specifically including any reporting requirement under such law. Neither the acceptance by the fiduciary of property or a failure by the fiduciary to inspect property will be deemed to create any inference as to whether or not there is or may be any liability under any environmental law with respect to such property.[87]

The Virginia law states that these protections may by reference be made applicable to a fiduciary of the estate of a decedent as well as to the trustee of an inter vivos or testamentary trust.[88]

West Virginia

Voluntary Cleanup Program

In 1995, West Virginia enacted the Voluntary Remediation and Redevelopment Act.[89] In that legislation, the legislature found that:

(1) there is property in West Virginia that is not being put to its highest productive use because it is contaminated or it is perceived to be contaminated as a result of past activity on the property;

(2) abandonment or underuse of contaminated or potentially contaminated industrial sites results in inefficient use of public facilities and services and increases the pressure for development of uncontaminated pristine land;[90]

(3) the existing legal structure creates uncertainty regarding the legal effect of remediation upon liability;[91]

---

[87]Of course, the fiduciary also has the power:

To resign as a fiduciary if the fiduciary reasonably believes that there is or may be a conflict of interest between it in its fiduciary capacity and in its individual capacity because of potential claims or liabilities which may be asserted against it on behalf of the trust or estate because of the type or condition of assets held therein.

Id. § 64.1-57(1)(u).

[88]For survey of Virginia law on wills, trusts, and estates for the year 1969-70, *see* 56 VA. L. REV. 1559 (1970); for the year 1971-72, *see* 58 VA. L. REV. 1363 (1972); for the year 1972-73, *see* 59 VA. L. REV. 1621 (1973); for the year 1973-74, see 60 VA. L. REV. 1632 (1974); for the year 1975-76, *see* 62 VA. L. REV. 1497 (1976); for the year 1979-80, *see* 67 VA. L. REV. 369 (1981); for the year 1989, *see* 23 U. RICH. L. REV. 859 (1989). For article on the Uniform Custodial Trust Act, *see* 24 U. RICH. L. REV. 65 (1989).

[89]W.VA. CODE § 22-22 *et seq.* (1996).

[90]Since existing industrial areas frequently have transportation networks, utilities, and an existing infrastructure, it can be less costly to society to redevelop existing industrial areas than to relocate amenities for industrial areas at pristine sites.

[91]Legal uncertainty serves as a further disincentive to productive redevelopment of brownfields. Therefore, incentives should be put in place to encourage voluntary redevelopment of contaminated or potentially contaminated sites.

(4) an administrative program should be established to encourage persons to voluntarily develop and implement remedial plans without the need for enforcement action by the division of environmental protection.

For that reason, the West Virginia law established (1) an administrative program to facilitate voluntary remediation activities and brownfield revitalization and (2) limitations on liability under environmental laws and rules for those persons who remediate sites in accordance with applicable standards established under state environmental law.[92] The law also provides financial incentives to entice investment at brownfield sites.[93]

For purposes of the West Virginia program, the term *brownfield* means any industrial or commercial property that is abandoned or not being actively used by the owner as of the July 1, 1996.[94] The term does not include any site subject to a unilateral enforcement order under CERCLA § 104 through § 106[95] or which has been listed or proposed to be listed by the U.S. EPA on the national priorities list.[96] The term also does not include a unilateral enforcement order under § 3008 and § 7003 of RCRA or West Virgina environmental law.[97]

Brownfield sites being remediated by persons who did not cause or contribute to the contamination of the site are eligible for consideration for remediation loans.[98]

Implementing regulations must be promulgated to get the program up and running by July 1, 1997. These rules must establish:

(1) an administrative program for both brownfield revitalization and voluntary remediation, including application procedures;
(2) procedures for the licensure of remediation specialists, including, but not limited to, establishing licensing fees, testing procedures, disciplinary procedures, and methods for revocation of licenses;
(3) procedures for community notification and involvement;
(4) risk-based standards for remediation;
(5) standards for the remediation of property;
(6) a risk protocol for conducting risk assessments and establishing risk-based standards.[99]

---

[92]W.Va. Code § 22-22-1 (1996).

[93]For information concerning the span, history, and subject matter of former articles one through thirteen, see generally Flannery, Beckett, and McThomas, *Consolidated Environmental Regulation in West Virginia,* 97 W.Va. L. Rev. 401 (1995).

[94]W.Va. Code § 22-22-2(b) (1996).

[95]Comprehensive Environmental Response, Compensation and Liability Act, 94 Stat. 2779, 42 U.S.C. §§ 9601, 9604-9606, as amended.

[96]W.Va. Code § 22-22-2 (1996).

[97]*Id.*

[98]W.Va. Code § 22-22-5(b) (1996).

[99]The risk protocol shall:

(1) require consideration of existing and reasonably anticipated future human exposures based on current and reasonably anticipated future land and water uses and significant adverse effects to ecological receptor health and viability;

(7) chemical and site-specific information, where appropriate for purposes of risk assessment.[100]
(8) criteria to evaluate and approve methods for the measurement of contaminants using the practical quantification level and related laboratory standards and practices to be used by certified laboratories;
(9) standards and procedures for the utilization of certificates of completion, land use covenants, and other legal documents necessary to effectuate the purposes of this article; and
(10) any other rules necessary to carry out the requirements and the legislative intent of the act.

Business Transfer Statute

West Virginia[101] enacted a statute affecting the transfer of hazardous waste sites and requiring notice and disclosure.

In West Virginia, a grantor in a deed or other transfer instrument or a lessor in a lease agreement must disclose in the document the fact that the real property or the subsurface thereof has been "used for the storage, treatment, or disposal of hazardous waste."[102] The West Virginia Hazardous Waste Management Act

---

(2) include, at a minimum, both central tendency and reasonable upper bound estimates of exposure;
(3) require risk assessments to consider, to the extent practicable, the range of probabilities of risks actually occurring, the range or size of populations likely to be exposed to risk, and quantitative and qualitative descriptions of uncertainties;
(4) establish criteria for what constitutes appropriate sources of toxicity information;
(5) address the use of probabilistic modeling;
(6) establish criteria for what constitutes appropriate criteria for the selection and application of fate and transport models;
(7) address the use of population risk estimates in addition to individual risk estimates;
(8) to the extent deemed appropriate and feasible by the director considering available scientific information, define appropriate approaches for addressing cumulative risks posed by multiple contaminants or multiple exposure pathways;
(9) establish appropriate sampling approaches and data quality requirements; and
(10) this protocol shall include public notification and involvement provisions so that the public can understand how remediation standards are applied to a site and provide for clear communication of site risk issues, including key risk assessment assumptions, uncertainties, populations considered, the context of site risks to other risks, and how the remedy will address site risks.

[100]Risk assessments should use chemical and site-specific data and analysis, such as toxicity, exposure, and fate and transport evaluations in preference to default assumptions. Where chemical and site-specific data are not available, a range and distribution of realistic and plausible assumptions should be employed.
[101]W. VA. CODE § 20-5E-20(a).
[102]W. VA. CODE § 20-5E-20(a). The code states:

(a) The grantor in any deed or other instrument of conveyance or any lessor in any lease or other instrument whereby any real property is let for a period of time shall disclose in such deed, lease or other instrument the fact that such property or the subsurface of such property, (whether or

(WVHWMA) only applies to a grantor or lessor who has an interest in the property at the time it is used for such purpose or who has actual knowledge of the previous use for such purpose.[103] Additionally, the grantee or the lessee must disclose, in writing and no later than the transfer or lease, to the grantor or the lessee any plan to use the property for the purpose of storage, treatment, or disposal of hazardous waste.[104]

RCRA

The West Virginia Hazardous Waste Management Act[105] applies to real property or subsurface[106] used for the storage, treatment, or disposal of hazardous waste.[107] The notice requirement of WVHWMA is triggered by conveyance or lease of such property.[108] If a grantor or lessor owned the property at the time treatment, storage, or disposal occurred, or has actual knowledge it was used for such purposes in the past,[109] he must disclose this knowledge in the deed, other instrument of conveyance, or lease.[110]

---

not the grantor or lessor is at the time of such conveyance or lease the owner of such subsurface) was used for the storage, treatment or disposal of hazardous waste. The provisions of this subsection shall only apply to those grantors or lessors who owned or had an interest in the real property when the same or the subsurface thereof was used for the purpose of storage, treatment or disposal of hazardous waste or who have actual knowledge that such real property or the subsurface thereof was used for such purpose or purposes at any time prior thereto.

(b) Any grantee of real estate or of any substrata underlying said real estate or any lessee for a term who intends to use the real estate conveyed or let or any substrata underlying the same for the purpose of storing, treating or disposing of hazardous waste shall disclose in writing at the time of such conveyance or lease or within thirty days prior thereto such fact to the grantor or lessor of such real estate or substrata. Such disclosure shall describe the proposed location upon said property of the site to be used for the storage, treatment or disposal of hazardous waste, the identity of such waste, the proposed method of storage, treatment or disposal to be used with respect to such waste and any and all other information required by rules and regulations of the director.

[103]*Id.*

[104]W.VA. CODE § 20-5E-20(b).

[105]*Id.*

[106]*Id.* at § -20(a). The requirements of WVHWMA apply to grantors or lessors of such property even if they do not own the property at the time of conveyance or lease. *Id.*

[107]*Id.* at §§ 20-5E-20(a), (b).

[108]*Id.* at § 20-5E-20(a).

[109]*Id.* at § 20-5E-20(a).

[110]*Id.* WVHWMA also imposes disclosure requirements on grantees and lessees. If a grantee or lessee intends to use the real estate or substrata for treatment, storage, or disposal, he must disclose this intent to the grantor or lessor in writing either at the time of the conveyance or lease or within thirty days prior to commencement of such acts. *Id.* at 20-5E-20(b). This provision is favorable to the grantor or lessor since CERCLA imposes liability on previous owners and present lessors. Thus, a statute requiring notice to a former owner or current lessor protects him from CERCLA liability if the grantee or tenant complies with such a provision.

CHAPTER 7

Region IV:
Alabama, Florida, Georgia, Kentucky, Mississippi, North Carolina, South Carolina, and Tennessee

The states in Region IV were the last to add voluntary cleanup programs to their various arsenals of brownfield redevelopment laws. All of the voluntary cleanup laws enacted by states in Region IV have been enacted since 1995.

The following chart describes the more recent programs developed in Region IV to encourage voluntary cleanup:[1]

States	Brownfields Financing Programs	VCP with Written Assurances	Incentives to Attract Private Development
Alabama	X	X	
Florida	X	X	X
Georgia	X	X	
Kentucky		X	
Mississippi			
North Carolina	X		
South Carolina	X		
Tennessee	X	X	

Alabama

Specialized Stakeholder Laws—Trustee

Alabama is one of the states that amended its state equivalent of CERCLA to include some specialized trustee liability protection.

On April 18, 1990, an Alabama law protecting fiduciaries from environmental liabilities went into effect.[2] Alabama law dictates that, in addition to powers, remedies, and rights that may be set forth in any will, trust agreement, or other document that is the source of authority, a trustee, executor, administrator, guardian, or one acting in any other fiduciary capacity, whether an individual, corporation,

---

[1]*See* Charles Bartsh, BROWNFIELDS STATE OF THE STATE REPORT: 50-STATE PROGRAM ROUNDUP (Northeast-Midwest Inst., January 28, 1998).

[2]ALA. CODE § 19-3-11 (powers of fiduciaries regarding environmental laws affecting property held by fiduciary).

or other entity ("fiduciary"), will have the following powers, rights, and remedies whether or not set forth in the will, trust agreement, or other document that is the source of authority:

(1) to inspect, investigate, or cause to be inspected and investigated property held by the fiduciary,[3]
(2) to take, on behalf of the estate or trust, any action necessary to prevent, abate, or otherwise remedy any actual or potential violation of any environmental law affecting property held by the fiduciary,[4]
(3) to refuse to accept property in trust,[5]
(4) to settle or compromise at any time any and all enviornmental claims against the trust or estate,[6]
(5) to disclaim any power granted by any document, statute, or rule of law that, in the sole discretion of the fiduciary, may cause the fiduciary to incur personal liability under any environmental law,
(6) to decline to serve as a fiduciary.[7]

The fiduciary will be entitled to charge the cost of any inspection, investigation, review, abatement, response, cleanup, or remedial action against the income or principal of the trust or estate. A fiduciary will not be personally liable to any beneficiary or other party for any decrease in value of assets in trust or in an estate by reason of the fiduciary's compliance or efforts to comply with any environmental law, specifically including any reporting requirement under such law. Neither the acceptance by the fiduciary of property or a failure by the fiduciary to inspect or

---

[3]For purpose of this law, the term "property" includes interests in sole proprietorships, partnerships, or corporations and any assets owned by any such business enterprise. The purpose of the investigation is limited to determining compliance with environmental law affecting such property and to respond to any actual or potential violation of any environmental law affecting property held by the fiduciary.

[4]These actions can be either before or after the initiation of an enforcement action by any governmental body.

[5]This rule is conditioned upon the fiduciary determining that any property to be donated or conveyed to the trust either is contaminated by any hazardous substance or is being used or has been used for any activity directly or indirectly involving any hazardous substance that could result in liability to the trust or otherwise impair the value of the assets held therein.

[6]This would include claims asserted by any governmental body or private party involving the alleged violation of any environmental law affecting property held in trust or in an estate.

[7]This ability is conditioned upon the fiduciary's reasonably belief that there is or may be a conflict of interest between the fiduciary in its or his fiduciary capacity and in its or his individual capacity because of potential claims or liabilities that may be asserted against the fiduciary on behalf of the trust or estate because of the type or condition of assets held therein.

investigate property will be deemed to create any inference as to whether or not there is or may be any liability under any environmental law[8] with respect to such property.

Finally, Alabama law states that "[a] fiduciary in its individual capacity shall not be considered an owner or operator of any property of the trust or estate for purposes of any environmental law."[9]

RCRA

Alabama has a state statute requiring that some form of general notice be placed on the public record or in the chain of title to alert prospective purchasers and the public of a site's status as formerly managing hazardous waste.[10]

Florida

Voluntary Cleanup Law

Florida enacted the Brownfields Redevelopment Act, which was signed into law on May 30, 1997.[11] Florida law defines *brownfield sites* as sites "that are generally abandoned, idled, or under-used industrial and commercial properties where expansion or redevelopment is complicated by actual or perceived environmental contamination."[12] Similarly, the law defines a *brownfield area* as "a contiguous area of one or more brownfield sites, some of which may not be contaminated, and which has been designated by a local government by resolution. Such areas may include all or portions of community redevelopment areas, enterprise zones, empowerment zones, other such designated economically deprived communities and areas, and environmental protection agency-designated brownfield pilot projects."[13]

---

[8]For purposes of this section, *environmental law* means any federal, state, or local law, rule, regulation, or ordinance relating to protection of the environment or human health. For purposes of this section, *hazardous substances* means any substance defined as hazardous or toxic or otherwise regulated by any environmental law. *Id.* §19-3-11(c).

[9]*Id.* §19-3-11(d). Note that Acts 1990, No. 90-476, § 2, provides:

> The provisions of this act shall apply to all fiduciary relations now existing or hereafter created, but only to the fiduciary actions or inactions occurring after the effective date hereof [April 18, 1990]. Nothing in this act shall change any standard of conduct otherwise applicable to the fiduciary.

[10]ALA. CODE § 22-30-9; ALA. ADMIN. CODE r. 14-5.07.

[11]*See* FLA. STAT. § 288.107 (1997) (brownfield redevelopment bonus refunds); FLA. STAT. § 376.77 *et seq.*

[12]FLA. STAT. § 367.79 (1997) (definitions).

[13]*Id.*

The Florida Act is predicated on certain legislative findings. First, the legislature determined that "the reduction of public health and environmental hazards on existing commercial and industrial sites is vital to their use and reuse as sources of employment, housing, recreation, and open-space areas."[14] The reuse of industrial land is an important component of sound land-use policy for productive urban purposes that will help prevent the premature development of farmland, open-space areas, and natural areas, and reduce public costs for installing new water, sewer, and highway infrastructure.[15]

Second, the legislature determined that the abandonment or underuse of brownfield sites also results in the inefficient use of public facilities and services, as well as land and other natural resources, and extends conditions of blight in local communities. It also contributes to concerns about environmental equity and the distribution of environmental risks across population groups.[16]

Third, it was determined that incentives should be put in place to encourage responsible persons to voluntarily develop and implement cleanup plans without the use of taxpayer funds or the need for enforcement actions by state and local governments.[17]

Fourth, Florida recognized that "environmental and public health hazards cannot be eliminated without clear, predictable remediation standards that provide for the protection of the environment and public health."[18]

Fifth, the state determined that "site rehabilitation should be based on the actual risk that contamination may pose to the environment and public health, taking into account current and future land and water use and the degree to which contamination may spread and place the public or the environment at risk."

Sixth, according to the statistical proximity study contained in the final report of the environmental equity and justice commission, minority and low-income communities are disproportionately impacted by targeted environmentally hazardous sites. The results indicate the need for the health and risk exposure assessments of minority and poverty populations around environmentally hazardous sites in this state. Redevelopment of hazardous sites should address questions relating to environmental and health consequences.[19]

Seventh, Florida legislated that "environmental justice considerations should be inherent in meaningful public participation elements of a brownfields redevelopment program."[20]

Eighth, "the existence of brownfields within a community may contribute to, or may be a symptom of, overall community decline, including issues of human

---

[14]Fla. Stat. § 367.78 (1997) (legislative intent).
[15]Id.
[16]Id.
[17]Id.
[18]Id.
[19]Id.
[20]Id.

disease and illness, crime, educational and employment opportunities, and infrastructure decay."[21] The environment is, of course, "an important element of quality of life in any community, along with economic opportunity, educational achievement, access to health care, housing quality and availability, provision of governmental services, and other socioeconomic factors. Brownfields redevelopment, properly done, can be a significant element in community revitalization."[22]

Ninth, cooperation among federal, state, and local agencies, local community development organizations, and current owners and prospective purchasers of brownfield sites is required to accomplish timely cleanup activities and the redevelopment or reuse of brownfield sites.[23]

Accordingly, the Act asks local governments in Florida to designate brownfields areas that should be targeted for redevelopment. Local governments participating in the Florida brownfields redevelopment program must establish an advisory committee to ensure adequate public participation in the redevelopment process.[24]

The person responsible for brownfield site rehabilitation must enter into a brownfield site rehabilitation agreement with the department or an approved local environmental program. The brownfield site rehabilitation agreement must include a commitment to:

> a brownfield site rehabilitation schedule, including milestones for completion of site rehabilitation tasks and submittal of technical reports and rehabilitation plans as agreed upon by the parties to the agreement;
>
> time frames for the department's review of technical reports and plans submitted in accordance with the agreement;[25]
>
> conduct site rehabilitation activities under the observation of professional engineers or geologists who are registered under Florida state law;[26]
>
> conduct site rehabilitation in accordance with an approved comprehensive quality assurance plan under department rules;

---

[21]*Id.*

[22]*Id.*

[23]*Id.*

[24]*Id.* § 376.80.

[25]The department must make every effort to adhere to established agency goals for reasonable time frames for review of such documents.

[26]*See* chapter 471 or chapter 492, respectively. Submittals provided by the person responsible for brownfield site rehabilitation must be signed and sealed by a professional engineer registered under chapter 471, or a professional geologist registered under chapter 492, certifying that the submittal and associated work comply with the law and rules of the department and those governing the profession. In addition, upon completion of the approved remedial action, the department shall require a professional engineer registered under chapter 471 or a professional geologist registered under chapter 492 to certify that the corrective action was, to the best of his or her knowledge, completed in substantial conformance with the plans and specifications approved by the department.

conduct site rehabilitation consistent with state, federal, and local laws and consistent with the brownfield site contamination cleanup criteria required under Florida law;[27]

secure site access for the department or approved local environmental program to all brownfield sites within the eligible brownfield area for activities associated with site rehabilitation;

consider appropriate pollution prevention measures and to implement those that the person determines are reasonable and cost-effective, taking into account the ultimate use or uses of the brownfield site;[28]

other provisions that the person responsible for brownfield site rehabilitation and the department agree upon, that are consistent with law[29] and that will improve or enhance the brownfield site rehabilitation process.

In addition, the agreement must contain the following additional provisions:

An agreement between the person responsible for site rehabilitation and the local government with jurisdiction over the brownfield, which agreement must contain terms for the redevelopment of the brownfield;

Any professional engineer or geologist providing professional services relating to site rehabilitation program tasks must carry professional liability insurance with a coverage limit of at least $1 million;

During the cleanup process, if the department or local program fails to complete review of a technical document within the timeframe specified in the brownfield site rehabilitation agreement, the person responsible for brownfield site rehabilitation may proceed to the next site rehabilitation task;[30]

If the person responsible for brownfield site rehabilitation fails to comply with the brownfield site rehabilitation agreement, the department must allow ninety days for the person responsible for brownfield site rehabilitation to return to compliance with the provision at issue or to negotiate a modification to the brownfield site rehabilitation agreement with the department for good cause shown;[31]

---

[27]*See* § 376.81 (including any applicable requirements for risk-based corrective action).

[28]Such measures may include improved inventory or production controls and procedures for preventing a loss, spills, and leaks of hazardous waste and materials, and include goals for the reduction of releases of toxic materials.

[29]§ § 376.77-376.84.

[30]Note, however, that the person responsible for brownfield site rehabilitation does so at its own risk and may be required by the department or local program to complete additional work on a previous task. Exceptions to this subsection include requests for "no further action," "monitoring only proposals," and feasibility studies, which must be approved prior to implementation.

[31]If an imminent hazard exists, the ninety-day grace period will not apply. If the project is not returned to compliance with the brownfield site rehabilitation agreement and a modification cannot be negotiated, the immunity provisions are revoked.

The department is specifically authorized and encouraged to enter into delegation agreements with local pollution control programs,[32]

Local governments are encouraged to use the full range of economic and tax incentives available to facilitate and promote the rehabilitation of brownfield areas, to help eliminate the public health and environmental hazards, and to promote the creation of jobs and economic development in these previously run-down, blighted, and underutilized areas.

Any contractor performing site rehabilitation program tasks must demonstrate to the department that the contractor:

meets all certification and license requirements imposed by law; and

has obtained approval for the comprehensive quality-assurance plan prepared under department rules.

The contractor must certify to the department that the contractor:

complies with applicable OSHA regulations;

maintains workers' compensation insurance for all employees as required by the Florida workers' compensation law;

maintains comprehensive general liability and comprehensive automobile liability insurance with minimum limits of at least $1 million per occurrence and $1 million annual aggregate, sufficient to protect it from claims for damage for personal injury, including accidental death, as well as claims for property damage that may arise from performance of work under the program, designating the state as an additional insured party;

maintains professional liability insurance of at least $1 million per occurrence and $1 million annual aggregate;

---

[32]Approved under section 403.182. To administer the brownfield program within their jurisdictions, thereby maximizing the integration of this process with the other local development processes needed to facilitate redevelopment of a brownfield area. When determining whether a delegation pursuant to this subsection of all or part of the brownfields program to a local pollution control program is appropriate, the department shall consider the following. The local pollution control program must:

(a) have and maintain the administrative organization, staff, financial, and other resources to effectively and efficiently implement and enforce the statutory requirements of the delegated brownfields program; and

(b) provide for the enforcement of the requirements of the delegated brownfields program, and for notice and a right to challenge governmental action, by appropriate administrative and judicial process, which shall be specified in the delegation. The local pollution control program shall not be delegated authority to take action on or to make decisions regarding any brownfield on land owned by the local government. Any delegation agreement entered into pursuant to this subsection shall contain such terms and conditions necessary to ensure the effective and efficient administration and enforcement of the statutory requirements of the brownfields program as established by the act and the relevant rules and other criteria of the department.

has the capacity to perform or directly supervise the majority of the work at a site in accordance with Florida law.[33]

The state is to promulgate implementing regulations dictating, among other things, the cleanup criteria to be used in brownfields sites. The rule must also include protocols for the use of natural attenuation and the issuance of "no further action" letters. The law only applies to parties that did not cause or contribute to the creation of the brownfield site. The law will become effective on July 1, 1997. With regard to brownfield redevelopment economic incentives:

> It is the intent of the legislature that brownfield redevelopment activities be viewed as opportunities to significantly improve the utilization, general condition, and appearance of these sites. Different standards than those in place for new development, as allowed under current state and local laws, should be used to the fullest extent to encourage the redevelopment of a brownfield. State and local governments are encouraged to offer redevelopment incentives for this purpose, as an ongoing public investment in infrastructure and services, to help eliminate the public health and environmental hazards, and to promote the creation of jobs in these areas. Such incentives may include financial, regulatory, and technical assistance to persons and businesses involved in the redevelopment of the brownfield pursuant to this act.

Accordingly the legislature has designated that the agency may wish to consider the incentives described in the following table:

Type of Incentive	Examples
Financial incentives and local incentives for redevelopment	(a) tax increment financing; (b) enterprise zone tax exemptions for businesses; (c) safe neighborhood improvement districts; (d) waiver, reduction, or limitation by line of business with respect to occupational license taxes; (e) tax exemption for historic properties; (f) residential electricity exemption of up to the first 500 kilowatts of use may be exempted from the municipal public service tax; (g) minority business enterprise programs; (h) electric and gas tax exemption; (i) economic development tax abatement; (j) grants, including community development block grants; (k) pledging of revenues to secure bonds. (l) low-interest revolving loans and zero-interest loan pools; (m) local grant programs for facade, storefront, signage, and other business improvements;

(continued)

---

[33]*See* § 489.113(9).

Type of Incentive	Examples
	(n) governmental coordination of loan programs with lenders, such as microloans, business reserve fund loans, letter of credit enhancements, gap financing, land lease and sublease loans, and private equity.
	(o) payment schedules over time for payment of fees, within criteria, and marginal cost pricing.
Regulatory incentives	(a) cities' absorption of developers' concurrency needs;
	(b) developers' performance of certain analyses;
	(c) exemptions and lessening of state and local review requirements;
	(d) water and sewer regulatory incentives;
	(e) waiver of transportation impact fees and permit fees;
	(f) zoning incentives to reduce review requirements for redevelopment changes in use and occupancy, establishment of code criteria for specific uses, and institution of credits for previous use within the area;
	(g) flexibility in parking standards and buffer zone standards;
	(h) environmental management through specific code criteria and conditions allowed by current law;
	(i) maintenance standards and activities by ordinance and otherwise, and increased security and crime prevention measures available through special assessments;
	(j) traffic-calming measures;
	(k) historic preservation ordinances, loan programs, and review and permitting procedures;
	(l) one-stop permitting and streamlined development and permitting process.
Technical assistance incentives	(a) expedited development applications;
	(b) formal and informal information on business incentives and financial programs;
	(c) site design assistance;
	(d) marketing and promotion of projects or areas.

Other Laws

Although not specifically a "brownfields" statute, the Florida Quality Developments Program was enacted to "encourage development which has been thoughtfully planned to take into consideration protection of Florida's natural amenities, the cost to local government of providing services to a growing community, and the high quality of life Floridians desire."[34] To this extent, the statute mandates "that the developer be provided, through a cooperative and coordinated effort, an expeditious and timely review by all agencies with jurisdiction over the project of his or her proposed development."[35]

Among the criteria listed as necessary for a developer to participate in the program is that the project "participate in a downtown reuse or redevelopment program to improve and rehabilitate a declining downtown area."[36]

---

[34]FLA. STAT. § 380.061(1)(1996).
[35]*Id.*
[36]*Id.*

In addition, the Florida state comprehensive plan[37] dictates that the state focus on "downtown revitalization." The legislature mandated that, "In recognition of the importance of Florida's developing and redeveloping downtowns to the state's ability to use existing infrastructure and to accommodate growth in an orderly, efficient, and environmentally acceptable manner, Florida shall encourage the centralization of commercial, governmental, retail, residential, and cultural activities within downtown areas." To do this, three policies were established:

(1) provide incentives to encourage private sector investment in the preservation and enhancement of downtown areas;
(2) assist local governments in the planning, financing, and implementation of development efforts aimed at revitalizing distressed downtown areas;
(3) promote state programs and investments that encourage redevelopment of downtown areas.

In addition, the Florida state comprehensive plan dictates that the state:

participate in a downtown reuse or redevelopment program to improve and rehabilitate a declining downtown area;
promote state programs, investments, and development and redevelopment activities that encourage efficient development and occur in areas that will have the capacity to service new population and commerce;
develop a system of incentives and disincentives that encourages a separation of urban and rural land uses while protecting water supplies, resource development, and fish and wildlife habitats;
enhance the livability and character of urban areas through the encouragement of an attractive and functional mix of living, working, shopping, and recreational activities;
"promote state programs and investments which encourage redevelopment of downtown areas;"
promote effective coordination among various modes of transportation in urban areas to assist urban development and redevelopment efforts.[38]

Kentucky

Voluntary Cleanup Legislation

Kentucky's newly proposed bill centers around the application for a No Further Remediation letter to be given by the state to entities who successfully and voluntarily cleanup facilities to the appropriate level.[39] The bill allows for the application

---

[37]*Id.*
[38]*Id.*
[39]Kentucky H.B. 725, introduced February 28, 1996.

by any party to the state for a No Further Remediation letter. The entity's application must include:

(1) a legal description of the property;
(2) a copy of the deed to the property;
(3) an environmental site assessment of the property sufficient to show the nature and extent of any hazardous waste at the site;
(4) a proposed plan to remediate the environmental contamination at the site; and
(5) the subsequent proposed use of the property after obtaining the No Further Remediation letter.[40]

The state has the ability to approve the application, to deny the application after public comment, or to enter into negotiations with the party regarding the proposed remediation activities.[41] If approval of the application is made, the applicant has the affirmative duty to follow the provisions of the approved plan to afford remediation.[42] The state will issue the No Further Remediation letter only after the successful completion of the remediation activities.[43]

The bill further states that the No Further Remediation letter shall be construed as prima facie evidence that the property does not constitute a threat to human health and the environment and, additionally, the property does not require further remediation activities.[44] Such a letter is to be limited to and include:

(1) acknowledgment that the requirement of the remediation plan was satisfied;
(2) legal description of the property location;
(3) any remediation objectives, including monitoring requirements or land use restrictions imposed;
(4) statement that the No Further Remediation letter constitutes completion of the remediation activities with no further cleanup efforts required, secures a release from further responsibilities under the laws, and stands as prima facie evidence of no threat to public health or the environment;
(5) a prohibition against uses that are inconsistent with the land use limitations imposed on the property, without additional remedial activities and conveyance of the parcel to a third party; and
(6) a description of any preventive, engineering, and institutional controls required in the remediation plan and notification that failure to manage and maintain the controls with the terms of the remediation plan may result in voidance of the No Further Remediation letter.[45]

---

[40]H.B. 725 § 2(1).
[41]*Id.* § 2(2).
[42]*Id.* § 2(3).
[43]*Id.* § 2(4).
[44]*Id.* § 3(1).
[45]*Id.* § 3(2).

The letter shall apply in favor of the applicant who was issued the letter, mortgagee, trustee, or successor in interest of a deed of trust, any successor in interest to the property, any transferee of the property through bankruptcy, partition, settlement, or adjudication of a civil action gift or bequest, and any financial institution gaining control over the property by and through foreclosure or other terms of security interest held on the property.[46] Finally, the bill states that the No Further Action letter may be voidable if the property is managed in contrary to the law.[47]

Business Transfer Statute

The Kentucky Waste Management Act (KWMA)[48] is the only statute that requires the state to record disclosure of hazardous waste disposal activities.[49] The KWMA applies to hazardous waste disposal sites.[50] If the state determines that disposal activities have occurred, the notice requirement is triggered.[51] Once triggered, the Natural Resources and Environmental Protection Cabinet (the Cabinet) must record a notice with the local county clerk.[52] The notice must contain a legal description of the property that discloses to any potential transferee the land was used to dispose hazardous waste and that further information on the site is available from the Cabinet.[53]

Specialized Stakeholder Laws—Lender and Trustee

Kentucky was one of the states that amended its state equivalent of CERCLA to include some specialized lender and trustee liability protection.

Kentucky law defines a *financial institution* as:

(1) a bank or trust company;[54]
(2) a savings and loan association;[55]
(3) a credit union;[56]
(4) a mortgage loan company or loan broker;[57]
(5) an insurer;[58] and

---

[46]*Id.* § 3(3).
[47]*Id.* § 3(4).
[48]KEN. REV. STAT. ANN. § 224.876(16).
[49]*Id.*
[50]*Id.*
[51]*Id.*
[52]*Id.*
[53]*Id.*
[54]Defined by *id.,* at Chapter 287.
[55]Defined by *id.,* at Chapter 289.
[56]Defined by *id.,* at Chapter 290.
[57]Defined by *id.,* at Chapter 294.
[58]Defined by *id.,* at Chapter 304.

(6) any other financial institution engaged in the business of lending money, the lending operations of which are subject to state or federal regulation.[59]

If a financial institution exempted from liability under the state Superfund law conveys the site it has acquired, then the state will have a lien against the site for the actual and necessary costs expended in response to a release or threatened release or an environmental emergency. The lien must be filed with the county clerk of the county in which the site is located.[60]

A financial institution[61] that acquired a site by foreclosure, by receiving an assignment, by deed in lieu of foreclosure, or by otherwise becoming the owner as a result of the enforcement of a mortgage, lien, or other security interest held by the financial institution will not be held liable if:

(1) the financial institution served only in an administrative, custodial, financial, or similar capacity with respect to the site before its acquisition;
(2) the financial institution did not control or direct the handling of the material causing the environmental emergency, or control or direct the handling of the hazardous substance, pollutant, or contaminants at the site before its acquisition;
(3) the financial institution did not participate in the day-to-day management of the site before its acquisition;
(4) the financial institution, at the time it acquired the site, did not know and had no reason to know that a hazardous substance, pollutant, or contaminant was disposed of at the site.

For purposes of this law, the financial institution must have undertaken, at the time of acquisition, all appropriate inquiries into the previous ownership and uses of the property consistent with good commercial or customary practice in an effort to

---

[59]A *fiduciary* is defined by Ky. Rev. Stat. Ann., Chapter 386.
[60]*See also* (23)(a):

The cabinet shall have a lien against the real and personal property of a person liable for the actual and necessary costs expended in response to a release or threatened release or an environmental emergency. The lien shall be filed with the county clerk of the county in which the property of the person is located.

[61]*Financial institution* is defined as:

1. a bank or trust company defined by KRS Chapter 287;
2. a savings and loan association defined by KRS Chapter 289;
3. a credit union defined by KRS Chapter 290;
4. a mortgage loan company or loan broker defined by KRS Chapter 294;
5. an insurer defined by KRS Chapter 304; and
6. any other financial institution engaged in the business of lending money, the lending operations of which are subject to state or federal regulation.

Id. § 224.01-400 (h).

minimize liability. What actions constitute all appropriate inquiries will be determined by taking into account any specialized knowledge or experience on the part of the financial institution, the relationship of the market value of the site to the value of the site if uncontaminated, commonly known or reasonably ascertainable information about the site, the obviousness of the presence or likely presence of contamination at the site, the ability to detect the contamination by appropriate inspection, and any other relevant factor.

Even more importantly, under Kentucky law, when a financial institution undertakes actions to protect or preserve the value of the site, it will not be held liable if it undertakes those actions in accordance with law and the financial institution, its employees, agents, and contractors did not cause or contribute to an environmental emergency or to a release or threatened release of a hazardous substance, pollutant, or contaminant. The financial institution must, however, comply with the release notification requirements of the state law in order to qualify for this exemption from liability.

A financial institution serving as a fiduciary[62] with respect to an estate or trust, the assets of which contain a site,[63] will not be liable if the financial institution:

(1) served only in an administrative, custodial, financial, or similar capacity with respect to the site before it became a fiduciary;

(2) did not control or direct the handling of the material causing the environmental[64] emergency[65] or control or direct the handling of the hazardous substance,[66]

---

[62]*Fiduciary* is defined the same for purposes of this statute as "a fiduciary as defined by KRS Chapter 386." *Id.* § 224.01-400 (i).

[63]*Site* is defined as "any building, structure, installation, equipment, pipe, or pipeline, including any pipe into a sewer or publicly-owned treatment works, well, pit, pond, lagoon, impoundment, ditch, landfill, storage containers, motor vehicles, rolling stock, or aircraft, or any other place or area where a release or threatened release has occurred." The term does not include any consumer product in consumer use. *Id.* § 224.01-400 (1)(a).

[64]*Environment* means "the waters of the Commonwealth, land surface, surface, and subsurface soils and strata, or ambient air within the Commonwealth or under the jurisdiction of the Commonwealth." *Id.* § 24.01-400 (1)(g).

[65]*Environmental emergency* is defined as "any release or threatened release of materials into the environment in such quantities or concentrations as cause or threaten to cause an imminent and substantial danger to human health or the environment; the term includes, but is not limited to, discharges of oil and hazardous substances prohibited by Section 311(b)(3) of the Federal Clean Water Act—(Public Law 92-500), as amended." *Id.* § 224.01-400 (1)(d).

[66]*Hazardous substance* is defined as "any substance or combination of substances including wastes of a solid, liquid, gaseous, or semi-solid form which, because of its quantity, concentration, or physical, chemical, or infectious characteristics may cause or significantly contribute to an increase in mortality or an increase in serious irreversible or incapacitating reversible illness, or pose a substantial present or potential hazard to human health or the environment. The substances may include but are not limited to those which are, according to criteria established by the cabinet, toxic, corrosive, ignitable, irritants, strong sensitizers, or explosive, except that the term "hazardous substance" shall not include petroleum, including crude oil or any fraction thereof which is not otherwise specifically listed or designated as a hazardous substance under this section, and shall not include natural gas, natural gas liquids, liquified natural gas, or synthetic gas usable for fuel, or mixtures of natural gas and synthetic gas." *Id.* § 224.01-400 (1)(a).

pollutant, or contaminants,[67] at the site before it became a fiduciary;

(3) did not participate in the day-to-day management of the site before it became a fiduciary;

(4) at the time it became a fiduciary, did not know and had no reason to know that a hazardous substance, pollutant, or contaminant was disposed at the site;

(5) when it undertakes actions to protect or preserve the value of the site, undertakes those actions in accordance with state law and regulations;

(6) its employees, agents, and contractors did not cause or contribute to an environmental emergency or to a release[68] or threatened release[69] of a hazardous substance, pollutant, or contaminant; and

(7) complies with the release notification requirements.

For purposes of this exemption, the financial institution also must have undertaken, at the time it became a fiduciary, all appropriate inquiries into the previous ownership and uses of the property consistent with good commercial or customary practice in an effort to minimize liability. What actions constitute all appropriate inquiries will be determined by taking into account any specialized knowledge or experience on the part of the financial institution, the relationship of the market value of the site to the value of the site if uncontaminated, commonly known or

---

[67]*Pollutant or contaminant* shall include, but not be limited to:

any element, substance, compound, or mixture, including disease-causing agents, which after release into the environment and upon exposure, ingestion, inhalation, or assimilation into any organism, either directly from the environment or indirectly by ingestion through food chains, will or may reasonably be anticipated to cause death, disease, behavioral abnormalities, cancer, genetic mutation, physiological malfunctions (including malfunctions in reproduction) or physical deformations, in such organisms or their offspring; except that the term "pollutant or contaminant" shall not include petroleum, including crude oil or any fraction thereof which is not otherwise specifically listed or designated as a hazardous substance under this section and shall not include natural gas, liquified natural gas, or synthetic gas of pipeline quality (or mixtures of natural gas and such synthetic gas).

Id. § 224.01-400 (1)(f).

[68]*Release* is defined as "any spilling, leaking, pumping, pouring, emitting, emptying, discharging, injecting, escaping, leaching, dumping, or disposing hazardous substances, pollutants, or contaminants into the environment, including the abandonment or discarding of barrels, containers, and other closed receptacles containing any hazardous substance, pollutant, or contaminant, but excludes emissions from the engine exhaust of a motor vehicle, rolling stock, aircraft, vessel, or pipeline pumping station engine; the release of source, by-product, or special nuclear material from a nuclear incident, as those terms are defined in the Atomic Energy Act of 1954, if the release is subject to requirements with respect to financial protection established by the Nuclear Regulatory Commission under Section 170 of the Act, or any release of source by-product, or special nuclear material from any processing site designated under Sections 102(a)(1) or 302(a) of the Uranium Mill Tailing Radiation Control Act of 1978; and the normal application of fertilizer." *Id.* § 224.01-400(1)(a).

[69]*Threatened release* means "a circumstance which presents a substantial threat of a release." *Id.* § 224.01-400(1)(e).

reasonably ascertainable information about the site, the obviousness of the presence or likely presence of contamination at the site, the ability to detect the contamination by appropriate inspection, and any other relevant factor.

Environmental Lien

Kentucky has enacted a nonsuperlien statute.[70]

RCRA

Kentucky has a state statute requiring that some form of general notice be placed on the public record or in the chain of title to alert prospective purchasers and the public of a site's status as formerly managing hazardous waste.[71]

Mississippi

Mississippi has no brownfields redevelopment program.

North Carolina

Voluntary Cleanup Program

Until 1997, North Carolina did not have any voluntary cleanup legislation, although interest in brownfields legislation was evident within the business community. As one commentator said:

> There may be a general perception in the state that North Carolina does not have a large number of brownfields compared to more industrialized states in the Northeast and Midwest, and because of the relative abundance of greenfield property, brownfields need not be an issue that warrants significant attention from either the public or private sector. However, most cities and towns are keenly aware of the loss of economic vitality of their downtowns and have public and private initiatives to encourage redevelopment of target areas; many of these areas contain contaminated properties that can affect redevelopment.[72]

Numerous changes in North Carolina regulatory programs aided brownfield site redevelopment even before the state enacted voluntary cleanup legislation. For

---

[70]Other states enacting similar legislation include: Alaska, Arkansas, Illinois, Indiana, Iowa, Kentucky, Maryland, Michigan, Minnesota, Montana, Ohio, Oregon, Pennsylvania, Tennessee, and Texas.

[71]KEN. REV. STAT. ANN. § 244.866 (4)401; KEN. ADMIN. REGS. § 35.070(10)(2).

[72]Kevin R. Boyer, *Regulators Look at Ways to Put Contaminated Property Back to Work,* 7 NORTH CAROLINA ENVIRONMENTAL LAW LETTER (August 1996).

example, in 1994, the North Carolina Inactive Hazardous Sites Branch developed a nonstatutory voluntary cleanup program (under its enforcement discretion authority) that allows responsible parties to clean up their sites without oversight by the branch until cleanup is completed. Under this program, responsible parties must retain professional firms and individuals who have been approved by the branch to conduct site cleanups. This private oversight is designed to expedite the site cleanup process.

In addition, North Carolina groundwater quality rules have been revised to allow consideration of natural processes that reduce contaminant concentration over time (known as natural attenuation or intrinsic bioremediation). These rules have reduced the need for active site cleanups, particularly in areas where there are no water supply wells, such as most urban downtowns.

Finally, the Department of Environment, Health, and Natural Resources (DEHNR) is currently developing guidance for requiring cleanup based on future land uses. This guidance would require consistent assessment of risk in determining cleanup levels for contaminated sites among its regulatory programs. The guidance would allow consideration of future land uses other than residential, which is not currently allowed under some of DEHNR's programs. If implemented, the guidance could allow less stringent and less costly cleanup of some sites, with land use restrictions to prevent unintended future exposures.

Brownfields Property Reuse Act of 1997[73]

In 1997, the North Carolina legislature enacted the Brownfields Property Reuse Act of 1997.[74] This law enables the state, at its discretion, to enter into brownfields agreements with a prospective developer provided that the liability protection gained by the developer is commensurate with the public benefit provided by the project.[75] Decisions on whether or not to enter into a brownfield agreement are, however, reviewable.[76]

After entering into an agreement, the brownfields developer must file a Notice of Brownfields Property in the land records to put subsequent landowners on notice of the cleanup efforts.[77] Liability protection provided under the cleanup law runs with the land, except that if a land use restriction set out in the Notice of Brownfields Property is violated, the owner at the time of violation and all the successors and assigns will be liable for remediate up to current standards.[78]

Developers that proposed a brownfields agreement must pay a non-refundable fee of one thousand dollars ($1,000) in order to obtain review by the state.[79]

---

[73]N.C. GEN. STAT. § 130A-310.31 *et seq.* (1997).
[74]N.C. GEN. STAT. § 130A-310.30 (1997).
[75]N.C. GEN. STAT. § 130A-310.32 (1997).
[76]*See* § 130A-310.36 (1997).
[77]*See* § 130A-310.35 (1997).
[78]*See* § 130A-310.33(c) (1997).
[79]*See* § 130A-310.39(a)(1) (1997).

When the developer submits his final report certifying the completion of remediation under the brownfield agreement, he must pay an additional fee of five hundred dollars.[80] The law establishes a Brownfields Property Reuse Act Implementation Account. Fees collected under the statute must be deposited into this account, along with any monies appropriated by the general assembly.[81]

It is considered a crime to provide any false information to the state in applying for a brownfield agreement.[82] Similarly, it is a crime to give false information to the state if that information is required pursuant to a brownfield agreement—even if that information is not otherwise required by law.[83]

Environmental Lien—Nuisance

North Carolina is among the growing number of states that have codified their nuisance laws to enable the state to undertake cleanup and attach a lien against the property for the cleanup costs.

In North Carolina, if the state or a local health director determines that a public health nuisance exists, then that official may issue an order of abatement directing the owner, lessee, operator, or other person in control of the property to take any action necessary to abate the public health nuisance.[84] If the person refuses to comply with the order, then the official may institute an action in the superior court of the county where the public health nuisance exists to enforce the order. The action must be calendared for trial within sixty days after service of the complaint upon the defendant.

The court may order the owner to abate the nuisance or direct the secretary or the local health director to abate the nuisance. If the secretary or the local health director is ordered to abate the nuisance, the department or the local health department must have a lien on the property for the costs of the abatement of the nuisance in the nature of a mechanic's and materialmen's lien[85] which lien may be enforced in the same manner as a mechanic's lien.[86]

If the secretary or a local health director determines that an imminent hazard exists, then that official may, after notice to or reasonable attempt to notify the owner, enter upon any property and take any action necessary to abate the imminent

---

[80]*See* § 130A-310.39(a)(2) (1997).

[81]*See* § 130A-310.38 (1997).

[82]N.C. Gen. Stat. § 130A-26.1 (5) (1997).

[83]N.C. Gen. Stat. § 130A-26.1 (6) (1997).

[84]N.C. Gen. Stat. § 130A-19 (1993) (abatement of public health nuisance).

[85]This is the same type of lien as provided in Chapter 44A of the General Statutes.

[86]EEE-ZZZ Lay Drain Co. v. North Carolina Dep't. of Human Resources, 422 S.E.2d 338 (N.C. Ct. App. 1992) (citing Warren County v. North Carolina, 528 F. Supp. 276 [E.D.N.C. 1981]). Use of property by the state may not be enjoined by the courts as a nuisance where the use of the property is in a manner authorized by valid legislative authority; *see also* Warren County v. North Carolina, 528 F. Supp. 276 [E.D.N.C. 1981] (decided under former § 130-20; because the power to initiate action to abate a public nuisance is vested in the local health director, a county may not proceed without him).

hazard.[87] The department or the local health department will have a lien on the property for the cost of the abatement of the imminent hazard in the nature of a mechanic's and materialmen's lien. The lien may be defeated by a showing that an imminent hazard did not exist at the time the official took the action.

Within thirty days after termination of the temporary management of the property containing the nuisance, the temporary manager must give the court a complete accounting of:

(1) all property of which the temporary manager took possession;
(2) all funds collected under this article;
(3) expenses of the temporary management; and
(4) all disbursements or transfers of facility funds or other assets made during the period of temporary management. On the same day the accounting is filed with the court, the temporary manager must serve on the respondent by registered mail a copy of this accounting.[88]

If the operating funds collected during the temporary management exceed the reasonable expenses of the temporary management, the court must order payment of the excess to the respondent, after reimbursement to the contingency fund. If the operating funds are insufficient to cover the reasonable expenses of the temporary management, the respondent must be liable for the deficiency, except as described in this section. If the respondent demonstrates to the court that repayment of amounts spent from the contingency fund would significantly impair the provision of appropriate care or services to residents, the court may order repayment over a period of time with or without interest or may order that the respondent be required to repay only part or none of the amount spent from the contingency fund. In reaching this decision, the court may consider all assets, revenues, debts, and other obligations of the long-term care facility, the likelihood of the sale of the long-term care facility where repayment forgiveness would result in unjust enrichment of the respondent, and must consider the impact of its determination on the provision of care to residents. The respondent may petition the court to determine the reasonableness of any expenses of the temporary management. The respondent must not be responsible for expenses in excess of amounts the court finds to be reasonable. Payment recovered from the respondent must be used to reimburse the contingency fund for amounts used by the temporary manager.[89]

The court may order that the department have a lien for any reasonable costs of the temporary management that are not covered by the operating funds collected by the temporary manager and for any funds paid out of the contingency fund during the temporary management upon any beneficial interest, direct or indirect, of any respondent in the following property:

---

[87]N.C. GEN. STAT. § 30A-20 (1993) (abatement of an imminent hazard).
[88]N.C. GEN. STAT. § 131E-244 (1993) (accounting lien for expenses).
[89]*Id.* § 131E-244(b).

(1) the building in which the long-term care facility is located;
(2) the land on which the long-term care facility is located;
(3) any fixtures, equipment, or goods used in the operation of the long-term care facility; or
(4) the proceeds from any conveyance of property used as a long-term care facility which were made by the respondent within one year prior to the filing of the petition for temporary management unless such transfers were made in good faith, in the ordinary course of business, and without intent to frustrate the cleanup and accounting efforts. Transfers made coincidental with serious deficiencies in resident care may be considered evidence of such an intent.[90]

To the extent permitted by law, the lien provided for will be considered superior to any lien or other interest that arises subsequent to the filing of the petition for temporary management, except for a construction or mechanic's lien arising out of work performed with the express consent of the temporary manager.[91] The clerk of court in the county in which the long-term care facility is located must record the filing of the petition for temporary management in the lien docket opposite the names of the respondents and licensees named in the petition.[92]

Within sixty days after termination of the temporary management, the temporary manager must file a notice of any lien created. If the lien is on real property, the notice must be filed with the clerk of court in the county where the long-term care facility is located and entered on the lien docket. If the lien is on personal property, the lien must be filed with the person against whom the lien is claimed and must state the name of the temporary manager, the date of the petition for temporary management, the date of the termination of temporary management, a description of the property involved, and the amount claimed. No lien will exist against any person, on any property, or for any amount not specified in the filed notice.[93]

Business Transfer Statute

The North Carolina Solid Waste Management Act (NCSWMA)[94] applies to inactive hazardous waste disposal sites.[95] The sale, lease, conveyance, or transfer of a disposal site required by official notification by the department that a "notice of inactive hazardous substance or waste disposal site" (the Notice) is required triggers NCSWMA disclosure provisions.[96]

---

[90]Id. § 131E-244(c).
[91]Id. § 131E-244(d).
[92]Id. § 131E-244(e).
[93]Id. § 131E-244(f).
[94]Id.
[95]N.C. Gen. Stat. § 130A-310.8(a).
[96]Id. Sites that are undergoing voluntary remedial action are excluded from the disclosure requirements. Id. at § (g).

Within 180 days after notification from the department, the owner must submit a survey plat of areas the department has designated in the notice.[97] The survey must be prepared and certified by a professional land surveyor and titled "notice of inactive hazardous substance or waste disposal site."[98] The Notice must include the size and location of the disposal area[99] and the type, location, and amount of hazardous waste disposed "to the best of the owner's knowledge."[100]

The owner must file a certified copy of the Notice in the register of deeds' office in the county where the land is located after the department approves and certifies the Notice.[101] The Notice is then filed in the grantor index under the owner's name.[102] The deed or other instrument of transfer must contain a statement that the property was used as a hazardous waste disposal site and reference the recordation of the Notice whenever an inactive site is sold, leased, conveyed, or transferred.[103]

If the owner does not submit or file the Notice, the secretary may do so and recover costs from the party responsible for contamination.[104] In addition, if the owner who submits and files the Notice is not the responsible party, he may recover the reasonable costs for the submission and filing.[105]

South Carolina

Business Transfer Statute

The South Carolina Hazardous Waste Management Act (SCHWMA)[106] applies to permitted hazardous waste storage and disposal sites.[107] SCHWMA is triggered when the property is "sold, leased, conveyed or transferred in any manner."[108] At the time of the sale, lease, conveyance or transfer, the deed must contain the statement: "The real property conveyed or transferred by this instrument has previously been used as a storage or treatment facility for hazardous wastes."[109]

---

[97]*Id.*
[98]*Id.*
[99]*Id.* at § (a)(1).
[100]*Id.* at § (a)(2).
[101]*Id.* at § (b).
[102]*Id.* at § (c).
[103]*Id.* at § (e).
[104]*Id.* at § (d).
[105]*Id.*
[106]*Id.*
[107]*Id.*
[108]*Id.*
[109]*Id.* The notice can be no smaller than the print in the main body of the document. *Id.*

Tennessee

Voluntary Cleanup Program

In 1994, the Tennessee legislature amended the Hazardous Waste Management Act of 1983[110] to create the Voluntary Cleanup and Assistance Program.[111] The Tennessee Voluntary Cleanup Program allows willing and able parties to conduct investigations and voluntary cleanups of inactive contaminated industrial sites.[112] Certain activities conducted under the Voluntary Cleanup Program must occur with state oversight.[113] Specifically, investigatory and remedial measures, including monitoring, are among the activities requiring state oversight. State oversight costs incurred during the voluntary remediation of a contaminated industrial site must be reimbursed by volunteers.[114] Additionally, a $5,000 participation fee must be paid by volunteers.[115] This fee is used to establish a Voluntary Cleanup and Oversight Assistance Fund.[116]

The criteria for selecting containment and cleanup actions, including monitoring and maintenance options to be followed under the Voluntary Cleanup and Oversight Assistance Program, are specified by law.[117] Determined on a site-specific basis, the following relevant factors must be considered by program participants for state approval of voluntary remedial activities:

> technological feasibility of each alternative;
> cost effectiveness of each alternative;
> the nature of the danger to the public health, safety, and the environment posed by existing site contamination; and
> the extent to which each alternative would achieve the goal of the Voluntary Cleanup and Assistance Program.[118]

Volunteers are required to evaluate reasonable alternatives and select those actions which adequately protect the public health, safety, and the environment from risks posed by hazardous substances.[119]

---

[110]TENN. CODE ANN. § 68-212-202 *et seq.* (West 1994).
[111]*Id.* § 68-212-224.
[112]*Id.* § 68-212-224(a).
[113]*Id.*
[114]*Id.*
[115]*Id.* § 68-212-224(b).
[116]*Id.* § 68-212-224(c)(1).
[117]*Id.* § 68-212-224(e).
[118]*Id.* § 68-212-206(d)(1)-(4).
[119]*Id.* § 68-212-206(d).

Specialized Stakeholder Laws—Trustee

Tennessee was one of the states that amended its state equivalent of CERCLA to include some specialized lender liability protection. In Tennessee,[120] a fiduciary is empowered to:

inspect and monitor property[121] for the purpose of determining compliance with environmental laws affecting the property;

respond or take any other action necessary to prevent, abate or clean up,[122] any actual or threatened violation of any environmental laws affecting property held by the fiduciary relating to hazardous substances[123] or environmental laws;[124]

refuse to accept property in trust;[125]

settle or compromise, at any time, any and all claims involving the alleged violation of any environmental laws affecting property held in the estate or trust;[126]

disclaim any power[127] that, in the sole discretion of the fiduciary, may cause the fiduciary to incur personal liability under any environmental laws;[128]

decline to serve as fiduciary.[129]

In addition, the fiduciary is entitled to charge the cost for any inspection, insurance, review, abatement, response, or cleanup, or any other remedial action against

---

[120]TENN. CODE ANN. § 35-50-110 (specifically enumerated fiduciary powers that may be incorporated by reference).

[121]The fiduciary may inspect only property to which the fiduciary takes legal title (including interests in sole proprietorships, partnerships, or corporations and any assets owned by such business enterprises).

[122]Such action can be either on behalf of the trust or estate as is necessary, before or after the initiation of enforcement action by any governmental body.

[123]For the purposes of this law, *hazardous substances* is defined as "any substance defined as hazardous or toxic or otherwise regulated by any federal, state or local law, rule or regulation relating to the protection of the environment or human health."

[124]TENN. CODE ANN. § 32(A)(i).

[125]*Id.* § 32(A)(ii). This refusal is conditioned upon the fiduciary determining that any property to be donated to a trust estate is contaminated by any hazardous substances, or such property is being used or has been used for any activities, directly or indirectly involving hazardous substances, which could result in liability to the trust or estate or otherwise impair the value of the assets held therein.

[126]*Id.* § 32(A)(iii). This is true if the claims against the estate or trust whether the claims are asserted by a governmental body or private party.

[127]Powers granted by any document or any statute or rule of law may all be disclaimed under this provision.

[128]*Id.* § 32(A)(iv).

[129]*Id.* 32(A)(v). This provision is conditioned, however, on the fiduciary having a reasonably belief that there is or may be a conflict of interest between it in its fiduciary capacity and in its individual capacity because of potential claims or liabilities which may be asserted against it on behalf of the estate or trust resulting from the type or condition of assets held therein.

the income or principal of the estate or trust. Just as important, the fiduciary will not be personally responsible for these. Nor will the fiduciary be personally liable to any beneficiary or any other party for any decrease in value or exhaustion of assets in the estate or trust by reason of the fiduciary's compliance with any environmental laws, specifically including any reporting requirements under such laws.

While acting in good faith and according to traditional fiduciary standards, the fiduciary similarly will not be considered an "owner," "operator," or other party otherwise liable for violation of environmental laws unless the fiduciary has actually caused or contributed to such violation.[130]

Environmental Lien

After the courts determined that environmental lien may be constitutionally suspect absent certain protections, the trend toward superliens faltered. Tennessee deleted superlien provisions from its cleanup statute and enacted a nonsuperlien statute.

---

[130]*See generally* TENNESSEE FORMS (ROBINSON, RAMSEY, AND HARWELL), Nos. 4-305-4-307, 4-501-4-506, 4-612; Herman E. Taylor, *The Family Trust in Estate Planning in Tennessee*, 16 No. 2 TENN. B.J. 32.

CHAPTER 8

Region V:
Illinois, Indiana, Michigan, Minnesota,
Ohio, and Wisconsin

"For a site to be an available candidate for industrial or commercial brownfields redevelopment it must possess three sets of necessary criteria for success."[1] These criteria are:

  cost effective environmental remediation;
  support from local, state and federal regulators;
  attractive real estate location.

The states in Region V have been among the most active in developing brownfields legislation. Ohio's and Michigan pioneered voluntary cleanup legislation for the region—and ultimately the nation. Borrowing heavily from the privatization concept of Massachusetts, Ohio's became one of the first voluntary cleanup programs to receive national attention. The Ohio legislation—and lessons learned from experiences with it—served as a template for many subsequent statutes.

The success of brownfields programs in the states in Region V is in large part due to the support EPA Region V has lent to these initiatives. EPA Region V was among the first of the EPA regions to enter into a Superfund Memorandum of Agreement (SMOA) and Superfund Memorandum of Understanding with the states. Most states in Region V have either executed a SMOA or are in the process of negotiating one. These formed a gentleman's agreement between the EPA regional administrator and the state that U.S. EPA would not prosecute under federal CERCLA or RCRA programs that had received state cleanup approval under a bona fide brownfields program.

The following table describes the more recent programs developed in Region V to encourage voluntary cleanup:[2]

---

[1] See NATIONAL BROWNFIELDS TRANSACTION CONFERENCE (EPA/SONREEL June 18, 1996) (quoting Lewis Norry, "Site Selection Criteria Used for Brownfields Redevelopment," REMEDIATION & REUSE, April/May 1996, vol. 2, Issue 3, at 1, 7, 8).

[2] See Charles Bartsh, BROWNFIELDS STATE OF THE STATE REPORT: 50-STATE PROGRAM ROUNDUP (Northeast-Midwest Inst., January 28, 1998).

States	Brownfields Financing Programs	VCP with Written Assurances	Incentives to Attract Private Development
Illinois	X	X	X
Indiana	X	X	X
Michigan	X	X	X
Minnesota	X	X	X
Ohio	X	X	X
Wisconsin	X	X	

As this table indicates, all the states in Region V have developed financial mechanisms to help attract brownfields redevelopments. Three states—Indiana, Michigan and Wisconsin—adopted the concept of establishing brownfields redevelopment zones as a means of facilitating financing of brownfield redevelopment. These three states have all codified the grant program for brownfield redevelopment zones.

Illinois

Voluntary Cleanup Program

In 1989, Illinois began its Voluntary Cleanup Program. The VCP provides for traditional review and evaluation of plans for site investigation and cleanup of hazardous substances and/or petroleum.[3] Once the nature and extent of contamination is determined, a remediation goal is pursued jointly by the Illinois EPA and the property owner.[4] Site-specific voluntary cleanup goals may be achieved through the use of engineering controls, institutional controls, future land use restrictions placed in the chain of title of the property, or numeric controls.[5]

Tier	Site	Description
Tier I	Residential	Consists of specific contaminant levels
Tier II	Residential	Utilizes the same as Tier I but allows site-specific contaminant level—often resulting in lower standards than Tier I
Tier III	Commercial and Industrial	Full risk assessment of site in order to determine site-specific pathways as well as contaminant levels

---

[3]ILL. ANN. STAT. ch. 415, para. 22.2b (Smith-Hurd 1994).

[4]*Id.* § 22b(a)(3). Under § 22(m)(1), the following provisions are deemed conditions precedent for Illinois EPA review and evaluation of proposed voluntary cleanup plans:

submission of remedial work plan for voluntary cleanup at the site;
proposed site visit or other site evaluation by the Illinois EPA;
agreement to perform the work plan as approved by Illinois EPA;
payment of reasonable costs incurred and documented by the Illinois EPA in providing such services; and
advance payment for anticipated services of Illinois EPA in the amount of $5,000 or one-half of the total anticipated costs, whichever is less.

[5]*Id.* at § 22b(d).

Following the satisfactory completion of voluntary remedial activities, the Illinois EPA would provide a release from liability to volunteers coextensive with the remedial actions taken.[6]

In 1997, the Illinois legislature modified the existing program by enacting the Site Investigation and Remedial Activities Program.[7] The new law (1) provides assistance in identifying and assessing properties potentially contaminated by commercial, industrial, or other uses; (2) promotes appropriate cleanup of commercial, industrial, or other properties through the Illinois Voluntary Cleanup Program; and (3) creates and administers the Brownfields Redevelopment Fund,[8] a fund administered by Illinois EPA containing funds collected pursuant to the program.[9] In many ways, the new law is symptomatic of the growing landowners rights movements in the various state legislatures.[10]

The enacted law was in part a reaction to H.B. 1153, a law addressing the issue of brownfields redevelopment from an owner's or operator's perspective. The bill would have created a rebuttable presumption against state claims and conclusive against private claims, that a defendant has made all appropriate inquiry at the time of acquisition of contaminated industrial property.[11] H.B. 1153 would have provided very detailed procedures for parties conducting Phase I and Phase II environmental site assessments (ESAs). A Phase I ESA that does not disclose the presence of contaminants would have been deemed to have met any legal presumption against public or private claims, thus releasing the volunteer from liability.[12] Similarly, a Phase II ESA demonstrating that the property meets or exceeds the applicable standards would have been deemed to have met all legal presumptions, relieving the volunteer of liability.[13]

The law enacted was less biased toward landowners, but retained the detailed procedures for carrying out remedial investigation and reports.[14] The new law

---

[6]*Id.* § 22b(a)(1).

[7]415 ILCS 5/58.3 (1997).

[8]415 ILCS 5/58 (1997).

[9]415 ILCS 5/58.3(b) (1997).

[10]*See generally* HERTHA L. LUND, *The Property Rights Movement and State Legislation. Cf.* Nancie G. Marzulla, *The Property Rights Movement: How It Began and Where It Is Headed,* in Bruce Yandle, LAND RIGHTS (1995).

[11]1995 IL H.B.1153, §§ 22.2(j)(1)(C) & 22.2(j)(6)(E)(i). To determine whether a party made "all appropriate inquiry" into the previous ownership and operation of the contaminated site consistent with good commercial or customary practice, the following factors will be taken into account:

any specialized knowledge of the party;
the relationship between the purchase price to the value of the property if uncontaminated;
the obviousness of the presence or likely presence of contamination at the property; and
the ability to detect such contamination.

Id. § 22.2(j)(6)(B).

[12]*Id.* § 22.2(j)(6)(E)((i)(I).

[13]*Id.* § 22.2(j)(6)(E)(i)(II).

[14]415 ILCS 5/58.6 (1997) (remedial investigations and reports).

establishes "a risk based system of remediation based on protection of human health and the environment relative to present and future uses of the site."[15] The law encourages private redevelopment of brownfields.[16] Accordingly, "any person", including PRPs may elect to proceed under the new law,[17] unless the site is listed on the NPL, subject to RCRA closure requirements, regulated under state or federal underground storage tank (UST) laws or subject to a federal EPA or court order concerning cleanup.[18] Parties participating in the program are subject to extensive review by Illinois EPA throughout the cleanup process.[19]

A party choosing to enter into an agreement with the state to perform a response action plan may upon successful completion be issued a No Further Remediation letter.[20] This document must be recorded in the land records.[21] The issuance of a No Further Remediation letter will be considered a release[22] from further responsibility for further preventive or corrective action at the site,[23] provided the site is used as intended. The site is forever precluded for any land use that is not specified in the land records as consistent with the land use limitation placed on the property.[24] If the land use changes, then further investigation and/or remedial action must be conducted to ensure that the remediation is consistent with the changing land use.[25] This further remediation must also be recorded in the land records.[26]

The protections afforded by the No Further Remediation letter will extend to, among others, "innocent" landlords, lenders, trustees, prospective and subsequent purchasers.[27] The law also excludes the state or any local government from cleanup liability for land acquired through tax delinquency, bankruptcy, abandonment, escheat or other involuntary action by the government.[28]

The new law also provided certain economic incentives. Specifically, the legislature established the Brownfields Redevelopment Grant Fund.[29] Adminis-

---

[15] 415 ILCS 5/58 (1) (1997). *See also id.* § 5/58.5 (1997) (risk-based remediation objectives).

[16] The term *brownfields* is defined under the statute at 415 ILCS 5/58.2 (1997) (definitions).

[17] 415 ILCS 5/58 (2) (1997).

[18] *Id.*

[19] *Id.* § 5/58.7 (b) (1997) (review and evaluation by agency).

[20] 415 ILCS 5/58.8 (1997). Similar to the current Voluntary Cleanup Program, a volunteer must submit a response action plan, arrange for a site visit or other site evaluation by Illinois EPA, agree to perform a response action plan, and pay any reasonable costs incurred through Illinois EPA oversight. *Id.* § 22.2(m)(1). Following the successful completion of an Illinois EPA-approved response action plan, the agency may issue a No Further Action letter applicable to the site. *Compare* H.B. 1153 § 22.2(m)(7)(A).

[21] 415 ILCS 5/58.8 (a) (1997).

[22] 415 ILCS 5/58.10 (1997).

[23] 415 ILCS 5/58.8 (1997). *Compare* H.B. 1153 22.2(m)(7)(A).

[24] 415 ILCS 5/58.8 (c) (1997).

[25] *Id.*

[26] *Id.*

[27] 415 ILCS 5/58. 9 (1997). *Compare* H.B. 1153 § 22.2(m)(B).

[28] 415 ILCS 5/58. 9 (C) (1997). *Compare* H.B. 1153 § 22.2(m)(B).

[29] 415 ILCS 5/58. 13 (1997).

tered by Illinois EPA, the new grant program is designed to assist municipalities in brownfield redevelopment efforts.[30]

Illinois Responsible Property Transfer Act of 1988

The Illinois Responsible Property Transfer Act of 1988[31] was the first brownfield law enacted in the state of Illinois. It was intended to see that parties involved in certain real property transactions know about existing environmental liabilities related to the properties as well as the prior use and environmental condition of the properties.[32] The Act applies to certain "transfers of real property . . . after January 1, 1990."[33] The Act defines a *transfer* as "any conveyance of an interest in real property" such as by deed; a lease of at least forty years; an assignment of at least 25 percent of the beneficial interest in a state land trust; or "a mortgage, trust deed, or collateral assignment of a beneficial interest in a" state land trust.[34]

The Act applies to transfers of "real property," including improvements thereon, located in Illinois and containing at least one "facility" subject to reporting requirements under the Emergency Planning and Community Right-to-Know Act of 1986[35] and regulations thereunder or a storage tank requiring notification under the Solid Waste Disposal Act.[36] Thus, the real property must be located in Illinois and be subject to reporting under the federal laws, the Emergency Planning and Community Right-to-Know Act, or the Solid Waste Disposal Act.

The Illinois Act requires the transferor, within thirty days after execution of any written contract and not later than thirty days before the transfer of real property, to deliver to the transferee and to any lender a disclosure statement in a form prescribed by Illinois.[37] A transferee or a lender may void a duty "to accept a transfer or finance a transfer . . . yet to be" consummated if the disclosure statement reveals environmental defects previously unknown to the transferee or lender or if the transferor fails to make the required disclosure, unless the transferee or lender waives the right.[38] However, the transferee or lender must act to void the transaction within ten days after requesting or receiving the disclosure statement.[39]

Both the transferor and the transferee are responsible for seeing that the disclosure statement is recorded in the county where the property is located within thirty days after transfer of the real property.[40] Both the transferor and the transferee

---

[30]415 ILCS 5/58. 13 (1) (1997).
[31]ILL. REV. STAT. ch. 30, para. 901.
[32]*Id.* at ch. 30, para. 902.
[33]*Id.* at ch. 30, para. 904, § 4(a).
[34]*Id.* at ch. 30, para. 903, § 3(g).
[35]42 U.S.C. § 11022.
[36]42 U.S.C. § 6991.
[37]ILL. REV. STAT. ch. 30, para. 904, § 4(a).
[38]*Id.* ch. 30, para. 904, § 4(c).
[39]*Id.*
[40]ILL. REV. STAT. ch. 30, para. 906, § 6.

may be required to file the disclosure statement with the Environmental Protection Agency.[41]

To enforce the Act, the Act imposes stiff civil penalties for violations. A person violating the duty to disclose is civilly liable for up to $1,000 for the violation and $1,000 for each day the violation continues.[42] A person who knowingly makes a false statement in the disclosure statement is civilly liable for up to $10,000 for the violation and $10,000 for each day the violation continues.[43] Failure to record a disclosure statement results in joint and several civil penalties of up to $10,000 against the transferor and the transferee.[44]

Business Transfer Statute—RCRA

In addition, Illinois has an additional statute affecting the transfer of hazardous waste sites and requiring notice and disclosure.[45] Under the Illinois Environmental Protection Act, a person may not transfer an interest in or use real property previously "used as a hazardous waste disposal site" unless the person notifies, in writing, the Illinois Environmental Protection Agency of the transfer and the transferee of any restrictions the agency imposes upon the use of the property.[46] The state is required to impose conditions "reasonably necessary to protect public health and the environment." These may include permanently prohibiting use of the property to prevent an "unreasonable risk of injury to human health or to the environment."[47]

Environmental Lien

Illinois enacted a nonsuperlien statute to enforce its state equivalent of CERCLA.

Indiana

Voluntary Cleanup Program

In 1993, Indiana began its Voluntary Cleanup Program. The program is administered under the Department of Environmental Management (IDEM). To date, at

---

[41]*Id.*

[42]*Id.* at ch. 30, para. 907, § 7 (Smith-Hurd 1992).

[43]ILL. REV. STAT. ch. 30, para. 907, § 7(b) (Smith-Hurd 1992).

[44]*Id.* at ch. 30, para. 907(c) (Smith-Hurd 1992).

[45]ILL. ANN. STAT. ch. 111½, paras. 1021 § 1021 (m), (n).

[46]*Id.*

[47]*Id.* at ch. 111 ¼, para. 1039(g) (Smith-Hurd 1988 & Supp. 1992). In addition, the Illinois Responsible Property Transfer Act, ILL. REV. STAT. ch. 30, ½ 901, requires that the transferor provide notice, in the form of a disclosure statement, to the transferee and lender. The transferee or lender is released from any obligation to the transfer prior to closing, if the transferor does not provide the disclosure statement.

least eighty-one sites have entered into the program with five sites receiving Certificates of Completion.[48]

The Indiana VCP requires program participants to provide notice to the state and submit work plans for the review and approval of the investigation and voluntary remediation of contaminated industrial property.[49] Any site owner or operator, or a prospective owner or operator, may participate in the VCP. An applicant will be rejected, however, if:

> a state or federal enforcement action concerning the proposed site cleanup is pending;
>
> a federal grant compels IDEM to take enforcement action; or
>
> conditions at the site are considered to be an imminent and substantial threat to human health or the environment.

Cleanups that have already been complete may be eligible for the program, provided proper documentation is included in the application.[50]

When applying, the volunteer must include a $1,000 application fee and submit an application that includes data from a completed site characterization. IDEM then has thirty days to make an eligibility decision.

Incomplete applications are grounds for rejection. Incomplete applications will, however, be returned within forty-five days. IDEM must state what information is missing. The application can then be resubmitted with the data required.

The Voluntary Remedial Agreement is a standard document that identifies the respective obligations of IDEM and the applicant. The agreement requires the applicant to:

> cooperate with IDEM;
>
> adhere to state cleanup rules and regulations;

---

[48]*See* http://www.epa.gov/r5b...successl.htm#voluntary.

[49]IND. CODE ANN. § 13-7-8.9-7 (West 1994). Specifically, an eligible party seeking to remediate used industrial property under the Indiana voluntary cleanup program must provide the following information:

> general information on the parties, the site to be remediated, and other requested background information;
>
> an environmental assessment of the actual or threatened release of the hazardous substance and/or petroleum release; and
>
> an application fee of $1,000.

Applicant eligibility should be determined within thirty days of receipt of an application and application fee. *Id.* § 13-7-8.9-9. Property subject to pending state or federal enforcement actions, having conditions creating an imminent and substantial threat to human health or the environment, or where a federal grant requires an enforcement action are ineligible for voluntary cleanup. *Id.* 13-7-8.9-10(a)(1)-(3).

[50]*See* NATIONAL BROWNFIELDS TRANSACTION CONFERENCE (EPA/SONREEL, June 18, 1996) (quoting Voluntary Remediation Program).

evaluate the nature and extent of environmental contamination;
recommend remediation strategies (including cleanup levels);
reimburse IDEM for oversight costs;
supply a timetable of completion (including milestone tasks).

IDEM will provide the applicant with an estimate of expected oversight costs. Once the agreement is executed, the applicant then prepares and submits a series of work plans and reports for IDEM's review. These include:

a Phase II investigation report;
a remediation work plan;
a remediation completion report.

The Remediation Work Plan is subject to a thirty-day public notice and comment period. During the investigatory phase of voluntary remedial activities, an environmental assessment (the contents of which are specified by law)[51] must be conducted.[52] The program established a three-tiered framework for cleanup standards:

Tier	Description
I	Reflects background concentrations
II	Similar to those used under CERCLA and RCRA corrective action programs
III	Site-specific risk assessment performed by the site owner or operator

Regardless of which tier is applicable, participants are required to reimburse the state for oversight costs associated with voluntary cleanups.[53]

Upon the satisfactory completion of voluntary cleanup activities, a Certificate of Completion would be issued.[54] Receipt of a Certificate of Completion is a condition precedent to the issuance of a Covenant Not to Sue by the governor of

---

[51]*Id.* § 13-7-8.9-8. The elements of an environmental assessment under the Indiana voluntary cleanup program include:

a legal description of the site;
a description of the physical characteristics of the site;
the operational history of the site to the extent known by the applicant;
information regarding the nature and extent of any relevant contamination and relevant releases; and
relevant information concerning the potential for human exposure to contamination at the site.

Id. § 13-7-8.9-8(1)-(5).

[52]*Id.* § 13-7-8.9-7(b)(2)(B).

[53]*Id.* § 13-7-8.9-13(a)(1). Volunteers are required to reimburse the state for all reasonable costs exceeding the application fee submitted by participating parties. Specifically, the state may recover costs incurred in the review and oversight of the work plan and in association with the voluntary remediation agreement

[54]*Id.* § 13-7-8.9-17(a). The issuance of a Certificate of Completion certifies the successful completion of an approved voluntary remediation work plan. The issuance of a Certificate of Completion will be deemed final agency action.

Indiana.[55] The covenant generally releases volunteer(s) from liability, including future liability, for a claim relating to contamination specifically redressed by an approved voluntary cleanup.[56] Further, Indiana Covenants Not to Sue are transferable and apply to both current and future owners of brownfield sites.[57]

Prior to approving a proposed voluntary remediation work plan, the state and the volunteer(s) must enter into a voluntary remediation agreement.[58] This agreement sets the terms and conditions for the evaluation and implementation of the work plan.[59] Once the voluntary remediation agreement is entered, the department may evaluate the proposed voluntary remediation work plan.[60] Voluntary remediation

---

[55]*Id.* § 13-7-8.9-18(a).

[56]*Id.*

[57]*Id.* § 13-7-8.9-18(b). A Covenant Not to Sue issued under this section should bar all public or private claims arising in connection with the release or threatened release of hazardous substances and/or petroleum that was the subject of an approved voluntary remediation work plan. *Id.* Protection under the Covenant Not to Sue specifically applies to persons receiving such covenants through a legal transfer of a Certificate of Completion by acquiring property to which the Certificate of Completion complies. *Id.*

[58]*Id.* §13-7-8.9-13.

Indiana offers volunteers three tiers of cleanup standards. Tier one concerns background levels of contamination. Background contamination has been defined as normally occurring pollution in the ambient environment. Background standards are generally recognized as being very stringent standards. Tier two cleanup standards are health-based standards. Specifically, property cleaned up for residential purposes must meet a 1×10^{-6} health-based standard. Non-residential uses of contaminated property must meet a lower standard of 1×10^{-5}. Tier three concerns site specific risk assessment procedures. These options allow volunteers to select a risk level that is comfortable and compatible with the intended use of a redeveloped facility. The more extensive the voluntary cleanup work plan is, the more extensive the protection for future liability received by program participants. Conversely, a more limited cleanup results in limited protection form future liability.

Conference on Brownfields Redevelopment and Environmental Risk Management in Business and Real Estate Transactions (March 22-23, 1995) (Statement of S. Richard Heymann).

[59]*Id.* A mechanism to resolve disputes arising from the evaluation, analysis, and oversight of the implement of the work plan, including arbitration, adjudication, or alternative dispute resolution procedures are required to be incorporated in the voluntary remediation agreement. *Id.* § 13-7-8.9-13(a)(2). Also required to be incorporated in the voluntary remediation agreement are provisions concerning the indemnification of parties, retention of record and timetables for the reasonable review, and evaluation of the adequacy of the voluntary work plan.

[60]*Id.* § 13-7-8.9-12. A proposed work plan must include the following criteria:

detailed documentation of the remedial investigation and a description of the work to be performed
    to determine the nature and extent of the actual or threatened release;
a proposed statement of goals to be accomplished by the work plan;
plans concerning:
    quality assurance measures for the implementation of the proposed remediation project;
    descriptions of sampling and analysis;
    health and safety considerations;
    community relations;
    data management and record keeping; and
    proposed work schedule concerning the implementation specific remedial activities.

Id. § 13-7-8.9-12(b).

work plans submitted for approval are subject to notice, comment, and, if requested, public hearings.[61] If approved, volunteers are notified in writing that remedial activities may be commenced, as well as the date by which the work must be completed.[62] Oversight and progress reporting duties regarding the implementation of the voluntary remediation work plans lies with the state.[63] Upon successful completion of an approved voluntary remediation work plan, volunteers should qualify for the issuance of a Certificate of Completion and a Covenant Not to Sue.[64] The protection afforded by the Covenant Not to Sue does not extend to:

(1) future liability for property previously remediated under a voluntary work plan;[65]
(2) contamination not known at the time the Certificate of Completion was issued;[66]
(3) liability to the federal government for claims based on federal law.[67]

Volunteers are, however, protected from public or private actions brought during the implementation of an approved voluntary remediation work plan.[68] Volunteers are protected from any cause of action relating to the release or threatened release of hazardous substances and/or petroleum.[69] Further, administrative actions brought by the state are barred during the pendency of the voluntary remediation work plan.[70] Similarly, private actions for legal or equitable relief brought by third-parties or citizens groups are barred during voluntary cleanup activities.[71]

Nothing under the Indiana voluntary cleanup program prevents volunteers from bringing an action or raising a claim, including a claim for contribution, against third parties responsible for existing site contamination.[72] Volunteers

---

[61]*Id.* § 13-7-8.9-15(b). Notice to local government units located in the county affected by the proposed remediation work plan is required. *Id.* A copy of the voluntary work plan must be provided to each municipality and placed in at least one public library. *Id.* Notice requesting comments concerning the proposed voluntary work plan must be published, followed by a thirty-day comment period. *Id.* Interested persons may submit written comments or request a public hearing. *Id.* § 13-7-8.9-15(b)(1)-(2). If at least one public comment is received, a public hearing may be held in the geographical area affected by the proposed voluntary remediation work plan on the issue of approval or denial of the plan itself. *Id.* § 13-7-8.9-15(b)(2).

[62]*Id.* § 13-7-8.9-15(d)(2)(A).

[63]*Id.* § 13-7-8.9-16.

[64]*Id.* §§ 13-7-8.9-17 and 13-7-8.9-18.

[65]*Id.* § 13-7-8.9-19(c).

[66]*Id.*

[67]*Id.* § 13-7-8.9-19(d).

[68]*Id.* § 13-7-8.9-19(e).

[69]*Id.*

[70]*Id.*

[71]*Id.*

[72]*Id.* § 13-7-8.9-20(a).

successfully completing voluntary cleanups are afforded contribution protection for matters specifically addressed in a work plan or Certificate of Completion.[73]

Financial Incentives

In 1997, the Indiana legislature enacted further financial incentives to encourage brownfields redevelopment. Specifically, the legislation took two forms:

the Brownfields Revitalization Zone Tax;[74] and
the Environmental Remediation Revolving Loan Program.[75]

Under the first program, the state may designate a property or area as a "brownfield"[76] or a brownfield "zone."[77] This designation may be done according to a mapping process conducted by the state[78] or after an individual has applied to the state.[79] To be eligible, the state must determine that there is a benefit to the community that the land in question be designated as a part of the program.[80] Properties so designated, after appropriate hearings and review by the state,[81] may be eligible to apply for "an assessed valuation deduction" for properties located in the brownfield revitalization zone.[82] In addition, parties improving the brownfields can apply for a tax abatement.[83]

The second program, the Environmental Remediation Revolving Loan Program, was established to "assist in the remediation of brownfields to encourage the rehabilitation, redevelopment, and reuse of real property by political subdividsion by providing loans or other financial assistance to political subdivisions" to conduct brownfield redevelopment activities.[84]

Business Transfer Statutes

The Indiana Environmental Hazardous Disclosure and Responsible Party Transfer Law[85] requires the transferor to complete and record a disclosure[86] statement

---

[73]*Id.* at § 13-7-8.9-20(b).
[74]IND. CODE ANN. § 6-1.1-42-1 *et seq.* (1997).
[75]IND. CODE ANN. § 13-19-5-1 (1997).
[76]IND. CODE ANN. § 6-1.1-42-1 (1997) (brownfield defined).
[77]IND. CODE ANN. § 6-1.1-42-4 (1997) (zone defined).
[78]IND. CODE ANN. § 6-1.1-42-8 (1997) (preparation of maps and plats of zone).
[79]IND. CODE ANN. § 6-1.1-42-5 (1997) (application as a brownfield revitalization zone).
[80]*See* IND. CODE ANN. § 6-1.1-42-18 (1997) (contents of statement of benefits).
[81]IND. CODE ANN. § 6-1.1-42-19 through 26 (1997).
[82]IND. CODE ANN. § 6-1.1-42-27 through 30 (1997).
[83]IND. CODE ANN. § 6-1.1-42-17 (1997).
[84]IND. CODE ANN. § 13-19-5-1 *et seq.*(1997).
[85]IND. CODE § 13-7-22.5-1 *et seq.*
[86]§ 13-7-22.5 -1 defines a *disclosure document* as "a document that sets forth certain information about a property that is to be transferred."

concerning the environmental[87] history of the property[88] upon transfer.[89] This disclosure statement must be provided to the transferee[90] and lender thirty days prior to the transfer.[91] The transferee or lender[92] is released from any obligation pertaining to the transfer if the statement discloses an environmental defect in the property.[93]

---

[87]§ 13-7-22.5 -1.5 of the Act defines an *environmental defect* as an environmentally related commission, omission, activity, or condition that:

(1) constitutes a material violation of an environmental statute, regulation, or ordinance;
(2) would require remedial activity under an environmental statute, regulation, or ordinance;
(3) presents a substantial endangerment to:
  (A) the public health;
  (B) the public welfare; or
  (C) the environment;
(4) would have a material, adverse effect on the market value of the property or of an abutting property; or
(5) would prevent or materially interfere with another party's ability to obtain a permit or license that is required under an environmental statute, regulation, or ordinance to operate the property or a facility or process on the property.

[88]§ 13-7-22.5 -6 of the statute defines *property* as a specific and identifiable parcel of real property, including any improvements:

(1) that contains one (1) or more facilities that are subject to reporting under Section 312 of the federal Emergency Planning and Community Right-to-Know Act of 1986 (42 U.S.C. 11022);
(2) is the site of one or more underground storage tanks for which notification is required under 42 U.S.C. 6991a and IC 13-7-20-13(8)(A); or
(3) is listed on the Comprehensive Environmental Response, Compensation, and Liability Information System (CERCLIS) in accordance with Section 116 of CERCLA (42 U.S.C. 9616).

The term does not include property that has been subject to bonding or other financial assurances released by the appropriate governmental agency after compliance with applicable state laws.

[89]§ 13-7-22.5 -7 defines *transfer* as "a conveyance of an interest in property by any of" the following:

(1) a deed or other instrument of conveyance of fee title to property;
(2) a lease whose term, if all options were exercised, would be more than forty years;
(3) an assignment of more than 25 percent of the beneficial interest in a land trust;
(4) a collateral assignment of a beneficial interest in a land trust;
(5) an installment contract for the sale of property;
(6) a mortgage or trust deed;
(7) a lease of any duration that includes an option to purchase.

The term does not include a conveyance of an interest in property by any of the following:

(1) a deed or trust document, which, without additional consideration, confirms, corrects, modifies, or supplements a deed or trust document that was previously recorded;
(2) a deed or trust document that, without additional consideration, changes title to property without changing beneficial interest;
(3) a tax deed or a deed from a county transferring property the county received under IC 6-1.1-25-5.5;
(4) an instrument of release of an interest in property that is security for a debt or other obligation;
(5) a deed of partition;

A buyer of property who finances the purchase of the property through a mortgage loan is not required to deliver a disclosure document to the mortgagee that provides that mortgage loan.[94] A person who lends money and takes a mortgage on property to secure that loan is not required to record a disclosure document concerning the property in the office of the recorder of the county in which the property is located or to file a copy of the disclosure document with the Department of Environmental Management.

---

(6) a conveyance occurring as a result of the foreclosure of a mortgage or other lien on real property;

(7) an easement;

(8) a conveyance of an interest in minerals, gas, or oil (including a lease);

(9) a conveyance by operation of law upon the death of a joint tenant with right of survivorship;

(10) an inheritance or devise;

(11) a deed in lieu of foreclosure;

(12) a Uniform Commercial Code sale or other foreclosure of a collateral assignment of a beneficial interest in a land trust;

(13) a deed that conveys fee title under an installment contract for the sale of property;

(14) a deed that conveys fee title under an exercise of an option to purchase contained in a lease of property.

[90]§ 13-7-22.5 -8(a) defines a *transferee* as:

(1) a buyer, mortgagee, grantee, or lessee of real property;

(2) an assignee of an interest of more than 25 percent in a land trust; or

(3) in the case of a transfer to the trustee of a land trust, the owners of the beneficial interest of the land trust.

The term includes a prospective transferee. § 13-7-22.5 -8(b).

[91]*Transferor* is defined as:

(1) a seller, grantor, mortgagor, or lessor of real property;

(2) an assignor of an interest of more than 25 percent in a land trust; or

(3) in the case of a transfer by the trustee of a land trust, the owner of the beneficial interest of the land trust.

Id. § 13-7-22.5 -9 (a). The term includes a prospective transferor. *Id.* § 13-7-22.5 -9 (b).

[92]§ 13-7-22.5 -3 defines a *lender* as "a person that provides loans secured by an interest in property or by an assignment of beneficial interest in a land trust." § 13-7-22.5 -2 defines a *land trust* as a trust that is established under terms providing that:

(1) the trustee holds legal or equitable title to property;

(2) the beneficiary has the power to manage the trust property, including the power to direct the trustee to sell the property; and

(3) the trustee may sell the trust property:

(A) only at the direction of the beneficiary or other person; or

(B) after a period of time stipulated in the terms of the trust.

[93]*Id. See generally Seller Beware: The Indiana Responsible Party Transfer Law,* 24 IND. L. REV. 761 (1991); *Indiana Environmental Law: An Examination of 1989 Legislation,* 23 IND. L. REV. 329 (1990); *Indiana Responsible Property Transfer Law,* 34 RES GESTAE 78 (1990); *Look Before You Leap Into Real Estate Deals: Indiana's Responsible Property Transfer Law,* 33 RES GESTAE 264 (1989).

[94]IND. CODE ANN. § 13-7-22.5 -19.5 (1993).

A party to a transfer of property may bring a civil action against another party to the transfer of property to recover consequential damages based upon a violation of this Act. The party may also recover reasonable costs and attorney's fees.[95]

It is unlawful for a transferor to knowingly make a false statement in a disclosure document.[96] Each day that the transferor knows of the falsity of the statement made in the disclosure document but fails to correct that statement through the filing, recording, and delivery of a corrected disclosure statement constitutes a separate infraction.[97]

A person who is responsible for filing a disclosure document in the office of the county recorder and who fails to record the disclosure document is also guilty of an unlawful act. However, the failure of a transferee to record a disclosure document within the period allowed is not an infraction under this section if the disclosure document:

(1) was not delivered to the transferee within the time allowed under; or
(2) contains one or more false statements about substantive matters.[98]

If a recorded disclosure document reports the existence of an environmental defect on a property, a person who has a financial interest in the property may record, in the same recorder's office in which the disclosure document is recorded, a document that reports that the environmental defect has been eliminated from the property. A document so filed must be certified by a professional registered engineer who does not have a financial interest in the property.[99]

The Indiana Environment Hazardous Disclosure and Responsible Party Transfer Law (RPLA)[100] covers the same property as the Illinois Responsible Party Transfer Act except that it adds real property listed on the Comprehensive Environmental Response, Compensation, and Liability Information System (CERCLIS).[101]

RPLA is triggered upon the transfer of any real property after December 31, 1989,[102] that meets the requirements above. Transfer is defined as the conveyance of any interest in real property and includes deeds, leases over forty years, assignments of more than 25 percent of the beneficial interest in a land trust, installation contracts, and leases with the option to purchase.[103] Foreclosures, deeds in lieu of foreclosure, easements, inheritance, and devise are some transactions specifically excluded from the definition of transfer.[104]

---

[95]IND. CODE ANN. § 13-7-22.5 -21 (1993).
[96]IND. CODE ANN. § 13-7-22.5 -18 (1993).
[97]Id.
[98]IND. CODE ANN. § 13-7-22.5 -19 (1993).
[99]IND. CODE ANN. § 13-7-22.5 -22 (1993).
[100]Id.
[101]IND. CODE ANN. § 13-7-22.5-6.
[102]IND. CODE ANN. § 13-7-22.5-6.
[103]Id. at § -22.5-7.
[104]Id.

Once a transaction triggers RPLA, a disclosure document must be completed and delivered to the transferee.[105] The transferor must disclose potential environmental liability and identify the property and its characteristics, the type of transfer, uses of the property by the transferor as well as previous owners and lessees of the transferor, and any release of hazardous substances or petroleum required to be reported by state or federal law.[106] The disclosure document must be delivered at least thirty days prior to the transfer[107] but is waivable if all parties execute written statements that they are aware of the purpose and intent of the disclosure document.[108] Even if the thirty-day period is waived, the disclosure document must still be delivered to all parties to the transfer on or before the date the transfer becomes final.[109]

If environmental defects are revealed in the disclosure document that were not known to the parties to the transfer, then any party, including the lender, may void the transfer before it becomes final.[110] In addition, if the transferor does not provide a disclosure document, then any party who did not receive the document may demand one.[111] If the party does not receive the disclosure within ten days of the demand, or receives it within ten days but it reveals environmental defects, the party may void the transfer or obligation to finance the transfer.[112] A party may not void the transaction due to a transferee's failure to respond to a demand or for environmental defects once the transfer is final.[113]

After delivery to the transferee and lender of the disclosure document, the disclosure must be recorded in the county where the property is located.[114] In addition, a copy of the document must be filed within thirty days after the transfer with the Indiana Department of Environmental Management.[115] The transferor and transferee are jointly responsible for recording the document.[116]

Numerous penalties are available under RPLA. Failure to prepare and provide the disclosure document is a class B offense punishable by up to a $1,000 civil

---

[105]*Id.* A *transferee* is defined as a buyer, mortgagee, grantee, lessee, assignee of more than 25 percent interest in a land trust or a trustee who is the owner of the beneficial interest of a land trust. *Id.* at § -22.5-8(a).

[106]IND. CODE ANN. § 13-7-22.5-15.

[107]*Id.*

[108]*Id.* at § -22.5-10(a).

[109]*Id.*

[110]*Id.* at § 13-7-22.5-11. An *environmental defect* is a commission, omission, activity, or condition that (1) is a material violation of an environmental statute, regulation, or ordinance; (2) would require remedial activity; (3) substantially endangers public health, welfare, or the environment; (4) would materially affect the market value of the property or adjacent property; or (5) would prevent or interfere with the ability of another party to receive a required permit or license. *See id.* at § -1.5.

[111]*Id.* at § -12.

[112]*Id.* The same remedy is available in the case of a transferor who waives the thirty-day requirement but fails to provide the disclosure document before the date the transfer is to become final. *Id.* at § -13.

[113]*Id.* at § -14.

[114]*Id.* at § -16(a).

[115]*Id.*

[116]*Id.* at § -16(c).

penalty and court costs.[117] Failure to record the disclosure document is a class A offense.[118] However, a transferee does not violate the recordation requirement if the disclosure document was not delivered during the required time period or contains false statements with respect to substantive matters.[119] Knowingly making a false statement in the disclosure document is also a class A offense.[120] The state is authorized to bring suit in the judicial circuit where the violation took place[121] and may impose a civil penalty of up to $10,000 as well as court costs.[122]

The primary strength of RPLA is its provision allowing the transferee or lender to void his obligations if the disclosure document reveals an "environmental defect." This is due to the fact that the only trigger that allows a party to void the transfer is that one of the defects exists.[123] Although most defects must be materially detrimental, the defect that would "require remedial activity" seems to mean any required level of remediation would be considered an environmental defect.[124] Thus, the broad environmental liabilities that encompass "environmental defects" gives a potential purchaser a substantial method by which to avoid acquiring contaminated property and its associated liabilities.

Another apparent strength of this statute is the ability of the transferee or lender to demand a disclosure document and to void the transfer if the demand is not met by the transferor. The problem is that, if the parties reach closing, the disclosure document is not delivered, and the parties proceed to close, the purchaser is left without a remedy under the statute.[125] The buyer's only recourse is to attempt to rescind the transfer pursuant to common law principles.[126] In addition, a transferee who goes through with a transfer without a disclosure statement would be potentially liable for CERCLA cleanup costs.

Specialized Stakeholder Laws—Lender and Trustee

Indiana is one of the first states to amend its state equivalent of CERCLA to include some specialized lender and trustee liability protection. It has been actively refining these protections ever since the problem of lender liability for environmental matters came to national attention. In 1997, the state enacted further legislation clarifying that lenders should not be liable under any "environmental management

---

[117]IND. CODE ANN. § 34-4-32-4.

[118]*Id.* at § -19.

[119]*Id.* at § -19.

[120]*Id.* at § -18.

[121]IND. CODE ANN. § 34-4-32-1.

[122]*Id.* at § -4.

[123]FREDRICK P. ANDES, THEODORE L. FREEMAN, AND JOHNATHAN LIPSON, MINIMIZING ENVIRONMENTAL LIABILITIES IN WINDING DOWN AND OPERATING SUBSIDIARY, at § 3.30. It does not matter how the defect occurred or whether it can be remediated. *Id.*

[124]*Id.* Even the common problem of overfilling an underground storage tank might be enough. *Id.*

[125]*Id.* at § 3.30.

[126]*Id.*

laws", except CERCLA § 107, unless the lender had actual and direct managerial control over the use, generation, treatment, storage or disposal of hazardous substance at the facility in question.[127]

Indiana also excludes "creditors" from its CERCLA equivalent, if the person (1) extended credit to an owner or operator of the facility, (2) has an interest in the facility to secure an extension of credit, or (3) acquired title or a right to title to the facility upon default, at foreclosure, or in lieu of foreclosure as a result of an extension of credit secured by an interest in the facility unless the extension of credit was solely for the purpose of avoiding environmental liability.[128]

A secured or unsecured creditor or a fiduciary is not liable under the Indiana CERCLA equivalent in connection with the release or threatened release of a hazardous substance from a facility unless the fiduciary or creditor exercised actual and direct managerial control over the:

(1) use;
(2) generation;
(3) treatment;
(4) storage; or
(5) disposal

of the hazardous substance at the facility.[129]

Indiana law also contains limitations on the liability of certain specialized types of creditors.[130] A secured or unsecured creditor will not be held liable with respect to a petroleum facility unless the creditor has exercised actual and direct managerial control over the petroleum facility. A person is a "creditor" with respect to a petroleum facility if the person:

(1) has extended credit to an owner or operator of the petroleum facility;
(2) has an interest in the facility to secure an extension of credit; or

---

[127]IND. CODE ANN. § 13-25-4-8 (1997).

[128]For the purposes of Indiana law, a person is considered a "fiduciary" if the person meets the definition set forth in:

(1) IC 29-1-1-3; or
(2) IC 30-2-4-1.

In addition, under Indiana law, the exceptions provided by Section 107(b) of CERCLA (42 U.S.C. 9607[b]) to liability otherwise imposed by Section 107(a) of CERCLA (42 U.S.C. 9607[a]) are equally applicable to any liability otherwise imposed under the state law. *Id.*

[129]The law states further that the liability of a fiduciary for a release or threatened release of a hazardous substance from a facility that is held by the fiduciary in its fiduciary capacity may be satisfied only from the assets held by the fiduciary in the same estate or trust as the facility that gives rise to the liability.

[130]IND. CODE ANN. § 13-7-20.1-14.

(3) has acquired title or a right to title to the petroleum facility upon default, at foreclosure, or in lieu of foreclosure as a result of an extension of credit secured by an interest in the petroleum facility unless the extension of credit was solely for the purpose of avoiding environmental liability.

Under Indiana law, the liability of a trustee is limited to the assets held by the fiduciary in the same estate or trust.[131]

Indiana law also limits the environmental liability of fiduciaries[132] in certain specialized circumstances. As of July 1, 1991, a fiduciary will not be liable with respect to a petroleum facility unless the fiduciary has exercised actual and direct managerial control over the petroleum facility.[133] The liability of a fiduciary with respect to a petroleum facility that is held by the fiduciary in its fiduciary capacity may be satisfied only from the assets held by the fiduciary in the same estate or trust as the petroleum facility that gives rise to liability.[134]

Environmental Lien

Indiana enacted a nonsuperlien statute to enforce its state equivalent of CERCLA.

Michigan

The Michigan Environmental Response Act (MERA) of 1982 established the means for the state to fund contaminated site cleanups and recover costs from responsible parties. In 1990, the Polluters Pay amendments to this act established strict joint and several liability for potentially responsible parties and provided for administrative orders and Covenants Not-to-Sue for use at brownfields. In 1995, the

---

[131]IND. CODE ANN. § 13-25-4-8 (1997).

[132]For the purposes of this section, a person is a "fiduciary" if the person meets the definition set forth in:

(1) IC 29-1-1-3; or
(2) IC 30-2-4-1.

Id. § 13-7-20.1-14(a).

[133]This provision also applies to secured and unsecured creditors. *See id.* § 13-7-20.1-14(c). A person will be considered a "creditor" with respect to a petroleum facility if the person:

(1) has extended credit to an owner or operator of the petroleum facility;
(2) has an interest in the facility to secure an extension of credit; or
(3) has acquired title or a right to title to the petroleum facility upon default, at foreclosure, or in lieu of foreclosure as a result of an extension of credit secured by an interest in the petroleum facility unless the extension of credit was solely for the purpose of avoiding environmental liability.

Id. § 13-7-20.1-14(a).
[134]*Id.* § 13-7-20.1-14(d).

state cleanup law was again amended and is now known as Part 201 of the Natural Resources and Environmental Protection Act. The amendment was specifically designed to encourage the reuse of brownfields by changing the liability standard to causation, allowing parties to purchase contaminated property without liability after completing a baseline environmental assessment, providing for due care in the reuse of the property, and providing for use-based cleanups as well as a lower cleanup standard.[135]

Voluntary Cleanup Statute

The Michigan State Legislature took affirmative steps to revise its former equivalent of "Superfund," or the Michigan Environmental Response Act, by enacting substantive amendments to the Act.[136] The amendments change significantly the liability scheme in MERA to completely eliminate liability for owners and operators of property contaminated with hazardous waste unless these persons caused the contamination.[137] In addition, persons who purchase or begin operations at property after June 5, 1995, will be exempt from liability for existing contamination if they conduct a baseline environmental assessment (BEA) of the property.[138] These new property owners/operators must also exercise "due care" with regard to any contamination that exists at the time of their purchase or occupancy.[139] The BEA must provide sufficient information on the site and the nature and extent of contamination to determine if, at some time in the future, the new owner has contributed to the contamination on the property. The party preparing the BEA may elect to limit the scope of the BEA if he limits the use of the hazardous substances at the facility in the future or if no activity will occur at the property during his ownership/operation.[140]

The new amendments also establish cleanup standards that are intended to protect the public health and environment yet also allow for a reduction in the cost of cleanups by allowing the cleanup standards to reflect the expected future use of the property.[141] The Michigan Department of Natural Resources (MDNR) is authorized to establish cleanup standards in land use-based categories.[142] The categories for cleanup criteria are residential, commercial, industrial, and recreational.[143] Voluntary cleanup of property will be determined in accordance with the correlating cleanup category and standards.[144]

---

[135]*See* http://www.epa.gov/r5b...successl.htm#voluntary.

[136]1995 P.A. 71, as amended. Part 201, effective June 5, 1995 (HB 4569); codified at MICH. COMP. LAWS ANN. §§ 20101-20133.

[137]*Id.* § 20126.

[138]*Id.* § 20126(1)(c).

[139]*Id.* § 20107a.

[140]*Id.* § 20101(1)(d).

[141]*Id.* § 20120.

[142]*Id.* § 20120a.

[143]*Id.* § 20120a(1)(a-e).

[144]*Id.*

In addition to the BEA, liability protection is offered to persons who clean up contaminated property in the form of Covenants Not to Sue[145] and contribution protection.[146] The amendments provide protection to lenders,[147] local governments,[148] and contractors.[149] Finally, the new voluntary program is fee based. The statutes establish a $750 fee for the nonliability letter, which is issued after the BEA has been accepted by the MDNR.[150]

Brownfield Redevelopment Board[151]

In 1997, the state legislature directed that the state establish a Brownfield Redevelopment Board within the state's environmental protection agency.

Brownfield Redevelopment Financing Act[152]

In 1997, the Michigan legislature authorized municipalities to create their own brownfield redevelopment authority to implement brownfield redevelopment plans in brownfield redevelopment zones. This act further endorses the brownfield redevelopment concept that remediation is at its heart a local land use issue. The Act is an effort of the state to decentralize the redevelopment efforts from the state capital to the local government where the brownfield is located.

Other Financing

Michigan is among the early states to develop grant programs to fund brownfields redevelopment. Michigan had two programs that were operating before the state legislature established the Brownfield Redevelopment Financing Act[153] in 1997. These two programs are:

> the Site Reclamation Grant; and
> the Site Assessment Grant.[154]

The Site Reclamation Grant provides funding to local governments to investigate and remediate sites known to be contaminated that are located in areas targeted

---

[145]*Id.* §§ 20132 (developers of property), 20133 (voluntary cleanup of property). These covenants are offered to property owners *prior* to cleanup of the property. *Id.*

[146]*Id.* §§ 20129(5), (6). The new amendments have also attempted to provide for such protection against the federal government. *Id.*

[147]*Id.* § 20101(a),(b).

[148]*Id.* § 20126(a),(b),(d).

[149]*Id.* § 20128.

[150]*Id.* § 20126(3).

[151]MICH. STAT. ANN. § 13A.20104a (1997).

[152]MICH. STAT. ANN. § 3.540(2651) *et seq.* (1997).

[153]*Id.*

[154]*See* Site Reclamation Program, State of Michigan Department of Environmental Quality (DEQ) web site at: http://www.deq.state.m...d/siterec/srpsumm.html (4/14/97 16:33:34).

for economic development projects. As of the end of 1995, the state had awarded twenty-one grants to Michigan local governments totaling $14,860,000. An additional $19 million remains in the fund, of which $16 million has yet to be appropriated by the Michigan legislature.

In order to be eligible, the local government must have sufficient environmental data for the site to be scored by the Department of Environmental Quality (DEQ). These grants are available to any local government, including cities, townships, villages, or counties—unless that local government is considered a PRP under state or federal environmental law. The grant, of up to $2 million per project, can be used for environmental investigation, interim response, or remediation[155]— provided the project will result in near term economic development. In other words, the project should result in job creation, private investment in the area, or a property tax increase for the local government. Applications for Site Reclamation Grants are evaluated by the state for the following characteristics:

> environmental benefit;
> economic benefit;
> preservation of undeveloped lands;
> utilization of existing infrastructure;
> potential for environmental contamination resulting from new greenfields development;
> utilization of public and private funding; and
> local support for the project.

The Site Assessment Grant program[156] is designed to provide local governments with funding to investigate properties that are identified as having redevelopment potential but about which the level of contamination is unknown. In 1993, $10 million was made available to Michigan local governments through this program. Grants totaling $7,680,000 were awarded for thirty-four projects in twenty different communities in 1994. Approximately $2.3 million remains in the program for future grants.

The program is available to communities that can demonstrate that the site has economic potential. It allows the local government to recover grants from PRPs and utilize the funds for response activities at the project site or at other properties, at the discretion of the locality.

The applications are evaluated based on how well they meet the objective of the program. These criteria include:

> targeting areawide redevelopment by providing funding for environmental assessment to investigate properties in areas with redevelopment potential;

---

[155]The limit is one per community per fiscal year, except that counties are not limited in the number of applications that can be submitted annually.

[156]*See* Site Reclamation Program, State of Michigan Department of Environmental Quality (DEQ) web site at: http://www.deq.state.m...d/siterec/srpsumm.html (4/14/97 16:33:34).

determining whether properties in eligible communities are suitable for reuse
and defining environmental response activities necessary for the planned
redevelopment;

encouraging marketing of properties to potential developers by providing
funding to projects that serve as developers' Baseline Environmental Assessment or due diligence by local units of government;[157]

encouraging businesses to reuse basic infrastructure rather than relocate to
greenfield space or move out of state;

providing funding to the projects with the greatest environmental and economic development benefits, taking into account demographic difference
throughout the state;

encouraging communities to demonstrate innovative approaches to public/
private partnerships to provide environmental improvements resulting in
eventual economic development.

Finally, Michigan is developing two new programs. The first is the State Revitalization Revolving Loan fund. This fund will allow local governments to obtain funds for site assessment and demolition activities needed prior to redeveloping sites with either known or suspected contamination.[158] "Loans are available to
help promote the economic development of properties. Potential projects do not
need to have an identified economic developer."[159] Applications for the program
became available in May 1997; however, the legislature did not appropriate any
funding for this program. As a result, the date funding will become available for
this program is not known.

In addition, the Brownfield Redevelopment Financing Act allows municipalities to establish Brownfield Redevelopment Zones.[160] The taxes generated from
these designated sites may be used for various response activities, including:

establishment of a local site remediation fund;

building demolition; and

repayment of revitalization loans.

The tax capture in this program is similar to the tax increment financing used by
some municipalities for downtown development in prior years.

Municipalities participating in the program work with DEQ to create Brownfield Redevelopment Authorities. These bodies then define local brownfield re-

---

[157]Site assessment grants also allow targeted marketing of properties to developers whose needs
match the condition of the property.

[158]*See* Site Reclamation Program, State of Michigan Department of Environmental Quality (DEQ)
web site at: http://www.deq.state.m...rd/glance/erdlpgr.html (4/14/97 16:34:01).

[159]*See* Site Reclamation Program, State of Michigan Department of Environmental Quality (DEQ)
web site at: http://www.deq.state.m...d/siterec/srpnews.html (4/14/97 16:34:26).

[160]*See* Site Reclamation Program, State of Michigan Department of Environmental Quality (DEQ)
web site at: http://www.deq.state.m...d/siterec/srpnews.html (4/14/97 16:34:26).

development zones, make plans for use of available financing tools, and target properties for cleanup and redevelopment.

Business Transfer Statutes

Michigan's Environmental Response Act[161] (MERA) requires written notification to the transferee of releases at the site. The transferor must provide the location, amount, and date of each release. MERA[162] is very broad because it covers "all real property." If a person transfers an interest in real property and has actual or constructive knowledge that there has been a release of a reportable quantity of a hazardous substance, the disclosure requirements of MERA are triggered.

In order to comply with MERA, prior to the transfer, the transferor must provide written notice to the transferee of the nature and extent of any such release. If the instrument of conveyance must be recorded, then the notice must also be recorded.[163] The act makes no provisions for remedies available for violations of these requirements.

Specialized Stakeholder Laws—Lender and Trustee

Until 1997, Michigan had no specialized stakeholder laws. The legislature enacted laws protecting lenders and trustees as part of its comprehensive brownfields legislation.[164]

Environmental Lien

Michigan enacted a nonsuperlien statute to enforce its state equivalent of CERCLA.

Minnesota

Voluntary Cleanup Program

The Minnesota Environmental Response and Liability Act (MERLA), as amended by the 1992 Land Recycling Act (LRA),[165] offers incentives to private parties conducting voluntary investigations and cleanups of contaminated industrial sites. Through Minnesota's Voluntary Investigation and Cleanup (VIC) Program, the LRA attempts to facilitate the transfer and development of property by providing

---

[161]MICH. COMP. LAWS ANN. § 299.610c.

[162]*Id.*

[163]*Id.*

[164]MICH. STAT. ANN. § 13A.20101b (1997) (liability of lender as fiduciary or representative for disabled person).

[165]MINN. STAT. ANN. § 115B.175 *et seq.* (West 1994).

resource and technical assistance at the request of program participants.[166] Moreover, program participants undertaking and completing cleanups in accordance with an approved voluntary cleanup plan, and who are not otherwise responsible for a release under MERLA, are eligible for protection from liability under the LRA.[167]

As amended in 1993, the LRA allows *responsible persons* to conduct cleanups and gain liability protection for associated parties who are not otherwise liable for the release, including lenders, developers, prospective purchasers, and municipalities.[168] Specifically, the LRA allows volunteers to choose from five separate types of written assurances for protection from environmental liability for parties successfully meeting the VIC program prerequisites:[169]

Technical Assistance Approval Letters;
No Action Letters;
Off-site Source Determination Letters;
Certificates of Completion; and
No Association Determination Letters.[170]

Upon successful completion of the VIC Program, volunteers will receive a No Action letter from the Minnesota Pollution Control Agency (MPCA).[171] No

---

[166]*Id.* § 115B.175. The LRA was specifically enacted to clarify the cleanup liability of specific parties and obtain statutory mechanisms to obtain liability protection. A number of these amendments offer incentives to promote the voluntary investigation and cleanup of hazardous substances and/or petroleum. These amendments also encourage and facilitate the transfer and development of property by providing additional resources and technical assistance regarding contaminated property.

[167]*Id.* § 115B.175 [subdiv. 1] (West 1994). Interestingly, the Minnesota Program started modestly with two staff personnel and twenty proposed projects waiting for approval when the doors opened. In 1995, the program has twenty staff members and has recently just received its five hundredth application for assistance. On average, the Minnesota Pollution Control Agency (MPCA) receives three to six applications per week. The program was designed to assist parties interested in developing contaminated land and determining the environmental status of land before purchasing the property. *Environmental Law Institute: Redeveloping Brownfields Workshop* (March 28, 1995) (Statement of Joseph Zachmann, project manager, Site Response Section, MPCA).

[168]MINN. STAT. ANN. § 115B.175 [subdiv. 6a] (West 1994). The benefit to responsible parties under the LRA stems from the release from liability for future, unanticipated cleanup of contamination specifically addressed under the VIC program. Thus, through remediating contaminated property under the VIC program, responsible persons preclude subsequent claims based on retroactive liability for site contamination.

[169]*Id.* § 115B.175 [subdiv. 6]. These assurances can be issued in a variety of forms, such as letters, agreements, or certificates, and can apply to technical, administrative, and liability matters.

[170]*Minnesota Pollution Control Agency: Voluntary Investigation and Cleanup,* Guidance Document No. 4, § 3.0 (revised January 1994). These five broad categories of written assurances may be issued by the MPCA commissioner or designated MPCA staff following approval of an investigation and a Response Action Plan (RAP). *Id.* Except for No Action letters or agreements, these assurances are issued under discretionary authority expressly provided under Minnesota Statutes. *Id.* No Action letters are also discretionary but are issued pursuant to the inherent authority of the MPCA to exercise discretion in its enforcement action. *Id. See* MINN. STAT. ANN. § 115B.175 [subdiv. 3] (West 1994).

[171]MINN. STAT. ANN. § 115B.175 [subdiv. 6] (West 1994).

Action letters provide qualified liability protection from future Superfund or administrative enforcement actions brought by the MPCA.[172] The assurances provided by No Action letters or agreements, however, apply only to parties expressly included within the VIC Program.[173] Unless explicitly stated, a No Action letter or agreement does not extend to other parties associated with properties, such as lenders and successors to the owner of the property.[174]

Cases where No Action letters may be issued include where:

no contamination is detected and an approved Phase I Investigation Report determines that the past and current use of the property would not have contributed to a release to soils or groundwater;[175]

contamination is detected but the levels are not significant as determined by MPCA staff, and as result no cleanup is required;[176]

the property has been remediated as approved by the MPCA and no additional cleanup is required.[177]

---

[172]*Minnesota Pollution Control Agency: Voluntary Investigation and Cleanup,* Guidance Document No. 4, § 3.2.

[173]*Id.*

[174]*Id.* It should be noted that lenders receive express statutory liability protection under the VIC Program. *See* MINN. STAT. ANN. § 115B.175 [subdiv. 6] (2) (West 1994).

[175]*See Minnesota Pollution Control Agency: Voluntary Investigation and Cleanup,* Guidance Document No. 8 (revised January 1994) (information on the procedures and criteria involving a Phase I Investigation Report).

[176]*Minnesota Pollution Control Agency: Voluntary Investigation and Cleanup Program,* Guidance Document No. 4, § 3.2. A No Action letter may be issued in this circumstance should state that a release of hazardous substances was detected, but the levels in groundwater were, for example, below the Minnesota Department of Health recommended allowable limits. A Phase I Investigation Report must also be reviewed and approved by the MPCA staff to qualify for this No Action letter.

[177]*Id.* Within the above three categories, a No Action letter may be tailored according to site specific conditions. *Id.* A Limited No Action letter may be issued depending upon the completeness of the investigation or cleanup performed, the specific boundaries of the investigation or cleanup, or an abbreviated list of contaminants analyzed. *Id.* Cases where Limited No Action letters have been issued include:

Sites where only soil is investigated and remediated. In such cases, a Limited No Action letter will address only the soil investigation and response actions and will contain disclaimers stating that groundwater was not investigated or remediated.

Sites where only a discrete area of the property is investigated and remediated. An example includes an area of a proposed building expansion investigated and cleaned up but other areas of the property were not addressed. As such, a Limited No Action letter would only address the expansion area and would contain disclaimers regarding other areas of the property that were not investigated or cleaned up.

Sites where only a limited number of contaminants were investigated or cleaned up. A Limited No Action letter would only address the contaminants investigated and would contain disclaimers regarding other contaminants that have been detected or may be present.

Id.

The No Action letter will state that the MPCA will refrain from referring the contaminated industrial property to CERCLIS, U.S. EPA's Superfund site tracking system, and preparing a Hazard Ranking System (HRS) score for the site.[178] Specific LRA requirements mandate the submission of a voluntary response plan and a report describing the results of an investigation of actual or threatened releases of hazardous substances and/or petroleum.[179] Program participants are required to meet cleanup standards protective of public health, welfare, and the environment.[180]

Since the Minnesota VIC Program was to create incentives for redevelopment and reuse of inactive or abandoned industrial sites,[181] specific statutory protection is included for lenders.[182] This protection compensates for the MPCA's position that No Action Letters or Agreements do not extend to the parties who have specifically applied to participate in the VIC program.[183] Specifically, the LRA's lender liability provision is aimed at facilitating the lending of funds to third parties undertaking a voluntary cleanup or using the land in a manner that does not affect the contamination already present.[184]

Under the LRA, parties responsible for site contamination cannot obtain liability protection for themselves from state or federal environmental enforcement actions.[185] Responsible parties may, however, obtain a Covenant Not to Sue for future cleanup liability when they conduct and successfully complete a voluntary response action.[186] Protection from liability takes effect when the completion of a voluntary cleanup is certified by the MPCA and a Certificate of Completion is issued.[187] The issuance of a Certificate of Completion must coincide with the approval of a voluntary Response Action Plan (RAP).[188] The LRA's liability protection assurances will only vest when a RAP is approved and implemented.[189] If

---

[178]*Minnesota Pollution Control Agency: Voluntary Investigation and Cleanup,* Guidance Document No. 4, § 3.2.

[179]MINN. STAT. ANN. § 115B.175 [subdiv. 3] (West 1994).

[180]*Id.* § 115B.175 [subdiv. 3] (c).

[181]*Environmental Law Institute: Redeveloping Brownfields Workshop* (Statement of Joseph Zachmann).

[182]MINN. STAT. ANN. § 115B.175 [subdiv. 6] (2) (West 1994).

[183]*Minnesota Pollution Control Agency: Voluntary Investigation and Cleanup,* Guidance Document No. 4, § 3.2 (revised January 1994).

[184]*Environmental Law Institute: Redeveloping Brownfields Workshop* (March 28, 1995) (Statement of Joseph Zachmann).

[185]MINN. STAT. ANN. § 115B.175 [subdiv. 6a] (West 1994).

[186]*Id.* § 115B.175 [subdiv. 6a] (b); *Minnesota Pollution Control Agency: Voluntary Investigation and Cleanup,* Guidance Document No 4, § 3.5 (revised January 1994).

[187]MINN. STAT. ANN. § 115B.175 [subdiv. 5] (West 1994).

[188]*Minnesota Pollution Control Agency: Voluntary Investigation and Cleanup,* Guidance Document No. 4, § 3.5.

[189]*Id.* A RAP is only required when the MPCA has determined that the property is the site of a release or threatened release that needs to be addressed to protect human health, welfare, or the environment.

a response action is not required to remedy site contamination, then the contaminated site is not eligible for assurances of liability protection under the LRA.[190]

Business Transfer Statutes

Minnesota[191] has a statute affecting the transfer of hazardous waste sites and requiring notice and disclosure. Minnesota requires pretransfer recording of disclosure statements at sites of hazardous disposal facilities.[192] If the owner of real property knows, or should have known, that property was used as a hazardous waste disposal facility site or was "subject to extensive contamination by release of a hazardous substance," the owner, before transferring ownership, is required to record an affidavit about the use and condition of the property.[193] The owner must record, in the county where the property is located, an affidavit informing the potential transferee of the prior use, contamination, and any restrictions upon the use of the property.[194] Failure to record the required "affidavit . . . does not affect or prevent any transfer of ownership of the property."[195] Any person who knowingly does not record the affidavit is, however, "liable . . . for any release or threatened release of any hazardous substance from a facility located on" the property.[196]

Specialized Stakeholder Laws

Minnesota is one of the states that amended its state equivalent of CERCLA to include some specialized lender liability protection.

Under the Minnesota equivalent of Superfund, a mortgagee is not a responsible person solely because the mortgagee becomes an owner of real property through foreclosure of the mortgage or by receipt of the deed to the mortgaged property in lieu of foreclosure. In addition, a mortgagee of real property where a facility is located or a holder of a security interest in facility assets or inventory is considered not to be an operator of the facility solely because the mortgagee or holder has a capacity to influence the operation of the facility to protect its security interest in the real property or assets.

Environmental Lien

Minnesota enacted a nonsuperlien statute under its state CERCLA-type statute. Although environmental liens are constitutionally suspect under *Reardon* analysis, Minnesota's is one of the few statutes that is almost certain to survive constitutional

---

[190]*Id.*

[191]MINN. STAT. ANN. § 115B.16, subd. 2.

[192]*See* § 115B.16.

[193]MINN. STAT. ANN. § 115B.16, subd. 2.

[194]*Id.*

[195]*Id.*

[196]MINN. STAT. ANN. § 115B.16, subd. 4(b).

scrutiny under the *Reardon* test. The Minnesota statute provides for a preattachment hearing before the state environmental protection agency to challenge the state's intended action of filing a lien.[197]

Ohio

Voluntary Cleanup Program

"One of the unfortunate by-products of Ohio's industrial heritage is the number of contaminated sites, commonly known as brownfields, scattered through the state."[198] For over a decade, "valuable land would sit idle because fears of immense liability and cleanup costs scared potential developers, businesses and banks."[199] In order to reverse this trend, "Ohio now has a new program that removes the environmental and legal barriers that have stalled redevelopment and reuse of brownfields."[200]

Signed into law on June 29, 1994, the Ohio Voluntary Action Program[201] (S.B. 221) will be implemented in two stages: (1) an Interim Voluntary Action Program,[202] which will sunset upon (2) the adoption of the Final Program by Ohio Environmental Protection Agency (Ohio EPA).[203] When enacting S.B. 221, the Ohio legislature intended to create a program that:

> allows private parties to voluntarily clean up property contaminated by hazardous and/or petroleum to standards meeting the actual use of the property;
> removes lender liability surrounding contaminated property;
> privatizes the cleanup of contaminated property;
> consolidates the permit process for remedial activities; and
> offers tax abatements and low interest loans to potential volunteers.[204]

Volunteers satisfactorily completing the Voluntary Action Program will receive a Covenant Not to Sue from the Ohio EPA.[205] Program participants are

---

[197]*See* MINN. STAT. ANN. §§ 115B.01-.37.

[198]Jennifer A. Kwasniewski, *Ohio's New Voluntary Action Program—A Solution That Benefits the Environment and the Economy,* reprinted in AMERICAN BAR ASSOCIATION, SECTION OF NATURAL RESOURCES, ENERGY, AND ENVIRONMENTAL LAW, 4TH ANNUAL FALL MEETING (Oct. 4, 1996, Boston, Massachusetts).

[199]*Id.*

[200]*Id.*

[201]OHIO REV. CODE ANN. § 3746.01 *et seq.* (Baldwin 1994).

[202]*Id.* § 3746.04.

[203]*Id.*

[204]*Conference on Brownfields Redevelopment and Environmental Risk Management in Business and Real Estate Transactions* (March 22-23, 1995) (statement of William Hinger, M. En., RELA, Env'tl. Spec., Huntington Technical Services, Huntington Bancshares Inc.).

[205]OHIO REV. CODE ANN. § 3746.12 (Baldwin 1994).

required to remit a program fee of $2,000 to receive a Covenant Not to Sue, plus additional fees if their cleanups involve a Consolidated Standards Permit or an "operation and maintenance agreement."[206] Conditioned upon continued compliance with applicable cleanup standards, the Covenants Not to Sue provide volunteers and subsequent purchasers with protection from civil liability for future, unanticipated cleanup work.[207]

Program participants must use state certified professionals and laboratories to verify the proper completion of cleanup work and determine that the property is remediated to acceptable standards.[208] All professionals must have at least five years of experience with investigating and remediating hazardous substances and three years of experience directly supervising remediation projects.[209] Certified professionals include engineers,[210] professional geologists,[211] certified industrial hygienists,[212] and toxicologists.[213] Certified professionals and laboratories are required pay a fee of $2,000 to participate in the Interim Program.[214]

Until Ohio EPA promulgates the Final Voluntary Action Program, the existence of groundwater contamination will render property ineligible for voluntary cleanup.[215] Moreover, properties listed on U.S. EPA's National Priorities List (NPL)[216] and properties subject to closure requirements under solid or hazardous waste laws are not eligible for voluntary cleanup under both the Interim and Final Programs.[217] Properties subject to Ohio's underground storage tank (UST) law[218]

---

[206]*Id.* § 3746.07(H).

[207]*Id.* § 3746.12(B). Releases from liability under Covenants Not to Sue remain in effect only for as long as the property or portion thereof subject to the covenant complies with the applicable standards issued as part of the covenant. *Id.* § 3746(B)(1). If a property falls out of compliance with the terms of a Covenant Not to Sue, a volunteer has thirty days, after being so notified, to bring the property in compliance and enter into a compliance schedule agreement with the Ohio EPA. *Id.* § 3746.12(B)(3). Failure to bring the property into timely compliance with terms set forth in the covenant will result in the revocation of Covenant Not to Sue. *Id.* § 3746.12(B)(4).

[208]*Id.* §§ 3746.04(B)(5) and 3746.07(D) and (E) (Baldwin 1994). "Acceptable standards" are determined on a case-by-case basis. Specifically, a volunteer may choose to remediate a site consistent with numeric standards promulgated by Ohio EPA or use site-specific risk assessment procedures depending upon the future proposed use of the facility.

[209]*See generally Id.* § 3746.07(D).

[210]*Id.* § 3746.07(D)(1). Specifically, § 3746.7(D)(1) requires that professional engineers must registered under OHIO REV. CODE ANN. § 4733 (Baldwin 1994).

[211]*Id.* § 3746.07(D)(2). Professional geologists participating in the Voluntary Action Program must be certified by the American Institute of Geologists. *Id.*

[212]*Id.* § 3746.07(D)(3). Industrial hygienists participating in the Voluntary Action Program are required to be certified by the American Board of Industrial Hygiene.

[213]*Id.* § 3746.07(D)(4). Toxicologists participating in the Voluntary Action Program must at least hold a master's degree in toxicology or a related field from an accredited school. *Id.*

[214]*Id.* § 3746.07(D)(4).

[215]*Id.* § 3746.07(A)(3).

[216]*Id.* § 3746.02(A)(1)(d).

[217]*Id.* § 3746.02(A)(2).

[218]*See* the Ohio Bureau of Underground Storage Tanks Regulations (BUSTR), OHIO REV. CODE ANN. §§ 3737.88, 3737.882 and 3737.889 (Baldwin 1994).

and sites where the director of the Ohio EPA has initiated an environmental enforcement action are also excluded.[219]

Carcinogenic hazardous substances must be cleaned up to a risk range of 1×10^{-4} to 1×10^{-6}.[220] Petroleum must be cleaned up to standards set by the Ohio Bureau of Underground Storage Tank Regulations (BUSTR). Parties participating in the Voluntary Action Program must conduct Phase I environmental site assessments (ESA) in accordance with ASTM standards (E 1527-94).[221] Primarily a document review process, a Phase I ESA includes an analysis of a property's history of environmental compliance, chain of title, previously conducted ESAs, and any aerial photographs taken of the site.[222] Employee interviews may also be conducted.[223]

If the Phase I ESA demonstrates that there is a reason to believe that contamination exists, a Phase II ESA must be performed following criteria set forth in the Ohio Voluntary Action Program.[224] Minimal procedural steps must be followed to demonstrate that any contamination present at a site does not exceed applicable standards warranting remedial activity or that remedial activities at the property have achieved compliance with applicable cleanup standards.[225] Phase II ESA procedures include a review of all documentation prepared in connection with

---

[219]*Id.* § 3746.02(A)(5).

[220]*Id.* § 3746.07(A)(2) (cancer-causing hazardous substances must be cleaned up to the point where one person in 10,000 to one person in 1,000,000 is at risk of cancer after a lifetime of exposure to hazardous contaminants). The cleanup standards relating to carcinogenic substances will be determined by the proposed future use of the facility, i.e., whether it is for industrial, commercial, or residential purposes.

[221]*Id.* § 3746.04(B)(3).

[222]*Id.* The standards for Phase I property assessments, at a minimum, require all of the listed requirements. Stated in greater detail, Phase I requirements specifically include:

  review and analysis of deeds, mortgages, easements of record, and similar documentation related to the chain of title that are publicly available;
  review and analysis of previous environmental assessments, property assessments, environmental or geological studies of the property, and any land within 2,000 feet of the boundaries of the contaminated site that are publicly available.
  review of current and past environmental histories of *persons* who owned or operated the property;
  review of aerial photographs of the property that indicate prior usage of the site;
  interviews with managers of activities conducted at the property who have knowledge of environmental conditions of the property;
  conducting a walkover inspection of the property;
  identifying the current and past uses of the property, adjoining tracts of land, and the area surrounding the property, including, without limitation, interviews with persons who reside or have resided, or who are or were employed, within the area surrounding the property regarding the current and past uses of the site and adjacent tracts of land.

Id. § 3746.04(B)(3)(a)-(g).

[223]*Id.*

[224]*Id.* § 3746.04(B)(3)(g).

[225]*Id.* § 3746.04(B)(4).

the Phase I ESA 180 days prior to commencing the Phase II, quality assurance procedures for samples or measurements taken, sampling procedures ensuring representative sampling of potentially contaminated media, analytical and data assessment procedures, and data objectives ensuring that samples collected are biased toward potentially contaminated areas.[226] The criteria set forth under the Ohio Voluntary Action Program allow for the succinct quantification of the nature and extent of hazardous substance and/or petroleum contamination at brownfield sites.

Upon adoption of the Voluntary Action Final Program by the Ohio EPA, the following additional provisions will be implemented:

> creation of a "multidisciplinary board" comprised of public, private, environmental, scientific, and business advocates to advise the director of the Ohio EPA regarding the adoption of rules for the final program;[227]
> numeric cleanup standards meeting the U.S. EPA Performance Standards;[228]
> criteria for certification of professionals[229] and laboratories,[230] as well as for revoking or suspending certification;
> information regarding the content of NFA letters in order to receive a Covenant Not to Sue;[231]

---

[226]*Id.*

[227]*Id.* § 3746.03.

[228]*Id.* § 3746.04 (B)(1). The generic numerical standards established under § 3746.04(B)(1) must be consistent and equivalent in scope, content, and coverage to any applicable standard established under federal environmental laws and regulations, including, without limitation:

Federal Water Pollution Control Act, as amended, 33 U.S.C. § 1251;
Resource Conservation and Recovery Act, as amended, 42 U.S.C. § 6921;
Toxic Substances Control Act, as amended, 15 U.S.C. § 2601 *et seq.*;
Comprehensive Environmental Response, Compensation, and Liability Act, as amended, 42 U.S.C. § 9601 *et seq.*; and
Safe Water Drinking Act, as amended, 42 U.S.C. § 300f *et seq.*

[229]OHIO REV. CODE ANN. § 3746.04(B)(5) (Baldwin 1994). Factors considered for certification of environmental professionals include evaluating previous performance records regarding remedial investigations and activities and environmental compliance histories. *Id.* § 3746.04(B)(5)(a)(i).

[230]*Id.* § 3746.04(B)(6). The criteria and procedures for issuance, denial, suspension, and revocation of laboratory certifications can be found at OHIO REV. CODE ANN. § 3745 (Baldwin 1992).

[231]*Id.* § 3746.04(B)(7). Specifically, the information included in the NFA letter should specify:

a summary report of remedial activities prepared by the certified professional responsible for preparing the NFA letter;
if performed in lieu of generic numeric cleanup standards, notification that a risk assessment was performed;
the contaminants addressed, their source (if known), and contaminant levels prior to remediation;
identification of the person(s) who performed the remedial activity in support of the NFA letter;
a list of data, information, records, and documents relied upon by the certified professional in preparing the NFA letter.

Id. § 3746.04(B)(7)(a)-(e).

fees to defray all direct and indirect costs of the program;[232]

procedures for the Ohio EPA to audit properties;[233]

criteria for addressing groundwater contamination, including provisions for public participation on a case-by-case basis;[234]

variance procedures for volunteers providing sufficient evidence for a variance decision by the director of the Ohio EPA.[235]

Other provisions impacting both the Interim and Final Brownfields Programs provide incentives for parties to undertake voluntary cleanups for the redevelopment and reuse of brownfields. These incentives include:

low interest loans from the Ohio Department of Development, Ohio Water Pollution Control Loan Fund, and the Ohio Water Development Authority;[236]

a ten-year automatic tax abatement on the increase in land value with provisions allowing local abatement on real and personal property taxes for development projects for up to ten years;[237]

liability protection for the state, local governments, and contractors involved in voluntary cleanup projects;[238]

lender/trustee/fiduciary liability protection for property held to protect a security interest;[239]

---

[232]*Id.* § 3746.04(b)(8).

[233]*Id.* § 3746.17. Ohio EPA should conduct audits in connection with the issuance of NFA letters to ascertain the following:

the property subject to voluntary action meets applicable cleanup standards;

assess the qualifications and work performed by certified professionals; and

review the quality of the work performed by certified laboratories in connection with the Voluntary Action Program.

Id. 3746.17(A)(1)-(3). Pursuant to § 3746.171, property subject to institutional controls will be audited by Ohio EPA to confirm continuing compliance with specified land use controls every five years.

[234]*Id.* § 3746.04(B)(12).

[235]*Id.* § 3746.09.

[236]*See* OHIO REV. CODE ANN. §§ 6111.036, 6123.032, 6123.041 (Baldwin 1994).

[237]*Id.* §§ 5709.87(C)(2) and 5709.883.

[238]*Id.* § 3746.24.

[239]*Id.* § 3746.26. Paralleling the CERCLA § 101(20)(A), 42 U.S.C. § 9601(20)(A), "secured creditor exemption," this provision states that any person, who, without participating in the management of a property, holds indicia of ownership in a property primarily to protect a security interest is protected from liability for any civil action brought in connection with the property subject to the Voluntary Action Program, for investigative or remedial costs whether or not such costs were incurred in connection with property subject to the voluntary action program, and for conducting or causing to be conducted all such nonremedial activity at the property in compliance with applicable law. *Id.* § 3746.26(A)(1). The term *indicia of ownership* is defined at § 3746.26(B).

Borrowing much of the basic structure and definitions from the U.S. EPA's 1992 lender liability rules, this exemption from environmental liability applies only if, after taking title to the property, the

Consolidated Standards Permit agreements;[240]

NFA letters must be prepared and submitted to the state by certified professionals;[241]

Covenants Not to Sue are conditioned upon the continued land use specified in the NFA and only protect volunteers for state civil actions, not federal or third-party actions;[242]

Covenants Not to Sue and NFAs must be recorded with county records and will run with the land to ensure the property is used in a manner consistent with the use specified in the NFA letter;[243]

Ohio EPA has complete access to all cleanup documents and properties for audit purposes;[244]

information generated during the cleanup is privileged and is not admissible in court for subsequent enforcement actions;[245] and

parties participating in voluntary cleanups may recover remedial costs from people who caused the pollution (except in natural gas and petroleum cleanups).

---

holder of a security interest conducts all activities occurring at the property in compliance with the applicable requirements of the Ohio Revised Code and the exemption will be lost if new violations occur after the holder acquires title to the property. Section 3746.23(G)(6)-(7) exempts holders of security interests, trustees, and fiduciaries from liability for costs of conducting a voluntary cleanup under Chapter 3746, provided the requirements of §§ 3746.26 and 3746.27 are met. *Conference on Brownfields Redevelopment and Environmental Risk Management* (March 22-23, 1995) (statement of William Hinger).

[240]*Id.* § 3746.15. If volunteers are required to obtain permits for remedial activities under the Federal Water Pollution Control Act, as amended, 33 U.S.C. § 1251, Resource Conservation and Recovery Act, as amended, 42 U.S.C. § 6921, or the Clean Air Act, as amended, 42 U.S.C. § 1857, Ohio EPA will issue a consolidated standards permit for the activities in connection with the voluntary action. *Id.* § 3746.15(A).

[241]*Id.* § 3746.10(A).

[242]*Id.* § 3746.12(B)(2). Compliance with the terms of the Covenant Not to Sue issued by Ohio EPA should allow a volunteer to raise a "permit shield" defense to subsequent state actions seeking additional cleanup of contaminants. However, this provision is silent with regard to protection from federal environmental enforcement and civil third-party actions. Although compliance with the terms of the Covenant Not to Sue issued under the Voluntary Action Program may be asserted as a mitigating factor in federal or third-party enforcement proceedings, it will not meet the legal requisites of the permit shield defense.

Generally, a permit shield defense states that compliance with the terms of a permit or other enforceable agreement is deemed to be in compliance with applicable law, even if it is subsequently modified. Thus, until the permit renewal or the agreement must be renegotiated, a party is deemed to be in compliance with the law as long as the party does not violate the express terms of the permit or agreement. Interestingly, the question arises whether federal sign-off on a Covenant Not to Sue issued by Ohio EPA under the Voluntary Action Program, via a Memorandum of Agreement (MOA), or a Memorandum of Understanding (MOU) could be used to assert a permit shield defense from subsequent federal environmental enforcement actions.

[243]OHIO REV. CODE ANN. § 3746.14 (Baldwin 1994).

[244]*Id.* § 3746.21.

[245]*Id.* § 3746.28(C).

Compliance with the Ohio Voluntary Action Program has been criticized as an overwhelming and burdensome task. Moreover, contrary to the intent of the drafters, the lending community has begun to insist that all properties eligible for participation in the program must participate and obtain a Covenant Not to Sue as a condition precedent to making a loan. The net effect of this innovative legislation is, thus, uncertain.

Wisconsin

Voluntary Cleanup Program

In 1994, Wisconsin enacted the Land Recycling Act.[246] The law became effective on May 13, 1994, and is administered under the Wisconsin Department of Natural Resources (DNR) and the Department of Economic Development (DED). The former administers matters concerning environmental cleanup and remediation; the latter administers matters concerning which sites receive funding from the program.

In 1997, the state legislature, encouraged by the progress of the 1994 law, created additional funding for brownfields redevelopment.[247] The legislature also dictated that in awarding grants, the state should give higher priority to brownfields redevelopment than to other projects.[248]

Program participants are responsible for paying DNR's oversight costs. The DNR requires a formal application from the interested party to determine whether both the site and the applicant are eligible for the program. Applicants must be considered "purchasers" or "innocent landowners" to be allowed in the Wisconsin program. The legal definition of eligible parties requires that the current owner or prospective purchaser:

> purchased the property in an arm's length, good faith transaction;
> did not own the site when the contamination occurred; and
> did not cause any original contamination.

If the applicant is accepted into the program, the DNR will require a Phase I and a Phase II environmental investigation be conducted. These investigations must be performed in accordance with state law and must receive DNR approval. The investigations need not be conducted by the applicant; in fact they may be conducted by a PRP (such as the current owner) under contract with the applicant, provided they are approved by DNR.

The following table describes the documents subject to DNR oversight.

---

[246]Sections 292.11-.21.
[247]1997 Wisc. AB 100 § 20.143(1)(br) and § 764.23.09(19)(a).
[248]1997 Wisc. AB 100 § 768.23.175.

Document	Description
Phase I assessment	Preliminary environmental site assessment
Phase II work plan	More detailed environmental assessment, including sampling protocol for contaminated and clean areas
Remediation work plan	Plan for site remediation based on data obtained in phase I & II assessments
Closeout report	Report detailing the cleanup's completion

If no environmental contamination is found or that which is found is deemed insignificant, then no further action is required of the applicant. The applicant will not, however, receive any manner of written or oral release of liability from the program.

If contamination is found, all subsequent remedial activities must be conducted by the applicant itself. Where remediation is necessary, DNR must approve the remediation work plan. Once the plan is approved and the work completed, the applicant must then submit a "closeout" report. This last report demonstrates to DNR that the work was done in accordance with all state soil and groundwater standards.

The Certificate of Completion granted by DNR covers only environmental liability under the Wisconsin equivalent of CERCLA, the Hazardous Substance Discharge Law. The release is not applicable to any other state statutes; and, of course, it cannot bind federal causes of action. The release is transferable to future landowners and is considered binding, unlike that of many other states, even if:

DNR cleanup standards change in the future;
additional contamination is found; or
the remedy fails.

Properties contaminated with petroleum or agricultural chemicals are eligible for financial assistance under the Wisconsin UST Fund and the Agricultural Chemicals Cleanup Fund.

Financing

Encouraged by the preliminary success of the voluntary cleanup program, the legislature created a brownfields grant program in 1997.[249] Like the programs in Indiana and Michigan, the Wisconsin program targets brownfields properties as well as properties located in a brownfield development zone.[250]

---

[249] 1997 Wisc. AB 100 § 20.143(1)(qm). See also § 763.23.09(19)(a) which defines the term "brownfields redevelopment."

[250] *Id.* § 71.07(2dx).

CHAPTER 9

Region VI:
Arkansas, Louisiana, New Mexico,
Oklahoma, and Texas

Texas is among the nation's leading hazardous waste generators. The petrochemical industry accounts for 60 percent of the state's wastes "and exerts a dominant influence on Texas waste management policy."[1] As such, it is no surprise that Texas was the first state in Region VI to enact voluntary cleanup legislation. More recently, Oklahoma has followed Texas' lead. Arkansas has begun flirting with brownfields restoration legislation. The majority of states in Region VI, however, have little or no legislation in place addressing brownfields issues.

"EPA Region 6 has established its Brownfields program to encourage and support the development of State and Tribal voluntary cleanup programs."[2] To this extent, "the Region has publicized its authority to facilitate redevelopment by reducing potential liability of prospective purchasers and redevelopers and has solicited requests for grants and cooperative agreements.[3]

The following table describes the more recent programs developed in Region VI to encourage voluntary cleanup:[4]

States	Brownfields Financing Programs	VCP with Written Assurances	Incentives to Attract Private Development
Arkansas	X	X	X
Louisiana	X		
New Mexico		X	
Oklahoma		X	X
Texas		X	X

Arkansas

Arkansas passed "an act to encourage long-term environmental projects."[5] The legislature explained that:

---

[1]Harold C. Barnett, Toxic Debts and the Superfund Dilemma at 91 (1994).

[2]U.S. EPA, OSWER, Brownfields Initiatives in Region 6 (July 1995), reprinted in Key Environmental Issues in U.S. EPA Region VI, American Bar Association Section of Natural Resources, Energy, and Environmental Law (May 15-16, 1996, Dallas, Texas).

[3]Id.

[4]See Charles Bartsh, Brownfields State of the State Report: 50-State Program Roundup (Northeast-Midwest Inst., January 28, 1998).

[5]Ark. Stat. Ann. § 8-5-902 (1997). Arkansas 81st regular session, Act 401, House Bill 1563, 1997 Ark. ALS 401; 1997 Ark. Acts H 401; 1997 Ark. HB.

many areas of the state would benefit from long-term environmental remediation projects that significantly improve the effects caused by industrial or extractive activities. However, commitments by private enterprise to remedy such damages are discouraged by the prospect of civil liability based upon rigid application of state water quality standards to the enterprise's activities.

The purpose of the legislation "is to preserve the state's approach to establishing water quality standards, while also encouraging private enterprises to make significant improvements to closed or abandoned sites that are of such magnitude that more than three years will be required to complete the project." The statute specially includes brownfield sites in the new program.[6]

Environmental Liens

Arkansas enacted a nonsuperlien statute as part of its state equivalent of CERCLA.

Louisiana

Business Transfer Statutes[7]

The Louisiana Solid Waste Management Laws (LSWML)[8] provisions apply to abandoned and inactive hazardous and solid waste disposal sites.[9] The LSWML is triggered when: (1) the owner has actual or constructive knowledge that the property has been used for disposal and that wastes remain on the property; and (2) the property has been identified as an abandoned or inactive disposal site.[10]

Once triggered, the owner must record a notice in the mortgage and conveyance records of the parish in which the property is located.[11] The notice must indicate the location of the waste site on the property.[12]

Environmental Liens

Louisiana enacted a superlien as part of its state equivalent of CERCLA.

---

[6]Brownfield sites are defined in Act 125 of 1995 or as may be amended. The definition also includes "hazardous substance sites listed on the national priority list (42 U.S.C. Section 9605), or state priority list (ARK. CODE § 8-7-509(e)), or as may be amended."

[7]LA. REV. STAT. ANN. § 30:2157 (1992).

[8]*Id.*

[9]*Id.*

[10]*Id.*

[11]*Id.*

[12]*Id.*

New Mexico

The state of New Mexico has no brownfields redevelopment program at this time.

Oklahoma

The Oklahoma Brownfields Voluntary Redevelopment Act[13] was designed to:

(1) provide for the establishment of a voluntary program by the Department of Environmental Quality (DEQ);
(2) foster the voluntary redevelopment and reuse of brownfields by limiting the liability of property owners, lenders, lessees, and successors and assigns from administrative penalties assessed by the department and civil liability with regard to the remedial actions taken by the applicant for environmental contamination caused by regulated substances, as required by a consent order,[14] if the remedial action is not performed in a reckless or negligent manner; and
(3) provide for a risk-based system for all applicable sites based on the proposed use of the site.[15]

The statute defines *brownfield* as meaning "an abandoned, idled or underused industrial or commercial facility or other real property at which expansion or redevelopment of the real property is complicated by environmental contamination caused by regulated substances."[16] The law applies only to applications made and/or consent orders issued after January 1, 1988.[17]

The law dictates that the Oklahoma Brownfields Voluntary Redevelopment Act will "not be construed to authorize or encourage any person or other legal entity to cause or increase environmental contamination, to avoid compliance with state and federal laws and regulations concerning environmental contamination or to in any manner escape responsibility for maintaining environmentally sound operations."[18]

The DEQ is charged with implementing the voluntary redevelopment program for brownfields. In administering the Oklahoma Brownfields Voluntary Redevelopment Act, the department must:

---

[13]27A Okl. St. § 2-15-101 (1996). *See also* KEY ENVIRONMENTAL ISSUES IN U.S. EPA REGION VI, American Bar Association Section of Natural Resources, Energy, and Environmental Law (May 12-13, 1997, Dallas, Texas).

[14]"Consent order" means "an order entered into by the Department of Environmental Quality and an applicant, binding an applicant and the department to specified authorizations, activities, duties, obligations, responsibilities, and other requirements." 27A Okl. St. § 2-15-103 (1996).

[15]27A Okl. St. § 2-15-102 (1996).
[16]27A Okl. St. § 2-15-103(2) (1996).
[17]27A Okl. St. § 2-15-110(D) (1996).
[18]27A Okl. St. § 2-15-102(B) (1996).

(1) approve site-specific remediation plans for each site as necessary, using a risk-based system,

(2) review and inspect site assessment and remediation activities and reports, and

(3) use risk-based remediation procedures as determined by the agency to establish cleanup levels.[19] "Risk-based remediation" means site assessment or site remediation, the timing, type, and degree of which are determined according to case-by-case consideration of actual or potential risk to human health and safety or the environment from environmental contamination caused by regulated substances of a brownfield site.[20]

These rules can also include "emergency rules necessary pursuant to the Administrative Procedures Act to implement the provisions of the Oklahoma Brownfields Voluntary Redevelopment Act."[21] Such rules could include provision for applications, consent orders, notice and public participation opportunities, brownfield remediation plans, and no action necessary determinations issued by the department.[22]

The law dictates that the program developed must be a "voluntary program."[23] For this purpose, the law dictates that "No state governmental entity regulating any person or institution shall require evidence of participation in the Oklahoma Brownfields Voluntary Redevelopment Act."[24] The law is, however, silent with regards to private requirements of participation in the program—for example, by a lender.

Only certain persons can participate in the new program. The law provides that the following persons cannot apply:

(1) any person responsible for taking corrective action on the real property pursuant to orders or agreements issued by the federal Environmental Protection Agency;

(2) any person not in substantial compliance with a final agency order or any final order or judgment of a court of record secured by any state or federal agency relating to the generation, storage, transportation, treatment, recycling, or disposal of regulated substances; or

(3) any person who has a demonstrated pattern of uncorrected noncompliance.[25]

For purposes of determining program eligibility, the law defines a "demonstrated pattern of uncorrected noncompliance" as meaning "a history of noncompliance

[19]27A Okl. St. § 2-15-104(A) (1996).
[20]27A Okl. St. § 2-15-103(8) (1996).
[21]27A Okl. St. § 2-15-104(E)(2) (1996).
[22]27A Okl. St. § 2-15-104(E)(3) (1996).
[23]27A Okl. St. § 2-15-104(B) (1996).
[24]27A Okl. St. § 2-15-104(C) (1996).
[25]27A Okl. St. § 2-15-104(D) (1996).

by the applicant with state or federal environmental laws or rules or regulations promulgated thereto, as evidenced by past operations clearly indicating a reckless disregard for the protection of human health and safety, or the environment."[26]

An applicant may apply to the DEQ for a consent order for risk-based remediation of a brownfield site or for a no action necessary determination.[27] The law defines *applicant*[28] as any person who or entity that:

(1) has acquired the ownership, operation, management, or control of a site through foreclosure or under the terms of a bona fide security interest in a mortgage or lien on, or an extension of credit for, a brownfields site and which forecloses on or receives an assignment or deed in lieu of foreclosure or other indicia of ownership and thereby becomes the owner of a brownfield;

(2) possesses a written expression of an interest to purchase a brownfield and the ability to implement a brownfield redevelopment proposal;

(3) is the legal owner in fee simple of a brownfield;

(4) is a tenant on or lessee of the brownfield site; or

(5) is undertaking the remediation of a brownfield site.

The application must, at a minimum, include in the application the following information:

Items	Brief Description
Site description	A description of: a. the brownfield that is the subject of the application pursuant to the Oklahoma Brownfields Voluntary Redevelopment Act, b. the concentrations of contaminants in the soils, surface water, or groundwater at the site, c. the air releases that may occur during remediation of the site, and d. any monitoring of the brownfield that is to occur after issuance of the Certificate of Completion or Certificate of No Action Necessary.
Site remediation plan	A remediation plan for remediating any contamination caused by regulated substances on the brownfield or a proposal that no action is necessary to remediate the brownfield considering the present levels of regulated substances at the site and the proposed future use of the property.
Groundwater use	The current and proposed use of groundwater on and near the site.
Operational history	The operational history of the site and the current use of areas contiguous to the site.
Site use	The present and proposed uses of the site.
Contamination	Information concerning the nature and extent of any contamination caused by regulated substances and releases of regulated substances that have occurred at the site and any possible impacts on areas contiguous to the site.

(continued)

---

[26]27A Okl. St. § 2-15-103(6) (1996).
[27]27A Okl. St. § 2-15-105 (1996).
[28]27A Okl. St. § 2-15-103(1) (1996).

Lab results	Any analytical results from a laboratory certified by the DEQ or other data that characterizes the soil, groundwater, or surface water on the site.
Pathway analysis	An analysis of the human and environmental pathways to exposure from contamination at the site based upon the property's future use as proposed by the applicant.

The term *remediation* means activities necessary to clean up, mitigate, correct, abate, minimize, eliminate, control, and contain environmental contamination caused by regulated substances in compliance with a consent order from the DEQ.[29]

Similarly, the law defines the various certificates that can be issued by the state as follows:

Certificate	Description
Certificate of Completion	"a document issued by the DEQ pursuant to Section 6 of this act upon a determination that an applicant has successfully completed agency-approved risk-based remediation."[30]
Certificate of No Action Necessary	"a document issued by the DEQ pursuant to Section 6 of this act upon a determination that no remediation is deemed necessary for the expansion or redevelopment of the property for a planned use."[31]

Remediation or proposal for a no action necessary determination will be based on the potential risk to human health and safety and to the environment posed by the environmental contamination caused by regulated substances at the site. The following factors will be considered:

(1) the proposed use of the brownfield;
(2) the possibility of movement of the regulated substances in a form and manner that would result in exposure to humans and to the surrounding environment at levels that exceed applicable standards or represent an unreasonable risk to human health and safety, or the environment as determined by the department; and
(3) the potential risks associated with the remediation proposal or no action necessary determination and the economic and technical feasibility and reliability of such proposal or determination.

For purposes of brownfields redevelopment, the DEQ is not authorized to hold any public meeting or hearing to require information, make any determination, or in any manner consider the zoning or rezoning for any proposed redevelopment of a site. The department shall assume that any proposed redevelopment of the site

---

[29]27A Okl. St. § 2-15-103(8) (1996).
[30]27A Okl. St. § 2-15-103(3) (1996).
[31]27A Okl. St. § 2-15-103(4) (1996).

meets or will meet any zoning requirements.[32] The department may reject or return an application if:

(1) a federal requirement precludes the eligibility of the site;
(2) the application is not complete and accurate; or
(3) the applicant is ineligible under the provisions of the Oklahoma Brownfields Voluntary Redevelopment Act or any rules promulgated pursuant thereto.[33]

The department may enter into a consent order with the applicant for remediation of a site if the department concludes that the remediation will:

(1) attain a degree of control of regulated substances pursuant to the Oklahoma Brownfields Voluntary Redevelopment Act, other applicable department rules and standards, and all applicable state and federal laws as determined by the department; and
(2) for constituents not governed by paragraph 1 of this subsection, reduce concentrations such that the property does not present an unreasonable risk, as determined by the department, to human health and safety or to the environment based upon the property's proposed use.[34]

The department may make a no action necessary determination if the application as required by the Oklahoma Brownfields Voluntary Redevelopment Act indicates the existence of contamination caused by regulated substances that, given the proposed use of the property, does not pose an unreasonable risk to human health and safety or to the environment as determined by the department.[35] The consent order and the no action determination apply only to conditions caused by contamination on the property, to applicable state or federal laws, and to applicable rules and standards promulgated by the Board of Environmental Quality that existed at the time of submission of the application.[36]

If an application is disapproved by the department, the department will promptly provide the applicant with a formal written statement of the reasons for such denial.[37]

If the department determines that the applicant has successfully completed the requirements specified by the consent order, the department will certify the completion by issuing to the applicant a Certificate of Completion. If the department determines that no remediation action is deemed necessary for the site, the

---

[32] 27A Okl. St. § 2-15-106(A) (1996).
[33] 27A Okl. St. § 2-15-106(B) (1996).
[34] 27A Okl. St. § 2-15-106(C) (1996).
[35] 27A Okl. St. § 2-15-106(D) (1996).
[36] 27A Okl. St. § 2-15-106(E) (1996).
[37] 27A Okl. St. § 2-15-106(F) (1996).

department will issue the applicant a Certificate of No Action Necessary.[38] In either case, the certificate will list the use specified in the consent order for the site.[39]

The certificate will also include provisions stating that:

Certificate of Completion	Certificate of No Action
a. the department will not pursue administrative penalties and civil actions against the applicant, lenders, lessees, and successors and assigns associated with actions taken to remediate the contamination caused by regulated substances that is the subject of the consent order,	a. the department will not pursue any administrative penalties or civil actions against the applicant, lenders, lessees, and successors and assigns associated with the determination that no action is necessary to remediate the contamination caused by regulated substances that is the subject of the certificate,
b. the applicant and all lenders, lessees, and successors and assigns will not be subject to civil liability with regard to the remedial actions taken by the applicant for environmental contamination caused by regulated substances, as required by the consent order if the remedial action is not performed in a reckless or negligent manner,	b. the applicant and all lenders, lessees, and successors and assigns will not be subject to civil liability with regard to the determination that no action is necessary to remediate the site,
c. no person responsible for contamination caused by regulated substances who has not participated in the voluntary remediation process will be released from any liability, and	c. no person responsible for contamination caused by regulated substances who has not participated in the application process for a no action necessary determination will be released from any liability,
d. the Certificate of Completion will remain effective as long as the property is in substantial compliance with the consent order.	d. the Certificate of No Action Necessary will remain effective as long as the site is in substantial compliance with the certificate as determined by the department, and
	e. the issuance of the Certificate of No Action Necessary will not be construed or relied upon in any manner as a determination by the department that the brownfield has not been or is not environmentally contaminated by regulated substances.

The department will keep and maintain a copy of the application, work plan, consent order, any other correspondence, record, authorization, and report received by the department, and an official copy of the Certificate of Completion or the Certificate of No Action Necessary pursuant to the provisions of the Oklahoma Brownfields Voluntary Redevelopment Act relating to the site in an accessible location.[40]

---

[38] 27A Okl. St. § 2-15-106(G)(2) (1996).

[39] 27A Okl. St. § 2-15-106(G)(1) (1996).

[40] 27A Okl. St. § 2-15-106(H) (1996). But note that Chapter 10A of Title 67 of the Oklahoma Statutes will not apply to any records or copies required to be kept and maintained pursuant to this section. 27A Okl. St. § 2-15-106(I) (1996).

The Certificate of Completion and the Certificate of No Action Necessary, issued by the DEQ, are called "land use disclosure" documents.[41] All land use disclosures will be filed in the land records by the applicant in the office of the county clerk where the site is located.[42] Within thirty days of receipt of the Certificate of Completion or the Certificate of No Action Necessary, the applicant will submit to the DEQ an official copy of the land use disclosure filed with the county clerk in the county in which the site is located.[43] Failure to record the land use disclosure with the county clerk and submit the official copy to the department as required by this section will render the Certificate of Completion or Certificate of No Action Necessary voidable.[44]

Whoever knowingly converts, develops, or uses a brownfield site in violation of an authorized use as specified in the land use disclosure will be deemed guilty of a misdemeanor and, upon conviction thereof, will be punishable by a fine of not more than $1,000, imprisonment in the county jail for not more than one year, or both such fine and imprisonment. Each day such violation continues will be considered a separate offense.[45]

The DEQ will not assess against an applicant administrative penalties or pursue civil actions associated with the contamination that is the subject of the consent order or no action necessary determination if:

(1) the applicant is in compliance with the consent order during remediation or with the Certificate of No Action Necessary, and
(2) the applicant is in compliance with any postcertification conditions or requirements specified in the consent order.

After issuance of the Certificate of Completion or Certificate of No Action Necessary, the department will not assess administrative penalties or pursue civil actions associated with the contamination that is the subject of the consent order or no action necessary determination against any lender, lessee, or successor or assign if the lender, lessee, or successor or assign is in compliance with any postcertification conditions or requirements as specified in the consent order or Certificate of No Action Necessary.[46]

Failure of the applicant and any lenders, lessees, or successors or assigns to materially comply with the consent order entered into pursuant to the Oklahoma Brownfields Voluntary Redevelopment Act will render the consent order or the

---

[41]27A Okl. St. § 2-15-103(7) (1996) ("land use disclosure" means "the Certificate of Completion or the Certificate of No Action Necessary, issued by the Department of Environmental Quality, which is required to be filed in the office of the county clerk of the county wherein the site is situated pursuant to Section 7 of this act").

[42]27A Okl. St. § 2-15-107(A)(1) (1996).

[43]27A Okl. St. § 2-15-107(A)(2) (1996).

[44]27A Okl. St. § 2-15-107(A)(3) (1996).

[45]27A Okl. St. § 2-15-107(B) (1996).

[46]27A Okl. St. § 2-15-108(A) (1996).

Certificate of Completion or the Certificate of No Action Necessary voidable. Similarly, submission of any false or materially misleading information by the applicant knowing such information to be false or misleading will render the consent order, Certificate of Completion, or Certificate of No Action Necessary voidable.[47]

An applicant to whom a Certificate of Completion or a Certificate of No Action Necessary has been issued pursuant to the Oklahoma Brownfields Voluntary Redevelopment Act and such applicant's lenders, lessees, or successors or assigns will not be subject to civil liability with regard to the remedial actions taken by the applicant for environmental contamination caused by regulated substances as required by the consent order if the remedial action is not performed in a reckless or negligent manner.[48] Nothing in the Oklahoma Brownfields Voluntary Redevelopment Act will be construed to limit or negate any other rights of any person from pursuing or receiving legal or equitable relief from the applicant or any other person or legal entity causing or contributing to the environmental contamination.[49] In those cases where an applicant conducts a voluntary remediation in conjunction with a party responsible for the contamination, the responsible party will also be released from liability to the same extent as the applicant.[50]

The release of liability from administrative penalties and any civil actions authorized by the Oklahoma Brownfields Voluntary Redevelopment Act will not apply to:

(1) any environmental contamination and consequences thereof that the applicant causes or has caused outside the scope of the consent order or the certificate issued by the department;
(2) any contamination caused or resulting from any subsequent redevelopment of the property;
(3) existing contamination caused by regulated substances not addressed prior to issuance of the Certificate of Completion or the Certificate of No Action Necessary; or
(4) any person responsible for contamination who has not participated in the voluntary remediation.[51]

The DEQ may require the applicant to reimburse the department for reasonable costs described in the consent order for the review and oversight of any remediation, reports, field activities, or other services or duties of the department pursuant to the Oklahoma Brownfields Voluntary Redevelopment Act that are performed by the department prior to the issuance of the Certificate of Completion or the Certificate of No Action Necessary, unless otherwise authorized by the consent order.[52]

[47]27A Okl. St. § 2-15-108(B) (1996).
[48]27A Okl. St. § 2-15-108 (C)(1) (1996).
[49]27A Okl. St. § 2-15-108 (C)(2) (1996).
[50]27A Okl. St. § 2-15-108 (C)(3) (1996).
[51]27A Okl. St. § 2-15-108 (D)(1996).
[52]27A Okl. St. § 2-15-109 (1996).

Any application for remediation of a site submitted to the DEQ prior to the effective date of the act that results in a consent order and any consent order issued by the department prior to the effective date of the law meeting the conditions and requirements established by the department or as otherwise determined by the department to be in compliance for such site is ratified by the new legislation.[53] To this extent, any person who has entered into a consent order with the department may continue to rely upon the consent order if the person has accepted the conditions of and in other respects complies with the requirements so established and with the provisions of the consent order as determined by the department.[54] Any benefits and releases of liability from administrative penalties and from civil action as provided by the Oklahoma Brownfields Voluntary Redevelopment Act will apply and be made part of the consent order,[55] provided they were issued after January 1, 1988.[56]

Texas

Voluntary Cleanup Program

In 1995, Texas enacted its voluntary cleanup program (VCP).[57] The program provides "administrative, technical, and legal incentives to encourage the cleanup of contaminated sites in Texas."[58] Any site not subject to a Texas Natural Resource Conservation Commission (TNRCC) order or permit is eligible to enter the VCP.

The program requires volunteers to remit a $1,000 application fee[59] and enter into an enforceable agreement with the state regarding the remediation of contaminated industrial property.[60] Moreover, the agreement between the state and a volunteer imposes a work schedule and precludes environmental enforcement activities while voluntary remedial activities are conducted.[61] Other features include

---

[53]27A Okl. St. § 2-15-110(A) (1996).

[54]27A Okl. St. § 2-15-110(B) (1996).

[55]27A Okl. St. § 2-15-110(D) (1996).

[56]27A Okl. St. § 2-15-110(D) (1996).

[57]1995 TX H.B. 2296; Tex. Health & Safety Code Ann. §§ 361.601-361.612 (West 1995).

[58]*See* Voluntary Cleanup Section, http://www.tnrcc.texas...ste/pcd/vcp/index.html (7/2/97 12:56: 16).

[59]*Id.* § 361.604(3) (West 1995); *see also* Richard Heymann, Esq., Foley and Lardner, Remarks from Conference on Brownfields Redevelopment and Environmental Risk Management in Business and Real Estate Transactions (March 22-23, 1995, Washington, D.C.).

[60]Tex. Health & Safety Code § 361.606 (West 1995). Under this program, specific requirements comprising the elements of voluntary cleanup work plans and reports submitted under § 361.608 are not articulated. In fact, the Texas program neither explicitly sets forth any cleanup standards nor references other state or federal provisions. *See Id.*

[61]*Id.* § 361.606(C). Specifically, the voluntary cleanup agreement must:

identify all applicable federal and state statutes and rules;
describe any work plan submitted, including a final report providing all information necessary to verify all work contemplated by the voluntary cleanup agreement has been completed;

reimbursement for state oversight costs, and expedited administrative processes for reviews and approvals of voluntary cleanups.[62] The extent of state oversight depends on the risk reduction standards selected by the volunteer.

Standard	Program Timetable
Background or health-based	Limited state oversight; volunteer must notify TNRCC of its intent to clean a site and its risk reductions standard to be used in cleanup; volunteer must then submit a final report upon completion of remediation; volunteers may request additional oversight.
Site specific	Greater state oversight; Preapproval of remediation project required; investigation work plan must then be approved by TNRCC; when accurate site assessment has been completed, volunteer submits baseline risk assessment for TNRCC approval; corrective action study plan must incorporate feasibility study and remedial work plan and be approved by TNRCC; written assurances may be given for completed sites.

Upon satisfactory completion of remedial activities, a Certificate of Completion will be issued by the state to volunteers.[63] The certificate states that "all lenders and future landowners who are not a responsible party are released from all liability to the state for cleanup of areas covered by the certificate."[64] Additionally, Texas provides releases from liability for volunteers, new owners, and lenders for the proper and timely completion of voluntary cleanup activities.[65]

---

include a schedule for the submission of required information;
identify the technical standards applied in evaluating work plans and reports referring to the proposed future use of the site.

Id. § 361.606(C)(1)-(4).
[62]*Id.* § 361.603(B)(1).
[63]*Id.* § 361.609. The Certificate of Completion must contain the following information:

acknowledgment of the release from liability as provided in § 361.610;
statement of the proposed future use of the site;
a legal description of the site.

Id. § 361.609(B)(1)-(3).
[64]*See* Voluntary Cleanup Section, http://www.tnrcc.texas...ste/pcd/vcp/index.html (7/2/97 12:56:16).
[65]*Id.* § 361.610. The release from liability under this section applies to a person who is not a "responsible party" as defined under TEX. HEALTH & SAFETY CODE ANN. §§ 361.271 and 361.275(G) (West 1995) as the person applies to perform a voluntary cleanup. *Id.* Moreover, the protection afforded extends to subsequent purchasers and lenders who make a loan secured by property subject to voluntary remediation. However, a release for liability under the Texas program will be void if the change in the land use increases the risk to human health and the environment. *Id.* § 3631.610(C).

In Texas voluntary cleanups may not be undertaken if a federal grant requires an enforcement action at the site.[66] Also, the Texas program fails to provide specific statutory protection for lenders securing commercial loans with impaired property. The Certificate of Completion issued by TNRCC covers only prospective purchasers and lenders. Even so, although Certificate of Completions[67] and releases from liability[68] address lenders, the program makes no effort to address under what circumstances a lender may have triggered liability as a responsible party under the Texas law.[69]

As of June 24, 1997, the TNRCC received 460 applications from volunteers applying to the program. One hundred sites have been issued certificates of completion.[70] Applicants include an assortment of industries: dry cleaners; manufacturing facilities, shopping centers, warehouses, auto-related businesses, and other commercial and industrial enterprises.[71]

The viability of the Texas program has been enhanced by the fact that EPA Region VI has entered into a SMOA with TNRCC.[72] In addition, the Texas legislature again endorsed the program in 1997 by amending the law to allow monies acquired under the federal brownfields initiatives to be used in the state cleanup program.[73]

Environmental Lien

Texas enacted a nonsuperlien statute as part of its state equivalent of CERCLA.

---

[66]*Id.* § 361.605(A)(2). In contrast to COLO. REV. STAT. ANN. § 25-16-309(2), which creates a nondiscretionary duty on the part of the state to petition U.S. EPA to withhold enforcement action, § 361.605(A)(2) explicitly subrogates the Texas program if a federal grant would require enforcement. Thus, volunteers under the Texas program are at the mercy of federal legislators and regulators.

[67] TEX. HEALTH & SAFETY CODE ANN. § 361.609 (West 1995).

[68]*Id.* § 361.610.

[69]*Compare and contrast*: OHIO REV. CODE ANN. § 3746.26(A)(1) (Anderson 1994); MICH. COMP. LAWS ANN. §§ 20101a & 20101b (West 1995); MINN. STAT. ANN. § 115B.175 [subdiv. 6] (2) (West 1992 & Supp. 1994); MONT. CODE. ANN. § 75-10-701(8)(b) (1995); 1995 NY S.B. § 19; VT. STAT. ANN. tit. 10, § 6615(C)(6)(A) (1995); and VA. CODE ANN. § 10.1-1429.1(A) (Michie 1995).

[70]Id; *Compare* 1997 Tex. ALS. 1452; 1997 Tex. Gen. Laws 1452; 1997 Tex. Ch. 1452; 1997 Tex. H.B. 1. (stating that at least 25 volunatry cleanups have been completed under the Texas program).

[71]Voluntary Cleanup Section, http://www.tnrcc.texas...ste/pcd/vcp/index.html (7/2/97 12:56:16).

[72]*See* http://www.tnrcc.texas...waste/pcd/vcp/moa.html (7/2/97 12:56:36).

[73]1997 Tex. ALS 855; 1997 Tex. Gen. Laws 855; 1997 Tex. Ch. 855; 1997 Tex. H.B. 1239.

CHAPTER 10

Region VII:
Iowa, Kansas, Missouri, and Nebraska

The states in Region VII have had relatively little brownfield activity. Although numerous states have enacted some brownfields legislation over the years, there is no clear pattern to the brownfields legislative enactment. The states in Region VII tend to enact legislation after other states have had some experience with the laws. Perhaps it is due to the vast amount of undeveloped land in this Region VII, that brownfields legislation exists, but does not seem to be an environmental priority.

The following table describes the more recent programs developed in Region VII to encourage voluntary cleanup:[1]

States	Brownfields Financing Programs	VCP with Written Assurances	Incentives to Attract Private Development
Iowa	X	X	X
Kansas		X	
Missouri	X	X	X
Nebraska		X	

Iowa

Land Recycling and Environmental Remediation Standards Act[2]

In 1997, the Iowa legislature enacted the Iowa Land Recycling and Environmental Remediation Standards Act[3] because it found that "[s]ome real property in Iowa is not put to its highest productive use because it is contaminated or it is perceived to be contaminated as a result of past activity on the property."[4] The legislature explained that the reuse of these brownfield sites "is an important component of a sound land-use policy that will prevent the needless development of

---

[1]*See* Charles Bartsh, BROWNFIELDS STATE OF THE STATE REPORT: 50-STATE PROGRAM ROUNDUP (Northeast-Midwest Inst., January 28, 1998).

[2]IOWA CODE § 455H.101 (1997).

[3]*Id.*

[4]§ 455H.104.

prime farmland and open-space and natural areas, and reduce public expenditures for installing new infrastructure."[5]

The new program created incentives to "encourage capable persons to voluntarily develop and implement cleanup plans."[6] The program does so by establishing environmental remediation standards that must be used for any response action or other site assessment or remediation that is conducted at a site enrolled in the program.[7] The legislature explained:

> The safe reuse of property should be encouraged through the adoption of environmental remediation standards developed through an open process which take into account the risks associated with any release at the site. Any remediation standards adopted by this state must provide for the protection of the public health and safety and the environment.[8]

The program is voluntary.[9] However, all participants must enter into an agreement with the department to reimburse the department for actual costs incurred by the department in reviewing documents submitted as a part of the enrollment of the site. This fee may not exceed $7,500 per enrolled site. The agreement entered must allow the department access to the enrolled site and must require a demonstration of the participant's ability to carry out a response action reasonably associated with the enrolled site.[10]

Certain sites are not eligible for participation.[11] These include: (1) sites requiring corrective action under the RCRA UST provisions due to petroleum spills, (2) NPL listed sites, and (3) certain animal feeding operations. Sites requiring

---

[5]*Id.*

[6]*Id.*

[7]§ 455H.102 These new remediation standards apply "notwithstanding provisions regarding water quality in chapter 455B, division III; hazardous conditions in chapter 455B, division IV, part 4; hazardous waste and substance management in chapter 455B, division IV, part 5; underground storage tanks, other than petroleum underground storage tanks, in chapter 455B, division IV, part 8; contaminated sites in chapter 455B, division VIII; and groundwater protection in chapter 455E." *Id.*

The program gave the state agencies broad rulemaking authority to implement the program. *See* § 455H.105 (duties of the commission), § 455H.106 (duties of the department) and § 455H.501 (rulemaking authority of the commission).

[8]§ 455H.104.

[9]§ 455H.107 dictates:

> A person may enroll property in the land recycling program pursuant to this chapter to carry out a response action in accordance with rules adopted by the commission which outline the eligibility for enrollment. The eligibility rules shall reasonably encourage the enrollment of all sites potentially eligible to participate under this chapter and shall not take into account any amounts the department may be reimbursed under this chapter.

[10]§ 455H.107 (2). The law contains a large number of definitions codified at § 455H.103.

[11]§ 455H.107 (3).

RCRA UST cleanups involving spills other than petroleum may participate in the program.

If the site cleanup assessment demonstrates that the release on the enrolled site has affected additional property, then all the property that is shown to be affected by the release must also be enrolled.[12] Participation in the program is not considered an admission of liability[13] or an acknowledgement of the conditions of the property. Protections provided pursuant to the voluntary cleanup program are afforded in addition to any other legal protections and are not considered as a legal substitute.[14]

Following enrollment of the property in the land recycling program, the participant must proceed on a timely basis to carry out response actions.[15] The participant must take such response actions as necessary to assure that conditions in the affected area comply with one of three standards: (1) background standards,[16] (2) statewide standards,[17] and (3) site-specific cleanup standards.[18] Any remediation standard which is applied must, however, provide for the protection of the public health and safety and the environment.[19] Once the participant has demonstrated the affected area is in compliance with the established remediation standards, the department must proceed on a timely basis and issue a No Further Action letter.[20]

The No Further Action letter must state that the participant and any protected party are not required to take any further action at the site related to any hazardous substance for which compliance with applicable standards is demonstrated by the participant in accordance with applicable standards, except for continuing requirements specified in the no further action letter. If the participant was a person having control over a hazardous substance, then the No Further Action letter may provide that a further response action may be required, where appropriate, to protect

---

[12] § 455H.107.

[13] § 455H.305.

[14] § 455H.306.

[15] § 455H.107.

[16] As established pursuant to § 455H.202.

[17] As established pursuant to § 455H.203.

[18] As established pursuant to § 455H.204.

[19] § 455H.201 A participant may use a combination of these standards to implement a site remediation plan and may propose to use the site-specific cleanup standards whether or not efforts have been made to comply with the background or statewide standards. *Id.*

Until rules setting out requirements for background standards, statewide standards, or site-specific cleanup standards are finally adopted by the commission and effective, participants may utilize site-specific cleanup standards for any hazardous substance utilizing the procedures set out in the department's rules implementing risk-based corrective action for underground storage tanks and, where relevant, the United States environmental protection agency's guidance regarding risk assessment for superfund sites. *Id.*

The standards may be complied with through a combination of response actions that may include, but are not limited to, treatment, removal, technological or institutional controls, and natural attenuation and other natural mechanisms, and can include the use of innovative or other demonstrated measures. *Id.*

[20] § 455H.107.

against an imminent and substantial threat to public health, safety, and welfare.[21] A protected party who was a person having control over a hazardous substance, at the time of the release, may be required by the department to conduct a further response action, where appropriate, to protect against an imminent and substantial threat to public health, safety, and welfare.

Protections afforded by the voluntary cleanup program will not relieve a person from liability (1) for a release of a hazardous substance occurring at the enrolled site after the issuance of a No Further Action letter or (2) from liability for any condition outside the affected area addressed in the cleanup plan and No Further Action letter.[22] Moreover, the liability protection and immunities extend only to liability or potential liability arising under state law. "It is not intended to provide any relief as to liability or potential liability arising under federal law."[23]

The No Further Action letter, Covenant Not to Sue, or any agreement authorized to be entered into and entered into under the voluntary cleanup program may be transferred by the participant or a later recipient to any other person by assignment or in conjunction with the acquisition of title to the enrolled site to which the document applies.[24] If a person transfers property to an affiliate in order for that person or the affiliate to obtain a benefit to which the transferor would not otherwise be eligible, the affiliate will be subject to the same obligations and obtain the same level of benefits as those available to the transferor under the law. The department must provide, upon request, a No Further Action letter as to the affected area to each protected party.[25] The No Further Action letter will, however, be conditioned upon compliance with any institutional or technological controls relied upon by the participant to demonstrate compliance with the applicable standards.[26] Further, the No Further Action letter must be recorded in the county real estate records.[27]

Upon issuance of a No Further Action letter,[28] a Covenant Not to Sue arises by operation of law.[29] The covenant releases the participant and each protected party from liability to the state, in the state's capacity as a regulator administering environmental programs, to perform additional environmental assessment, remedial activity, or response action with regard to the release of a hazardous substance for which the participant and each protected party has complied with the requirements of the voluntary cleanup program.

---

[21] § 455H.301 (2).

[22] § 455H.307.

[23] § 455H.308 ("this section shall not be construed as precluding any agreement with a federal agency by which it agrees to provide liability protection based on participation and completion of a cleanup plan under this chapter").

[24] § 455H.504.

[25] § 455H.301 (3).

[26] § 455H.301 (4).

[27] § 455H.301 (5). Chapter 558 dictates the recording requirements.

[28] *See* § 455H.301.

[29] § 455H.302.

The No Further Action letter will be void if the department demonstrates by clear, satisfactory, and convincing evidence that any approval was obtained by fraud or material misrepresentation, knowing failure to disclose material information, or false certification to the department.[30] Under these circumstances, the Covenant Not to Sue will no longer apply.

The participant may withdraw the enrolled site from further participation in the land recycling program at any time upon written notice to the department.[31] Any participant who withdraws an enrolled site from further participation in the program will not be entitled to any refund or credit for the enrollment fee paid and will be liable for any costs actually incurred by the department (up to the $7,500 cap). The department or court may determine that a participant who withdraws prior to completion of all response actions identified for the enrolled site forfeits all benefits and immunities provided by the law. If it is deemed necessary and appropriate by the department, a participant who withdraws may be required to stabilize the enrolled site in accordance with a plan approved by the department.

A participant may apply to the department for a variance from any provision of the new law. The department may issue a variance from applicable standards only if the participant demonstrates that: (1) it is either technically infeasible to comply with the applicable standards, or the cost of complying with the applicable standards exceeds the benefits, (2) the proposed alternative standard will result in an improvement of environmental conditions in the affected area and ensure that the public health and safety will be protected, and (3) the establishment of and compliance with the alternative standard or set of standards in the terms and conditions is necessary to promote, protect, preserve, or enhance employment opportunities or the reuse of the enrolled site. If requested by a participant, the department may issue a variance from any other provision of the new law if the department determines that the variance would be consistent with the declaration of policy and is reasonable under the circumstances.[32]

State Registry

The state must maintain a record of the affected areas or portion of affected areas for which No Further Action letters were issued[33] and that involve institutional or technological controls that restrict the use of any of the enrolled sites to comply with applicable standards.[34] The records pertaining to those sites must indicate the applicable use restrictions.

If the site has no institutional controls, an enrolled site listed on the registry of confirmed hazardous waste or hazardous substance disposal sites[35] that has

---

[30]§ 455H.301 (2).
[31]§ 455H.201.
[32]§ 455H.205.
[33]*See* § 455H.301.
[34]§ 455H.503.
[35]Established pursuant to section 455B.426.

completed a response action as to the conditions which led to its original listing on the registry, must be removed from the registry listing, once a No Further Action letter has been issued.[36]

Incremental Property Taxes

To encourage economic development and the recycling of contaminated land, cities and counties may provide by ordinance that the costs of carrying out response actions under the voluntary cleanup program are to be reimbursed, in whole or in part, by incremental property taxes over a six-year period. A city or county which implements incremental property taxes must provide that taxes levied on property enrolled in the land recycling program must be divided[37] in the same manner as if the enrolled property was taxable property in an urban renewal project. Incremental property taxes collected must be placed in a special fund of the city or county. A participant must be reimbursed with moneys from the special fund for costs associated with carrying out a response action.[38] Beginning in the fourth of the six years of collecting incremental property taxes, the city or county must begin decreasing by 25 percent each year the amount of incremental property taxes computed.[39]

Land Recycling Fund

The legislature created a land recycling fund within the state treasury under the control of the commission.[40] Moneys received from fees, general revenue, federal funds, gifts, bequests, donations, or other moneys so designated must be deposited in the fund.[41] The state may use the land recycling fund to provide for: (1) financial assistance to political subdivisions of the state for activities related to an enrolled site, (2) financial assistance and incentives for qualifying enrolled sites, and (3) funding for any other purpose consistent with the law.[42]

Federal Delegation

The Iowa legislature was concerned about federal oversight. The law expressly states that the provisions of the voluntary cleanup program:

> . . . shall not prevent the department from enforcing both specific numerical cleanup standards and monitoring of compliance requirements specifically

---

[36]*See* § 455H.301.

[37]*See* § 403.19, subsections 1 and 2.

[38]This will be done in accordance with rules adopted by the commission.

[39]§ 455H.309.

[40]§ 455H.401.

[41]*Id.* Any unexpended balance in the land recycling fund at the end of each fiscal year shall be retained in the fund, notwithstanding section 8.33.

[42]§ 455H.401.

required to be enforced by the federal government as a condition of the receipt of program authorization, delegation, primacy, or federal funds.[43]

In addition, the law requires that "any rules or standards established pursuant to this chapter shall be no more stringent than those required under any comparable federal law or regulation."[44]

Business Transfer Statute

Iowa law provides that "a person shall not substantially change the manner in which a hazardous waste or hazardous substance disposal site on the registry pursuant to section 455B.426 is used without the written approval of the director."[45] In addition, a person can "not sell, convey, or transfer title to a hazardous waste or hazardous substance disposal site which is on the registry pursuant to section 455B.426 without the written approval of the director," who must respond to a request for a change of ownership within thirty days of its receipt. Decisions of the state concerning the use or transfer of a hazardous waste or hazardous substance disposal site may be appealed.

If the state has reason to believe that a person has not complied with this requirement, or that there is an imminent danger that the statute will be violated, "the director may institute a civil action in district court for injunctive relief to prevent the violation and for the assessment of a civil penalty, not to exceed one thousand dollars per day for each day of violation."[46]

Immediately upon the listing of real property in the registry of hazardous waste or hazardous substance disposal sites, a person liable for cleanup costs must submit to the state a report consisting of documentation of the responsible person's liabilities and assets, including, if filed, a copy of the annual report submitted to the secretary of state pursuant to chapter 490. A subsequent report must be submitted annually on April 15 for the period the site remains on the registry.[47]

---

[43]§ 455H.510.

[44]§ 455H.511.

[45]Iowa Code § 455B.430.

[46]*Id.* Moneys collected under this subsection must, however, be deposited in the remedial fund.

[47]The Iowa code dictates:

With each declaration of value submitted to the county recorder under chapter 428A, there shall also be submitted a statement that no known wells are situated on the property, or if known wells are situated on the property, the statement must state the approximate location of each known well and its status with respect to section 159.29 or 455B.190. The statement shall also state that no known disposal site for solid waste, as defined in section 455B.301, which has been deemed to be potentially hazardous by the department of natural resources, exists on the property, or if such a known disposal site does exist, the location of the site on the property. The statement shall additionally state that no known underground storage tank, as defined in section 455B.471, subsection 11, exists on the property, or if a known underground storage tank does exist, the type and size of the tank, and any known substance in the tank. The statement shall also state that no known hazardous waste as defined in section 455B.411, subsection 3, or listed by the department pursuant to section 455B.412,

Kansas

Ironically, the state of Kansas had no brownfields redevelopment legislation until 1997,[48] despite the fact that one of the earliest and most successful brownfields redevelopment projects took place in the downtown sector of Wichita, Kansas.

In 1990, the city of Wichita began to develop an urban renewal plan for the city center. During the planning, "a six-square-mile lake of underground pollution was discovered beneath the business center."[49] Moreover, the "blob" of pollution was determined through environmental testing to be moving at a rate of one foot per day. As the press aptly noted, "The discovery had a chilling effect on all of downtown as banks stopped making loans, fearful that they would be held liable for environmental damages on properties they financed."[50]

The city looked for solutions in the experiences of other municipalities. In the early 1990s, however, there were no models for Wichita to adapt to its own. Most cities seemed to either be doing nothing[51] or turning to the federal Superfund for aid. Doing nothing was not an option in Wichita, because the contaminated area included almost the entire downtown—and its huge tax base. Similarly, the city thought it unwise to have the site listed on the federal NPL.

---

subsection 2, or section 455B.464, exists on the property, or if known hazardous waste does exist, that the waste is being managed in accordance with rules adopted by the department of natural resources. The statement shall be signed by at least one of the sellers or their agents. The county recorder shall refuse to record any deed, instrument, or writing for which a declaration of value is required under chapter 428A unless the statement required by this section has been submitted to the county recorder. A buyer of property shall be provided with a copy of the statement submitted, and, following the fulfillment of this provision, if the statement submitted reveals no well, disposal site, underground storage tank, or hazardous waste on the property, the county recorder may destroy the statement. The land application of sludges or soils resulting from the remediation of underground storage tank releases accomplished in compliance with department of natural resources rules without a permit is not required to be reported as the disposal of solid waste or hazardous waste.

If a declaration of value is not required, the above information shall be submitted on a separate form. The director of the department of natural resources shall prescribe the form of the statement and the separate form to be supplied by each county recorder in the state. The county recorder shall transmit the statements to the department of natural resources at times directed by the director of the department.

The owner of the property is responsible for the accuracy of the information submitted on the form. The owner's agent shall not be liable for the accuracy of information provided by the owner of the property. The provisions of this paragraph do not limit liability which may be imposed under a contract or under any other law.

[48]KAN. STAT. ANN. § 65-34,161 *et seq.* (1997).

[49]*Wichita Rescues Redevelopment By Paying for Toxic Cleanup, Banks Had Balked at Financing the Rebuilding of Downtown Because of Liability Fears. Now a Smaller Plan Is Moving Ahead*, LOS ANGELES TIMES (November 6, 1991).

[50]*Id.*

[51]*Id.* ("We tried to figure out what other cities were doing in this situation," Glaser said. "We didn't find much in the way of a model that we could copy. In most cases we found that they're not doing anything").

Instead, "City officials determined that the only way to save downtown was for the city to assume responsibility for the . . . cleanup."[52] Accordingly, the city established a tax increment financing district. Part of the property taxes levied from the affected area were used to pay cleanup costs that could not be recovered from PRPs. "The move not only eliminated the fear of federal government red tape and foot-dragging, but it also relieved banks and innocent property owners of liability for the pollution cleanup."[53]

"It was the first time a local government put itself between the EPA and state regulators and the property owners," said Marvin L. Wynn, chief operating officer of the Wichita/Sedgwick County Partnership for Growth, an organization of business leaders that sponsored the development plan. If Wichita "hadn't done what they did, it would've been tied up in litigation and development would have been stopped for years and years, maybe decades," said David Burke, one of the developers in a plan to turn a dilapidated downtown warehouse district into a vibrant residential, retail, and entertainment area called Old Town Marketplace.[54]

The Wichita blob was, thus, cleaned up through a creative partnership between private enterprise and local governments. "The city coming up with an innovative approach really has solved the problem. You're seeing property change hands again and there is lending in the area. Property values will start going back up again."[55]

Despite the success of the Wichita local government, the state of Kansas did not deem it necessary to pass legislation codifying the remediation process until 1997.

Voluntary Cleanup and Property Redevelopment Act of 1997[56]

The Kansas Voluntary Cleanup and Property Redevelopment Act applies to real property where environmental cleanup may be needed.[57] The program is voluntary[58] and may be initiated by submission of an application to the department for properties where investigation and remediation may be necessary to protect human health or the environment based upon the current or proposed future use or redevelopment of the property.[59] Certain properties may not take advantage of the program. These include:

---

[52]*Id.* The cleanup was expected to cost $20 million over twenty years, but, as Glaser noted: "The annual property taxes at risk were worth that much."

[53]*Id.*

[54]*Id.*

[55]*Id.*

[56]KAN. STAT. ANN. § 65-34,161 *et seq.* (1997).

[57]KAN. STAT. ANN. § 65-34,161 (1997).

[58]*Id.* § 65-34,164 (a) (Voluntary application; application of other laws; eligible property).

[59]*Id.* Note, however, that "property which may be eligible for reimbursement from trust funds established in the Kansas storage tank act, K.S.A. 65-34,100 et seq., and amendments thereto, or the Kansas dry cleaner environmental response act, K.S.A. 1997 Supp. 65-34,141 et seq., and amendments thereto, shall meet all of the requirements of the respective act." Id. § 65-34,164 (b) (1997).

property that is listed or proposed for listing on the CERCLA NPL;

property the contaminated portion of which is the subject of: (1) Enforcement action issued pursuant to city, county, state or federal environmental laws; or (2) environmental orders or agreements with city, county, state or federal governmental agencies;

a facility which has or should have a RCRA permit, which contains a corrective action component;

oil and gas activities regulated by the state corporation commission;

property that presents an immediate and significant risk of harm to human health or the environment; or

property that the department determines to be a substantial threat to public or private drinking water wells.[60]

Within these confines, the legislature gave the state agency broad discretion to adopt rules and regulations necessary to define, administer, and enforce the provisions of the new law.[61]

Application Process

Each application or reapplication for participation in the voluntary program must be accompanied by a nonrefundable application fee of $200 to cover processing costs.[62] The department must review and approve or deny all applications. Thereafter, the department must notify the applicant in writing, whether the application is approved or denied. If the application is denied, the notification must state the reason for the denial.

Following departmental approval of an application, a voluntary agreement must be executed between the participant and the department. The department may not commence oversight and review activities until the voluntary agreement is executed.

As part of the voluntary agreement, the department must require the applicant to post a deposit not to exceed $5,000. The deposit must be used to cover all direct and indirect costs of the department in administration of the program,

---

[60]*Id.* § 65-34,164 (c).

[61]*Id.* § 65-34,163. The legislature defined the following terms for purposes of the new program:

(a) *Contaminant* means such alteration of the physical, chemical or biological properties of any soils and waters of the state as will or is likely to create a nuisance or render such soils or waters potentially harmful, or injurious to public health, safety or welfare, or to the plant, animal or aquatic life of the state.

(b) *Department* means the department of health and environment.

(c) *Secretary* means the secretary of health and environment.

Id. § 65-34,162.

[62]*Id.* § 65-34,165 (application; fee; action on; agreement; deposit; access to property; termination of agreement; fund, use and disposition of).

including but not limited to providing technical review, oversight, and guidance in relation to the property covered in the application. If the costs of the department exceed the initial deposit, an additional amount agreed upon by the department and the applicant will be required prior to proceeding with any voluntary work under the program. Timely remittance of reimbursements to the department is a condition of continuing participation. After the mutual termination of the voluntary agreement, the department must refund any remaining balance within sixty days.[63]

During the time allocated for review of applications, assessments, other investigative activities and remedial activities, the department, upon reasonable notice to the applicant, must have access at all reasonable times to the subject real property.

The department must review reports, including any environmental assessments and investigations submitted by the applicant, and make a determination as to any required actions. If the department determines that no remedial action is necessary, the department may issue a no further action determination. If the department determines that further investigation or remediation is required, then the applicant must submit to the department a voluntary cleanup plan that follows the scope of work prepared by the department for voluntary investigation or remediation and includes the actions necessary to address the contamination.[64]

Remedial alternatives must be based on the actual risk to human health and the environment currently posed by contaminants on the property. In so determining the state must consider:

the present and proposed future uses of the property and surrounding properties;

the ability of the contaminants to move in a form and manner that would result in exposure to humans and the surrounding environment at levels that exceed applicable state standards and guidelines, or the results of a risk analysis if such standards and guidelines are not available; and

the potential risks associated with proposed cleanup alternatives and the reliability and economic and technical feasibility of such alternatives.[65]

The applicant may unilaterally terminate the voluntary agreement prior to completion of investigative and remedial activities if the applicant leaves the site in no worse condition from a human health and environmental perspective than when the applicant initiated voluntary activities. The applicant must, however, notify the department in writing of the intention to terminate the voluntary agreement. The department will cease billing for review of any submittal under the voluntary agreement upon receipt of notification. Within ninety days after receipt of notification for termination, the department must provide a final bill for services provided. If the applicant requests termination of the voluntary agreement under this subsection, initial deposits are not refundable. In the event the department has

---

[63]*Id.*

[64]*Id.* § 65-34,166 (remedial action; determination whether required; plan).

[65]*Id.* § 65-34,167 (Same; alternatives; factors considered).

costs in excess of the initial deposit, the applicant must remit full payment of those costs. Upon payment of all costs, the department must notify the applicant in writing that the voluntary agreement has been terminated.

In addition, to granting the applicant a right of termination, the new law gives the department the right to terminate the voluntary agreement if the applicant:

> violates any terms or conditions of the voluntary agreement or fails to fulfill any obligations of the voluntary agreement; or
>
> fails to address an immediate and significant risk of harm to public health and the environment in an effective and timely manner.

If the state elects to terminate the agreement, then the department must notify the applicant in writing of the intention to terminate the voluntary agreement and include a summary of the costs of the department. The notification must state the reason or reasons for the termination.[66]

The department must provide formal written notification to the applicant that a voluntary cleanup plan has been approved or disapproved within sixty days of submittal of the voluntary cleanup plan by the applicant unless the department extends the time for review to a certain date.[67] The department must approve a voluntary cleanup plan if the department concludes that the plan will attain a degree of cleanup and control of contaminants that complies with all applicable statutes and rules and regulations.[68] The approval of a voluntary cleanup plan by the department, however, applies only to those contaminants and conditions identified on the property based upon the statutes and rules and regulations that exist when the application is submitted.[69]

Upon determination by the department that a voluntary cleanup plan is acceptable, the department must publish a notice of the determination in a local newspaper of general circulation in the area affected and make the voluntary cleanup plan available to the public. The public must have fifteen days from the date of publication during which any person may submit to the department written comments regarding the voluntary cleanup plan. After fifteen days have elapsed, the department may hold a public information meeting if, in the department's judgment, the comments submitted warrant such a meeting or if the applicant requests such a meeting. Upon completion of the public notification and participation process, the department must make a determination to approve the plan in accordance with this section.[70]

---

[66]*Id.*

[67]§ 65-34,168 (a).

[68]§ 65-34,168 (b).

[69]§ 65-34,168 (d).

[70]*Id.* Approval is not absolute. The departmental approval of a voluntary cleanup plan must be void upon:

(1) Failure of an applicant to comply with the approved voluntary cleanup plan;

(2) willful submission of false, inaccurate or misleading information by the applicant in the context of the voluntary cleanup plan; or

If a voluntary cleanup plan is not approved by the department, then the department must promptly provide the applicant with a written statement of the reasons for denial. If the disapproval is based upon the applicant's failure to submit the information required, then the department must notify the applicant of the deficiencies in the information submitted.[71]

Within forty-five days after the completion of the voluntary cleanup described in the approved voluntary cleanup plan, the applicant must provide to the department assurance that the plan has been fully implemented. A verification sampling program must be required by the department to confirm that the property has been cleaned up as described in the voluntary cleanup plan.[72]

No Further Action

After an applicant completes the requirements of the act, the department may determine that no further remedial action is required.[73] Within sixty days after such completion, unless the applicant and the department agree to an extension of the time for review, the department must provide written notification that a no further action determination has been made.[74] The department must provide written notification of a no further action determination.[75]

In issuing the determination, the department may consider many factors.[76] These include whether (1) the contamination (or a release of contamination) originates from a source on adjacent property upon which the necessary action is or will be taken, (2) the action protects human health and the environment, (3) the person undertaking the action is viable and financially capable person or entity, and (4) the person may or may not be legally responsible for the source of contamination.[77]

The issuance of a no further action determination by the department applies only to identified conditions on the property. It is based upon applicable statutes and rules and regulations that exist as of the time of completion of the requirements. Then no further action determination may be voided by the department if:

> there is any evidence of fraudulent representation, false assurances, concealment, or misrepresentation of the data in any document to be submitted to the department under this act;

---

(3) failure to initiate the plan within six months after approval by the department, or failure to complete the plan within twenty-four months after approval by the department, unless the department grants an extension of time.

§ 65-34,168 (e).
[71]§ 65-34,168 (c).
[72]§ 65-34,168 (h).
[73]§ 65-34,169 (b) (1).
[74]§ 65-34,169 (a).
[75]§ 65-34,169 (b).
[76]§ 65-34,169 (c).
[77]*Id.*

the applicant agrees to perform any action approved by the department and
   fails to perform such action;
the applicant's willful and wanton conduct contributes to known environmental contamination; or
the applicant fails to complete the voluntary actions required in the voluntary
   cleanup plan.[78]

If a no further action determination is not issued by the department, the department must promptly provide the applicant with a written statement of the reasons for denial.[79]

EPA Involvement

Nothing in the voluntary cleanup act absolves any person from obligations under
any other law or rule and regulation, including any requirement to obtain permits
or approvals for work performed under a voluntary cleanup plan.[80] Moreover, if
U.S. EPA indicates that it is investigating a property that is the subject of an approved voluntary cleanup plan, the department must attempt to obtain agreement
with the EPA that the property be addressed under the appropriate state program
or, in the case of property being addressed through a voluntary cleanup plan, that
no further federal action be taken with respect to the property at least until the voluntary cleanup plan is completely implemented.

Enforcement

Voluntary cleanup plans are not enforceable against an applicant unless the department can demonstrate that an applicant who initiated a voluntary cleanup under an
approved plan has failed to fully implement that plan. In that case, the department
may require further action if such action is authorized by other state statutes administered by the department or rules and regulations of the department.[81]
   Information provided by an applicant to support a voluntary cleanup plan
must not provide the department with an independent basis to seek penalties from
the applicant pursuant to applicable statutes or rules and regulations. If, pursuant
to other applicable statutes or rules and regulations, the department initiates an enforcement action against the applicant subsequent to the submission of a voluntary
cleanup plan regarding the contamination addressed in the plan, the voluntary disclosure of the information in the plan must be considered by the enforcing authority to mitigate penalties which could be assessed to the applicant.[82]

---

[78]§ 65-34,169 (c).
[79]§ 65-34,169 (d).
[80]§ 65-34,171.
[81]§ 65-34,172 (a).
[82]§ 65-34,172 (b).

Voluntary Cleanup Fund

The new law established a fund in the state treasury entitled the Voluntary Cleanup Fund. The revenue in the fund consists of:

moneys collected for application fees;

moneys collected as deposits for costs associated with administration of the act, including technical review, oversight, and guidance;

moneys received by the secretary in the form of gifts, grants, reimbursements, or appropriations from any source intended to be used for purposes of the fund; and

interest attributable to the investment of moneys in the fund.

The state can use moneys in the Voluntary Cleanup Fund for limited purposes. These include:

review of applications;

technical review, oversight, guidance, and other activities necessary to carry out the provisions of the new act;

activities performed by the department to address immediate or emergency threats to human health and the environment related to a property under the act; and

administration and enforcement of the provisions of the act.[83]

Environmental Assessments, Preparer

A novel aspect of the Kansas law is the requirement to retain state qualified environmental consultants. The law requires that, the department may accept only environmental assessments prepared by a qualified environmental professional, as defined by rules and regulations adopted by the secretary.[84] Kansas had no such state certification of environmental experts prior to the enactment of this law.

---

[83]*Id.* The law further dictates that:

On or before the tenth of each month following the month in which moneys are first credited to the voluntary cleanup fund, and monthly thereafter on or before the tenth of each month, the director of accounts and reports must transfer from the state general fund to the Voluntary Cleanup Fund interest earnings based on:

(1) The average daily balance of moneys in the voluntary cleanup fund for the preceding month; and (2) the net earnings rate of the pooled money investment portfolio for the preceding month.

Id. In addition, the law dictates that "all expenditures from the fund must be made in accordance with appropriation acts upon warrants of the director of accounts and reports issued pursuant to vouchers approved by the secretary for the purposes set forth in this section." *Id.*

[84]§ 65-34,170.

Annual Report

The department must publish annually in the Kansas Register a summary of the number of applicants, the general categories of those applicants and the number of cleanups completed pursuant to the new voluntary cleanup law.[85]

Missouri

Voluntary Cleanup Program

The Missouri Voluntary Cleanup Program[86] affords state oversight of any and all voluntary cleanup measures commenced by private parties. Any person, including persons acquiring, disposing of, or in current possession of property, may apply to the Missouri Department of Natural Resources (MDNR) to commence voluntary remediation of real property.[87] All applications must be submitted with a $200 application fee.

The MDNR will review the application of the applicant, and shall approve or deny the application within ninety days of submission.[88] If the application is approved, the applicant is to submit copies of all reports prepared in preparation of remediation activities.[89] The MDNR has the duty to review this material, and to comment on any additional site assessments that will be needed, within 180 days of the original submission of these assessment documents.[90] In addition, the applicant will be required at the time of the submission of these documents to place a deposit with the MDNR of not more than $5,000, which will be utilized to cover the costs of the department's oversight.[91] Unused money is refunded. Additional oversight costs are billed quarterly at a rate of twenty-five times the actual hourly wage of the MDNR staff involved in the project.

After the approval of assessment documents, the applicant has the obligation to submit for approval his proposed remedial action plan.[92] This proposed plan is to include all work plans, safety plans, testing protocols, and appropriate monitoring plans for postremediation activities.[93] If the remediation plans are acceptable,

---

[85]§ 65-34,173.

[86]MO. ANN. STAT. §§260.565-260.575 (Vernon Supp. 1994).

[87]*Id.* § 260.567(1).

[88]*Id.* § 260.567(2).

[89]*Id.* § 260.567(3). These reports can and shall include any site assessment results, investigations, sample collections, and analyses, including Phase I audits. *Id.*

[90]*Id.*

[91]*Id.* In addition, the department may seek reimbursement for any costs incurred in the administration and oversight of voluntary cleanup activities. *Id.* § 260.569(1). These costs shall be at the rate of the lesser of the actual costs to the department, or one hundred dollars per hour. *Id.*

[92]*Id.* § 260.567(6).

[93]*Id.*

the applicant may commence operations to restore the property,[94] but he must file quarterly progress reports of his activities with the department.[95]

At the completion of the remedial activities, the person(s) conducting the remediation of the property are eligible for a No Further Action letter to be promulgated by the state.[96] If the department determines that the remedial activity was performed according to the original plans, and the applicant has met the appropriate requirements under the Act, such a letter can be drafted.[97] However, a restriction may be placed in the letter requiring the participant(s) to perform future monitoring activities and to report any additional contamination that is discovered.[98]

Finally, the Act provides for the promulgation of specific guidance and regulations pertaining to cleanup protocol, standards and cost administration to be applied by the department during voluntary cleanup activities.[99]

Specialized Stakeholder Laws

Missouri law[100] states that a person will not be deemed an owner-operator of real or personal property or a person having control over hazardous substances who, without participating in the management of such real or personal property, holds indicia of ownership primarily to protect a security or lienhold interest in the subject real or personal property or in the property in which such real or personal property is located. No lender-owner or representative will, by virtue of becoming the owner of real or personal property, be liable for any cleanup costs, response costs, or third-party liability arising from contamination or pollution of or from said property prior to the date that title vests in the lender-owner or representative including contamination or pollution that continues thereafter without the lender-owner knowingly or recklessly causing such contamination.

A foreclosing lender-owner or representative will not, by virtue of becoming the owner of real or personal property, be liable for any cleanup costs, response costs, or third-party liability arising from contamination or pollution of or from such property during the period of ownership so long as, and to the extent that, he does not knowingly or recklessly cause new contamination or pollution or does not knowingly or recklessly allow others to cause new contamination or pollution. The lender-owner must make reasonable efforts to resell the real or personal property to fit within this exemption.[101]

---

[94]*Id.* § 260.567(7).
[95]*Id.* § 260.567(8).
[96]*Id.* § 260.573.
[97]*Id.*
[98]*Id.*
[99]*Id.* § 260.571.
[100]Mo. Rev. Stat. § 427.031.
[101]In addition, the state law provides that:

No representative shall be personally liable to any beneficiary for diminution in the value of property held in its fiduciary capacity due to compliance with environmental laws.

Business Transfer Statute

Missouri law dictates that "[n]o person may substantially change the manner in which an abandoned or uncontrolled hazardous waste disposal site on the registry prepared and maintained by the department pursuant to section 260.440 is used without the written approval of the director."[102] The law also states that "[n]o person may sell, convey or transfer title to an abandoned or uncontrolled hazardous waste disposal site which is on the registry prepared and maintained by the department pursuant to section 260.440 without disclosing to the buyer early in the negotiation process that the site is on the registry, specifying applicable use restrictions and providing all registry information for the site."[103]

The seller is obligated under the statute to notify the buyer that he may be assuming liability for any remedial action at the site, provided, however, the sale, conveyance, or transfer of property shall not absolve any person responsible for site contamination, including the seller, of liability for any remedial action at the site. The seller must also notify the department of the transfer of ownership within thirty days after the transfer.

The transfer does not, however, relieve any person responsible for the environmental damage of liability for remedial measures.[104] If the director believes that a violation has occurred, or may occur, the director may seek a civil injunction to prevent such violation and a civil penalty of up to $1,000 per day for each day the violation continues.[105]

Decisions of the director concerning the use of an abandoned or uncontrolled hazardous waste site may be appealed to the commission. In addition, if the department has reason to believe that the notice provisions of the section have been violated, or are in imminent danger of being violated, it may institute a civil action in any court of competent jurisdiction for injunctive relief to prevent such violation and for the assessment of a civil penalty not to exceed $1,000 per day for each day of violation.[106]

Nebraska

The Nebraska Remedial Action Plan Monitoring Act took effect as of January 1, 1995.[107] The Act is administered by the Nebraska Department of Environmental Quality (NDEQ).

---

[102]*Id.*

[103]*Id.*

[104]*Id.*

[105]Mo. ANN. STAT. § 260.475(4).

[106]*Id.*

[107]NEB. REV. STAT. §§ 81-15, 181-188 (1996). *See generally* Todd S. Davis & Kevin D. Margolis, BROWNFIELDS: A COMPREHENSIVE GUIDE TO REDEVELOPING CONTAMINATED PROPERTY at 511-517 (1997).

Unlike many states, the Nebraska program has no statutory restriction on who can participate in the program. Unlike Ohio, the program includes sites with contaminated groundwater. NPL listed sites are eligible and have been included in the program. Finally, parties that caused or contributed to the pollution may also enter into the voluntary cleanup program.[108]

In addition, to a $5,000 application fee and a $5,000 participation fee, participants in the program must pay the oversight costs of NDEQ associated with site cleanup. Receipt of a No Further Action letter is expressly conditioned on:

(1) completion of site remediation;
(2) meeting the terms set by NDEQ; and
(3) payment of all oversight costs.[109]

---

[108]NEB. REV. STAT. §§ 81-15, 181-188 (1996).
[109]*Id.*

CHAPTER 11

Region VIII:
Colorado, Montana, North Dakota, South Dakota, Utah, and Wyoming

Of the states in Region VIII, only Colorado, Montana, and Utah have well-defined brownfields programs. Utah and Montana only enacted their programs in 1996 and 1997, respectively.

In contrast, the Colorado legislation reflects the typical trend evidenced by early states concerned with brownfields. First the state enacted CERCLA style legislation aimed at large hazardous substance sites. In Colorado, the classic site example would be Rocky Flats. Next, the state enacted business transfer legislation and legislation protecting lenders and trustees. These business transfer statutes required an audit of property containing hazardous wastes or hazardous substances prior to transfer. The lender protection statute, in turn, provided incentives for otherwise deep pockets to provide loans to people purchasing slightly or mildly contaminated property. Both statutes aimed at sites that were known to be or suspected to be environmentally problematic but did not rise to the level of needing state or federal attention. Finally, the state enacted voluntary cleanup legislation. This final law attempts to allow private parties to clean up very small sites, properties that are almost certainly outside the regulatory scope of the environmental enforcement authority.

The following table describes the more recent programs developed in Region VIII to encourage voluntary cleanup:[1]

States	Brownfields Financing Programs	VCP with Written Assurances	Incentives to Attract Private Development
Colorado		X	
Montana	X	X	
North Dakota			
South Dakota			
Utah		X	
Wyoming			

---

[1] *See* Charles Bartsh, BROWNFIELDS STATE OF THE STATE REPORT: 50-STATE PROGRAM ROUNDUP (Northeast-Midwest Inst., January 28, 1998).

Colorado

Voluntary Cleanup Program

Colorado's Voluntary Cleanup Program provides private parties contemplating the development of existing industrial sites with a mechanism for determining what the cleanup responsibilities will be when planning the actual reuse of existing sites.[2] The Colorado program endeavors to eliminate impediments regarding sale or redevelopment of brownfields, encourages prompt cleanup activities, and minimizes the administrative process.[3] Brownfield sites listed on the NPL, subject to state or federal corrective action, or subject to a consent order issued by the Colorado Water Quality Control Division are ineligible for voluntary cleanups.[4] RCRA treatment, storage, and disposal (TSD) facilities are also expressly excluded from voluntary cleanup.[5]

In Colorado, voluntary cleanup plans must provide for the cleanup and control of hazardous substances and/or petroleum consistent with state cleanup standards.[6] Under the Colorado program, voluntary cleanups must be conducted under a state-approved voluntary cleanup plan. Generally, the plan must include an environmental assessment of contaminated property, a proposal to remediate any contamination or condition that has or could lead to a release that poses an unacceptable risk to human health and the environment, and a description of applicable state standards establishing acceptable concentrations of contaminants in soils, surface water, or groundwater.[7] If state standards do not exist for the contaminants located at the site, the Colorado program requires volunteers to undertake risk assessment procedures to determine proposed cleanup levels and any current risk to human health or the environment. The reduction of contaminants must meet concentration levels that do not present unacceptable risks to health and the environment based upon the current and proposed uses of the site.[8]

---

[2]COLO. REV. STAT. ANN. § 25-16-301 *et seq.* (West 1994). *See also* http://governor.state. . . . /gov_dir/cdphe_dir/hm/.

[3]*Id.* § 25-16-302.

[4]*Id.* § 25-16-303(3)(b)(I)-(II).

[5]*Id.* § 25-16-303(3)(b)(III)-(IV).

[6]*Id.* § 25-16-304(2)(c).

[7]*Id.* § 25-16-304.

[8]*Id.*

Voluntary cleanup plans require the submission of an ESA,[9] a remediation plan proposal,[10] and remedial alternatives.[11] The ESA must be prepared by a qualified professional,[12] which is defined as a person with education, training, and experience in preparing environmental studies and assessments.[13]

A No Action determination may be submitted by a volunteer to the state under two circumstances:[14]

(1) at the successful completion of voluntary remedial activity; or
(2) upon determination that existing site contamination migrated onto the property from an off-site source.[15]

---

[9]*Id.* 25-16-302(2)(a). An environmental assessment for voluntary response actions should include:

a legal description and map of the site;

the physical characteristics of the site and contiguous areas, including any existing surface water bodies or groundwater aquifers;

location and description of the use of any wells on the site or on areas within a one-half mile radius of the site;

current and proposed use of on-site groundwater;

operational history of the site, including the current use of areas contiguous to the site;

the present and proposed uses of the site;

the nature and extent of any contamination and releases of hazardous substances and/or petroleum, including impacts on areas contiguous to the site;

the results of any sampling or other data characterizing the condition of the soil, groundwater or surface water on the site; and

a description of human and environmental exposure to contamination at the site based on the property's current and future use.

Id. § 25-16-308(2).
[10]*Id.* § 25-16-304.
[11]*Id.* § 25-16-305. Based on the actual risk to human health and the environment, remediation alternatives must consider the following factors:

present or proposed uses of the site;

potential pathways of migration for contaminants exceeding applicable state standards resulting in exposure to humans and the surrounding environment or, in the absence of state standards, representing an unacceptable risk to human health or the environment; and

potential risks associated with the proposed cleanup alternatives, the economic and technical feasibility, and the reliability of such alternatives.

[12]*Id.* § 25-16-308(1).
[13]*Id.*
[14]*Id.* § 25-16-307.
[15]*Id.*

This determination is appropriate when a qualified professional indicates that the existence of contaminants does not exceed applicable state standards or the contamination does not pose an unacceptable risk to health or the environment.[16]

Upon satisfactory completion of voluntary remedial activities, the state of Colorado must actively pursue a determination by U.S. EPA that the property subject to the voluntary plan should not be subject to further federal action.[17] It is unclear whether a private cause of action may lie in mandamus for failure to "actively" pursue a determination by U.S. EPA to take no further federal action. U.S. EPA is not, however, bound by legal authority to obey Colorado's petition. Any such action taken by a volunteer could, however, be asserted as a defense to or evidence of mitigating factors to federal environmental enforcement proceedings.

Although Colorado's VCP has been in place since August of 1994, according to commentators, its "enacting legislation provides it with little authority and few funds."[18] The program involves "very limited oversight, and its assurances are vague."[19] Reforms are being planned to make the program more acceptable to U.S. EPA headquarters and Region VIII, which have been very critical of many Colorado environmental programs since the state passed one of the first environmental audit privilege laws in the country.

Business Transfer Statute

The Colorado Hazardous Waste Act[20] covers hazardous waste disposal sites.[21] The Act is triggered by a "substantial change" in ownership, design, or operation of a hazardous waste disposal site.[22] Change in ownership includes assignment or transfer of the certificate of designation.[23]

---

[16]*Id.* § 25-16-307(2)(a). Generally, a No Action determination states that no further action is required to assure that the property, when used for the purposes identified in the no action petition, is protective of existing and proposed uses and does not pose an unacceptable risk to human health or the environment. *Id.* § 25-16-307(2)(b).

[17]*Id.* § 25-16-309(2). In pertinent part, the relevant provision states:

If the United States [E]nvironmental [P]rotection [A]gency indicates that it is investigating a site which is the subject of an approved voluntary cleanup plan or no action petition, the department *shall actively pursue* a determination by the United States . . . that the property not be addressed under the federal act, that no further federal action be taken with respect to the property at least until the voluntary cleanup plan is completely implemented.

[18]*See* Charles Bartsch & Elizabeth Collaton, COMING CLEAN FOR ECONOMIC DEVELOPMENT at 107.
[19]*Id.*
[20]*Id.* § 25-16-309(2).
[21]*Id.* at § 206(1).
[22]*Id.*
[23]*Id.*

When the owner intends to make such a "substantial change" he must submit an application[24] to the Board of County Commissioners or governing body of the municipality and request approval of the change before it becomes effective.[25] Approval of a substantial change may be made only if all of the factors in section 25-15-203 are satisfied.[26]

There are several weaknesses in this act. First, it makes no provisions for penalties for violations of these requirements. Second, like many other state statutes, the Colorado law does not provide notice to potential purchasers. Finally, the act only applies to hazardous waste disposal sites and not to the generators and handlers of hazardous waste, who are just as likely to have a release resulting in contamination.

Weaknesses in this law, in part, led to the passage of the Colorado VCP. Under the newer law, a broader array of parties can obtain state oversight in the cleanup process.

Specialized Stakeholder Laws

Colorado is one of the states that amended its state equivalent of CERCLA to include some specialized lender liability protection.

Colorado law dictates that, except as preempted by federal law, no person or entity will be deemed to be an owner or operator of real or personal property who, without participating in the management of the subject real or personal property, holds indicia of ownership primarily to protect a security or lienhold interest in the subject real or personal property or in the property in which the subject real or personal property is located. This language closely tracks the statutory language in the secured creditor exemption of the federal Superfund statute.

In addition, however, Colorado law requires that "no lender-owner or representative shall, by virtue of becoming the owner of real or personal property, be liable for any third-party liability arising from contamination or pollution emanating from said property prior to the date that title vests in the lender-owner or representative."[27] Moreover, "no lender-owner or representative shall, by virtue of becoming the owner of real or personal property, be liable for any third-party liability arising from contamination or pollution emanating from said property during the period of ownership so long as, and to the extent that, it does not knowingly or recklessly cause new contamination or pollution or does not knowingly or recklessly allow others to cause new contamination or pollution if lender-owner

---

[24]The application must be accompanied by a fee not to exceed $10,000, which may be partially or totally refunded. *Id.* at § 206(3). The purpose of the fee is to assist in funding the review of the substantial change. *Id.*

[25]*Id.* at § 206(1). However, if a hazardous waste disposal site has been designated by the council pursuant to section 25-15-217, any substantial change is to be approved by the council until January 1, 1986, and changes thereafter are to be approved by the department. *Id.*

[26]*Id.* at § 206(2).

[27]*Id.* §13-20-703(2).

has caused an environmental professional to conduct a visual inspection of the property and a record search of the recorded chain of title documents regarding the real property for the prior fifty years to determine the presence and condition of hazardous waste or substances, obvious contamination, or pollution and, if found by the enforcing agency to be in noncompliance with federal or state laws, takes steps to assure compliance with applicable laws."[28]

This applies to the lender-owner as long as it makes reasonable efforts to re-sell the property.[29]

Montana

Voluntary Cleanup and Redevelopment Act of 1996

The state of Montana had no brownfields redevelopment program prior to 1996, when the state legislature enacted the Voluntary Cleanup and Redevelopment Act of 1996. This law offers No Further Action letters to participants that successfully complete the state program. The program can be used to address all or only a portion of a site.

Controlled Apportionment Liability Act and Orphan Share Fund

In addition, in 1996, Montana created a fund that offers reimbursement for the cleanup of orphan shares. Reimbursement is dependent upon the level of the state fund.

Loans from the State Board of Investments

The state also encourages brownfields redevelopment by making loans available through the State Board of Investments. These loans are not specifically designed for brownfields, but may be applied to brownfield redevelopment projects.

Utah

Voluntary Environmental Cleanup Program

The Utah legislature created a voluntary cleanup program in 1997. It will be administered by the Department of Environmental Quality.[30] It will be funded by

---

[28]*Id.* § 13-20-703(3).

[29]*Id.* The law reserves that: "this section shall not affect any liability expressly created under federal or state health or environmental statutes, regulations, permits, or orders." *Id.*

[30]UTAH CODE ANN. § 19-8-104 (1997).

application fees and collection of oversight costs.[31] Fees so collected must be deposited in the General Fund in the Environmental Voluntary Cleanup Account.[32]

In Utah, any site is eligible for participation in the voluntary cleanup program except: (1) a RCRA TSD facility,[33] (2) that portion of a site that is listed on the NPL; or (3) that portion of a site for which an administrative, state, or federal enforcement action is existing or pending against the applicant (for remediation of the contaminants described in the application).[34]

Before the state may evaluate any detailed plan or report regarding the remediation goals and proposed methods of remediation, the applicant and the executive director must have entered into a voluntary cleanup agreement.[35] An agreement established under the new law may be terminated by the executive director or the applicant by giving fifteen days prior notice, in writing, to the other party.[36] The executive director must provide to the applicant written and itemized notification of all oversight costs within ninety days after the date the agreement is terminated. The state can enforce the contract to pay oversight costs (plus reasonable attorney fees) in court.

After the applicant and the executive director have signed the voluntary cleanup agreement, the applicant must prepare and submit the appropriate work plans and reports to the executive director as provided in the agreement. The executive director must review and evaluate the work plans and reports for accuracy, quality, and completeness. The executive director may approve a voluntary cleanup work plan or report, or if he does not approve the work plan or a report, he must notify the applicant in writing concerning additional information or commitments necessary to obtain approval.

At any time during the evaluation of a work plan or report, the executive director may request the applicant to submit additional or corrected information.

After considering the proposed future use of the property that is the subject of the agreement, the executive director may approve work plans and reports submitted under this section that do not require removal or remedy of all discharges, releases, and threatened releases on the property if the applicant's response actions under the agreement:

---

[31]§ 19-8-104(c). For applications submitted on or after May 5, 1997 through June 30, 1998, the application fee under this chapter is $2,000.

Regarding applications submitted on and after July 1, 1998, the executive director shall annually calculate the costs to administer the voluntary cleanup program under this chapter and shall establish the fees for the program. § 19-8-117.

[32]§ 19-8-107(5). The fund was created under § 19-8-103.

[33]*See* 42 U.S.C. § 6901 *et seq.*

[34]§ 19-8-105 (eligibility and exceptions—grounds for application rejection by executive director).

[35]§ 19-8-108(a)(1) (voluntary agreement—procedure and establishment).

[36]§ 19-8-109 (termination of agreement—cost recovery).

will be completed in a manner that protects human health and the environment;
will not cause, contribute to, or exacerbate discharges, releases, or threatened
  releases on the property that are not required to be removed or remedied
  under the work plan; and
will not interfere with or substantially increase the costs of response actions
  to address any remaining discharges, releases, or threatened releases resulting from releases initially generated on the property.

If the executive director determines that an applicant has completed a voluntary cleanup in accordance with the agreement entered into under this chapter, the
executive director must within thirty days after the determination issue to the applicant a certificate of completion. The certificate of completion must:

acknowledge the protection from liability provided by the voluntary cleanup
  law;[37]
state the proposed future use of the property as defined under the voluntary
  cleanup agreement;
provide a legal description of the property and identify the owner of the property; and
state the applicant has complied with the voluntary cleanup agreement and
  has met the requirements of the voluntary cleanup law.[38]

The executive director must upon issuance of the certificate of completion record
a copy in the real property records of the county in which the site is located.[39]
  If the executive director determines the applicant has not successfully completed a voluntary cleanup in accordance with an agreement entered into under this
chapter, the executive director must:

notify the applicant and the current owner of the property that is the subject
  of the agreement of the denial of a certificate of completion; and
provide to the applicant a list in writing of the reasons for the denial.

The applicant may appeal the determination of the executive director.[40]

Grandfather Clause

A person who has completed a response action prior to May 5, 1997, at a site that
would have been eligible for participation in the program, may now submit an
application[41] to the executive director[42] for a certificate of completion.[43] The

---

[37]See § 19-8-113.

[38]See § § 19-8-108 and 19-8-110.

[39]§ 19-8-111.

[40]§ 19-8-112. The appeal must be undertaken as provided in Title 63, Chapter 46b, Administrative
Procedures Act.

[41]§ 19-8-118 (cleanups conducted prior to May 5, 1997).

[42]See § 19-8-107.

[43]See § 19-8-111.

application must include information required by department rules concerning the property addressed by the application and the response action conducted at the site.

The executive director and applicant must identify in the voluntary agreement any necessary studies to be conducted by the applicant to demonstrate the cleanup has been completed as provided by law.[44]

Specialized Stakeholder Laws

Utah was one of the states that amended its state equivalent of CERCLA to include some specialized lender liability protection.

In Utah,[45] the legislature passed protections for secured lenders that became effective on July 1, 1993. Under this new law, a secured lender[46] may select between the following when the real property security is environmentally impaired[47] and the borrower's[48] obligations to the secured lender are in default. First, the

---

[44]*See* § 19-8-110.

[45]*Id.* § 78-37-1.5 (environmental impairment to real property security interest—remedies of lender).

[46]*Secured lender* is defined under the statute as meaning:

(i)   the trustee, the beneficiary, or both under a deed of trust against the real property security;

(ii) the mortgagee under a mortgage against the real property security; and

(iii) any successor in interest of the trustee, beneficiary, or mortgagee under the deed of trust or mortgage.

For purposes of this section, the statute defines:

(a) Estimated costs to clean up and remediate the contamination caused by the release include only those costs that would be incurred reasonably and in good faith.

(b) Fair market value is determined without giving consideration to the release and is exclusive of the amount of all liens and encumbrances against the security that are senior in priority to the lien of the secured lender.

(c) Any real property security for any loan or extension of credit secured by a single parcel of real property is considered environmentally impaired if the property is:

(i)   included in or proposed for the National Priorities List under Section 42 U.S.C. 9605;

(ii) on any list identifying leaking underground storage tanks under 42 U.S.C. 6991 *et seq.*; or

(iii) on any list published by the Department of Environmental Quality under Section 19-6-311.

Id. § 78-37-1.5(1)(f).

[47]*Environmentally impaired* is defined in that statute as meaning:

the estimated costs to clean up and remediate a past or present release of any hazardous matter into, onto, beneath, or from the real property security exceed 25 percent of the higher of the aggregate fair market value of all security for the loan or extension of credit at the time: (i) of the making of the loan or extension of credit; (ii) of the discovery of the release or threatened release by the secured lender; or (iii) an action is brought under this section.

Id. § 78-37-1.5(1)(b).

[48]For purposes of the statute, *borrower* means:

(i) the trustor under a deed of trust, or a mortgagor under a mortgage, when the deed of trust or mortgage encumbers real property security and secures the performance of the trustor or mortgagor lender may waive its lien.[49] This waiver could be against any parcel of real property security[50] or any portion of that parcel that is environmentally impaired all or any portion of the fixtures and personal property attached to the parcels. The secured lender may, in the alternate, exercise the rights and remedies of an unsecured creditor, including reduction of its claim against the borrower to judgment and any other rights and remedies permitted by law.[51] As another remedy, the lender may choose to exercise the rights and remedies of a creditor secured by a deed of trust or mortgage and, if applicable, a lien against fixtures or personal property attached to the real property security and any other rights and remedies permitted by law, including the right to obtain a deficiency judgment.[52]

In order to avail itself of these rights, however, the secured lender must have satisfied the following conditions:[53]

> In conjunction with and at the time of the making, renewal, or modification of the loan, extension of credit, guaranty, or other obligation secured by the real property security, the secured lender: (1) did not know or have reason to know of a release[54] of any hazardous matter[55] into, onto, beneath, or from the real property security; and (2) undertook all appropriate inquiry into the previous ownership and uses of the real property security consistent with good commercial or customary practice in an effort to minimize liability.

For the purposes of evaluating whether a secured lender has satisfied these conditions, the legislature directed the court to take into account:

---

under a loan, extension of credit, guaranty, or other obligation; and (ii) includes any successor-in-interest of the trustor or mortgagor to the real property security before the deed of trust or mortgage has been discharged, reconveyed, or foreclosed upon.

Id. § 78-37-1.5(1)(a).
[49]UTAH CODE ANN. § 78-37-1.5(3)(a).
[50]*Real property security* means any real property and improvements other than real property that contains only one but not more than four dwelling units, and is solely used for either:

(i) residential purposes; or
(ii) if reasonably contemplated by the parties to the deed of trust or mortgage, residential purposes as well as limited agricultural or commercial purposes incidental to the residential purposes.

Id. § 78-37-1.5(1)(d).
[51]*Id.* § 78-37-1.5(3)(b).
[52]*Id.* § 78-37-1.5(3)(c).
[53]*Id.* § 78-37-1.5(4).
[54]*Release* has the same meaning as in Section 19-6-302.
Id. § 78-37-1.5(1)(e).
[55]*Hazardous matter* is defined as:

(i) any hazardous substance or hazardous material as defined in Section 19-6-302; or
(ii) any waste or pollutant as defined in Section 19-5-102.

UTAH CODE ANN. § 78-37-1.5(1)(c).

(1) any specialized knowledge or experience of the secured lender;
(2) the relationship of the purchase price to the value of the real property security if uncontaminated;
(3) commonly known or reasonably ascertainable information about the real property security;
(4) the obviousness of the presence or likely presence of contamination at the real property security; and
(5) the ability to detect the contamination by appropriate inspection.

Before the secured lender may waive its lien against any real property security on the basis of environmental impairment the secured lender must: provide written notice of the default to the borrower;[56] bring a valuation and confirmation action against the borrower in a court of competent jurisdiction and obtain an order establishing the value of the subject real property security.

A secured lender's complaint may include causes of action for a money judgment for all or part of the secured obligation; but in such a case, the waiver of the secured lender's liens may result only if a final money judgment is obtained against the borrower.

If a secured lender selects to waive its lien[57] and the borrower's obligations are also secured by other real property security, fixtures, or personal property, the secured lender must first foreclose against the additional collateral to the extent required by applicable law.[58] In such a case, the amount of the judgment of the secured lender would be limited to the remaining balance of the borrower's obligations after the application of the proceeds of the additional collateral.[59] The borrower may waive or modify the foreclosure requirements provided by law if the waiver or modification is in writing and signed by the borrower after default.[60]

The Utah legislature expressly provided, however, that this law would "not affect any rights or obligations arising under contracts existing before July 1, 1993, and applies only to loans, extensions of credit, guaranties, or other obligations secured by real property security made, renewed, or modified on or after July 1, 1993."[61]

Wyoming

The state of Wyoming has no brownfields redevelopment program at this time.

---

[56]UTAH CODE ANN. § 78-37-1.5(5).
[57]*Id.* § 78-37-1.5(6).
[58]*Id.* § 78-37-1.5(6)(a).
[59]*Id.* § 78-37-1.5(6)(b).
[60]*Id.* §78-37-1.5(6)(c).
[61]*Id.* § 78-37-1.5(7).

CHAPTER 12

Region IX:
Arizona, California, Guam, Hawaii, and Nevada

California, like Texas, is among the nation's leading hazardous waste generators.[1] As such, it is no surprise that California was the first state in Region IX with well-developed brownfields legislation. This is due to the stage of development in California. It is overpopulated and desperately in need of water. As population density increases in California, the brownfields redevelopment efforts in these states have, thus, become increasingly important. Although the state is not as old and the manufacturing history not as long, the pressure to increase brownfields redevelopment in California is magnified in the state due to the severe water shortage.[2]

The following table describes the more recent programs developed in Region IX to encourage voluntary cleanup:[3]

States	Brownfields Financing Programs	VCP with Written Assurances	Incentives to Attract Private Development
Arizona		X	
California	X	X	X
Guam			
Hawaii		X	
Nevada			

Arizona

Voluntary Cleanup Program

Arizona does not have a voluntary cleanup statute but operates its voluntary cleanup programs under the enforcement discretion of three state statutes. These statutes include:

---

[1]Harold C. Barnett, TOXIC DEBTS AND THE SUPERFUND DILEMMA at 91 (1994).

[2]*See generally* KEY ENVIRONMENTAL ISSUES IN U.S. EPA REGION IX, American Bar Association, Section of Natural Resources, Energy, and Environmental Law (May 2-3, 1996, San Francisco, California).

[3]*See* Charles Bartsh, BROWNFIELDS STATE OF THE STATE REPORT: 50-STATE PROGRAM ROUNDUP (Northeast-Midwest Inst., January 28, 1998).

Arizona's Water Quality Assurance Revolving Fund (WQARF);
the Underground Storage Tank program; and
the emergency interim soil remediation standards.

The three programs developed to date by the state of Arizona are, therefore, based on certain media, as the following table describes:

Program	Description
WQARF Voluntary Program	Allows any person who undertakes a remedial action to request that the ADEQ approve the remedial action before, during, or after remediation.
	Parties undertaking a preapproved remedial action must file a RAP with ADEQ.
UST Voluntary Program	Allows a person who comes into possession or control of a property where a UST is located to voluntarily remove or otherwise close the UST in a safe and secure manner that prevents release of regulated substances.
Voluntary Environmental Mitigation Use Restriction (VEMUR) Program	A person conducting a remediation project based on predetermined remediation standards or background concentrations must provide initial notice of intent to remediate to ADEQ. The notice must set forth site location, site characteristics, and a statement that the property will not be used for residential purposes. The notice must also explain the reason for the selected remediation levels and technologies to be used at the site. A final report is submitted once site remediation is complete.

In addition to these established programs and the Covenant Not to Sue obtained working through them, the state of Arizona is willing to consider entering into prospective purchaser agreements under the state equivalent of CERCLA.

California

Voluntary Cleanup Program

The California Voluntary Expedited Remedial Action Reform Act of 1994,[4] which was enacted on September 7, 1994, applies only to sites selected by the Site

---

[4]For example, to expedite the conversion of property into productive use and to provide funds for response activities, CEPA may approve a site owner's request to modify the boundaries of a selected site for response action only if the following criteria have been met:

a remedial plan has been approved by CEPA;
the holder of the first deed of trust, if any, approves;
portions of the site proposed to be removed by the boundary modification do not require any response action;
the boundary modification will not result in an unacceptable risk of human exposure to hazardous substances at the site;

Designation Committee of the California Environmental Protection Agency (CEPA) for expedited remediation. Once selected, CEPA should notify all known potentially responsible parties for a selected site and enter into an enforceable agreement with a responsible party(ies) to undertake response actions at the site. The Act specifies procedures for, and the consequences of, filing a Certificate of Completion for a response action. The Act also requires the apportionment of liability for response costs and specifies criteria for such apportionment. An Expedited Site Remediation Trust Fund has been created to carry out CEPA's statutory mandate of expedited remediation of selected sites. Certain land use controls on sites subject to expedited remedial procedures must be recorded.[5] For example, to expedite the conversion of property into productive use and to provide funds for response activities, CEPA may approve a site owner's request to modify the boundaries of a selected site for response action only if the following criteria have been met:

a remedial plan has been approved by CEPA;
the holder of the first deed of trust, if any, approves;
portions of the site proposed to be removed by the boundary modification do not require any response action;
the boundary modification will not result in an unacceptable risk of human exposure to hazardous substances at the site;
hazardous substances have not migrated, and are not expected to migrate, onto portions of the site proposed for boundary redesignation; and
the modification of boundaries of the site will not significantly interfere with the response action at the site.[6]

The California program precludes state enforcement actions if program requirements are met.[7] Participants satisfying program requirements are protected from contribution actions brought by third parties for cleanup costs.[8] Moreover, parties entering into a valid voluntary remediation agreement and satisfying program requisites are entitled to seek contribution from other responsible parties.[9]

---

hazardous substances have not migrated, and are not expected to migrate, onto portions of the site proposed for boundary redesignation; and
the modification of boundaries of the site will not significantly interfere with the response action at the site.

CAL. HEALTH & SAFETY CODE § 25398.5(a)(1)(A)-(F) (West 1994).

[5] 1993 CA S.B. 923; CAL. HEALTH & SAFETY CODE § 25261 *et seq.* (West 1994).

[6] CAL. HEALTH & SAFETY CODE § 25398.5(a)(1)(A)-(F) (West 1994).

[7] *Id.* at § 25398.2(b)(1)(C). Responsible parties are deemed liable to CEPA for oversight response costs and other costs incurred by the state, which are allocated to responsible parties in a final liability apportionment. *Id.* at § 25399(a).

[8] *Id.* at § 25398.6(d). A responsible person settling with CEPA and paid any response cost as a part of the settlement agreement is not liable for claims of contribution from any other party for costs of the response action.

[9] *Id.* at § 25398.17(a).

Volunteers may sue recalcitrant responsible parties failing or refusing to comply with any order or agreement relating to a site selected by CEPA for expedited remediation for treble damages.[10]

The Voluntary Expedited Remedial Action Reform Act establishes a program whereby volunteers enter the program,[11] become subject to state oversight through micromanagement of remedial activities,[12] and receive a Certificate of Completion[13] upon satisfactory completion of the program. Specifically, property designated for expedited cleanup must employ remedies that are:

protective of human health and the environment;
provide long-term reliability at a reasonable cost;
provide reasonable protection to the waters of the state; and
leave the site in a condition that allows for permanent use free of any significant risk to human health or any potential for future environmental damage.[14]

A response action may achieve applicable cleanup standards through the use of:

engineering and appropriate land use controls applicable to the sites future use;
treatment reducing the toxicity, mobility, or volume of hazardous substances;
removal techniques for hazardous substances;
any combination of the above; or
other approved methods of protection.[15]

The California program was not specifically designed to encourage economic development of urban industrial areas. The California program merely resembles an expedited state Superfund program. For instance, the Expedited Remedial program specifies criteria for the apportionment of liability between potentially responsible parties for response costs.[16] During the apportionment phase, CEPA will weigh relevant equitable factors considered appropriate under the circumstances for which the remedial action was initiated.[17] In addition, the Expedited Remedial program requires that certain land use controls implemented at a site be recorded

---

[10]*Id.* One-half of any recovery of treble damages shall be deposited in the California Expedited Site Remediation Trust Fund. *Id.* § 25398.17(b).

[11]*Id.* at § 25396.6.

[12]*Id.* at § 25398.6.

[13]*Id.* at §§ 25398.14 and 25398.15.

[14]*Id.* at § 25398.6(a).

[15]*Id.* at § 25398.6(b)(1)-(5).

[16]*Id.* at § 25398.8(b). Under § 25398.7(c), CEPA will weigh equitable factors considered appropriate under the circumstances of the release for which the remedial action was initiated.

[17]*Id.*

in the site's deed and other chain of title documents to provide notice to subsequent purchasers and users.[18]

Other Voluntary Cleanup Programs in California

The Department of Toxic Substances Control (DTSC) created the Voluntary Cleanup Program in 1993.[19] The VCP permits parties, known as project proponents, to investigate and remediate sites after paying DTSC's oversight fees and entering into an agreement with DTSC to assess a site, prepare plans, and conduct specific remedial or removal actions.[20] Following the cleanup, when DTSC reviews and approves of the action, it may issue either a No Further Action (NFA) letter or Remedial Action Certification.[21] Unlike the Expedited Remedial Action Program agreement, however, the VCP does not provide a Covenant Not to Sue. Project proponents can, however, walk away from the program at any time without penalty.[22]

The following is a breakdown of the actual steps one must follow to participate in the Voluntary Cleanup Program. First, a project proponent applies to DTSC to participate in the VCP. If the application is accepted, the applicant then negotiates the terms and services of the VCP with the DTSC. Based on the estimated cleanup costs, either a Remedial Action Work Plan (RAW) or Remedial Action Plan (RAP) must be prepared. A RAW must be prepared for projects where the estimated total cleanup cost is less than $1 million.[23] A RAP is prepared for projects estimated to cost more than $1 million.[24]

The plan is put through notice and comment. It is then subject to final approval from DTSC. Finally, cleanup activities commence, under DTSC's supervision and per the requirements of the Hazardous Substance Account Act and the National Contingency Plan.[25]

Sites that are ineligible for the Voluntary Cleanup Program include the following:

DTSC's Annual Work Plan sites;
NPL sites;
federal facilities;
sites containing petroleum products or waste or otherwise outside of DTSC's jurisdiction.

---

[18]*Id.* at § 25398.7(a).

[19]CAL. HEALTH & SAFETY CODE, ch. 6.8 §§ 25300 *et seq.*

[20]Jane Kroesche, California, at 3, citing DTSC, *Policy and Procedure for Managing Voluntary Site Mitigation Projects, The Voluntary Cleanup Program,* #E0-95-006-PP, at 1-2 (Sept. 20, 1995) (unpublished manuscript, on file with author).

[21]*Id.* at 4, citing Barbara Coler and Steve Koyasako, Redevelopment and Revitalization of Brownfields, Department of Toxic Substance Control Initiatives, at 2 (Oct. 1995).

[22]*See* Charles Bartsch & Elizabeth Collaton, COMING CLEAN FOR ECONOMIC DEVELOPMENT at 108.

[23]CAL. HEALTH & SAFETY CODE §§ 25356.1(h), 25323.1.

[24]*Id.*

[25]*Id.* § 25356.1(d).

Project proponents must provide an advance payment of half the remediation project's estimated costs to DTSC. These estimated costs include oversight costs. DTSC has the discretion to adjust both the payment and schedule on a case by case basis. The cleanup standards used to estimate the site remediation costs are based on those used for the state CERCLA equivalent.[26]

At the conclusion of the process, DTSC may issue an NFA letter. NFAs are generally reserved for sites that are cleaned up to a standard that allows some contamination to remain on the property but which should not cause health hazards or problems for the environment. A site Certificate of Completion may be awarded to sites that proceed through assessment and more stringent remediation. This certificate reduces liability exposure, but like other states the certificate will not remove the contingent liability in its entirety. Numerous standard reopeners protect the state.[27]

The Unified Agency Review Law

The California Unified Agency Review Law[28] provides overview, final approval, and certification by an Administering Agency to responsible parties that successfully undertake investigation and cleanups.[29] In addition to providing qualified immunity to the responsible party, the certification binds state and local agencies.[30] The law is designed to enable participants to rely on the decisions of one lead agency within the California system.

The Private Site Management Program

The Private Site Management Program[31] allows private site managers to investigate sites and provide cleanup recommendations to DTSC.[32] If cleanup is recommended, either a RAW or RAP must be prepared, and remediation of the site occurs.[33] The private site manager will later request a Certificate of Completion that details the cleanup activities and confirms the site remediation.[34] DTSC's certification of the site must be made known to the public and may call for continued site maintenance.[35]

---

[26]*See* Charles Bartsch & Elizabeth Collaton, COMING CLEAN FOR ECONOMIC DEVELOPMENT at 110.

[27]*Id.*

[28]CAL. HEALTH & SAFETY CODE § 25260 *et seq.*

[29]*Id.* §§ 25262, 25264(a)(b).

[30]*Id.* § 25264(b), (c)(1-6).

[31]CAL. HEALTH & SAFETY CODE §§ 25395.1 *et seq.*

[32]Kroesche, *supra* note 20, at 20; CAL. HEALTH & SAFETY CODE § 25395.2.

[33]CAL. HEALTH & SAFETY CODE §§ 25395.5, 25395.6.

[34]*Id.* § 25395.8(a); Kroesche, *supra* note 20, at 21.

[35]CAL. HEALTH & SAFETY CODE, § 25395.8(d); Kroesche, *supra* note 20, at 21.

Business Transfer Statute

The California Health and Safety Code[36] requires that "any owner of nonresidential real property who knows, or has reasonable cause to believe, that any release of hazardous substance has come to be located on or beneath that real property must, prior to the sale, lease, or rental of the real property by that owner, give written notice of that condition to the buyer, lessee, or renter of the real property."[37] Failure of the owner to provide written notice to the buyer, lessee, or renter shall subject the owner to actual damages and any other remedies provided by law. In addition, where the owner has actual knowledge of the presence of any release of a material amount of a hazardous substance and knowingly and willfully fails to provide written notice to the buyer, lessee, or renter, the owner will be held liable for a civil penalty not to exceed $5,000 for each separate violation.[38]

In addition, the California code requires that any lessee or renter of real property who knows or has reasonable cause to believe that any release of a hazardous substance has come or will come to be located on or beneath that real property will, within a reasonable period of time, either prior to the release or following the discovery by the lessee or renter of the presence or believed presence of the hazardous substance release, give written notice of that condition to the owner of the real property or to the lessor under the lessee's or renter's lease or rental agreement. A lessee or renter who fails to provide written notice, when required, to the owner or lessor may be subject to actual damages as well as any other remedy provided by law. If the lessee or renter has knowledge of the presence of a release of a material amount of a hazardous substance, or of a hazardous substance release that is required to be reported to a state or local agency pursuant to law, on or under the real property leased or rented by the lessee or renter and knowingly and willfully fails to provide written notice, when required, to the owner or lessor, both of the following apply:

owner's or lessor's written notice to the lessee or renter, under the lessee's or renter's lease or rental agreement, except this does not apply to lessees and renters of property used exclusively for residential purposes; and

the lessee or renter is liable for a civil penalty not to exceed $5,000 for each separate violation.

A lessee or renter may cure a default under the lessee's or renter's lease or rental agreement by promptly commencing and completing the removal of, or taking other appropriate remedial action with respect to, the hazardous substance release. The removal or remedial action must automatically be conducted in

---

[36]CAL HEALTH & SAFETY CODE § 25359.7.
[37]*Id.* § 25359.7(a).
[38]*Id.*

accordance with all applicable laws and regulations and in a manner that is reasonably acceptable to, and approved in writing by, the owner or lessor. Curing the default will not, however, relieve the lessee or renter of any liability for actual damages or for any civil penalty for a violation.[39]

Specialized Stakeholder Statutes—Lender

California is one of the states that amended its state equivalent of CERCLA to include some specialized lender liability protection.

Under California law, a secured lender may select between the following where the real property security is environmentally impaired and the borrower's obligations to the secured lender are in default:

| Waiver of its lien against (i) any parcel of real property security that is environmentally impaired or is an affected parcel, and (ii) all or any portion of the fixtures and personal property attached to the parcels; and exercise of (i) the rights and remedies of an unsecured creditor, including reduction of its claim against the borrower to judgment, and (ii) any other rights and remedies permitted by law; | or | Exercise of (i) the rights and remedies of a creditor secured by a deed of trust or mortgage and, if applicable, a lien against fixtures or personal property attached to the real property security, and (ii) any other rights and remedies permitted by law.[40] |

Before the secured lender may waive its lien against any parcel of real property security on the basis of the environmental impairment, the secured lender must provide written notice of the default to the borrower and the value of the subject real property security must be established and its environmentally impaired status must be confirmed by an order of a court. The secured lender's complaint against the borrower for a valuation and confirmation action may include causes of action for a money judgment for all or part of the secured obligation, in which case the waiver of the secured lender's liens must result only if and when a final money judgment is obtained against the borrower.

If the secured lender elects the first option and the borrower's obligations are also secured by other real property security, fixtures, or personal property, the

---

[39]For further discussion of the California rules, see Handling Real Property Sales Transactions, CEB Action Guide; Minimizing Toxics Liability When Buying Real Property and Businesses, CEB Action Guide, Spring 1992; Taking a Closer Look: Significant New California Legislation Enacted in 1988. 12 CEB REAL PROP. L. REP. 55 (1988); Taking a Closer Look: Significant New California Legislation Relating to Real Property Enacted in 1987, 11 CEB REAL PROP. L. REP. 37 (1987).

[40]CAL. CIV. PROC. CODE § 726.5 (election between waiver of lien and exercise of specified rights and remedies). *See generally The Remediation of Hazardous Lending,* 10 CAL. REAL PROP. J. No. 3, p. 25.

secured lender must first foreclose against the additional collateral to the extent required by applicable law, in which case the amount of the judgment of the secured lender will be limited to the extent section 580a or 580d, or subdivision (b) of section 726 apply to the foreclosures of additional real property security. The borrower may waive or modify the foreclosure requirements of this law, provided that the waiver or modification is in writing and signed by the borrower after default.

The first option will, however, be inapplicable if the release or threatened release was not knowingly or negligently caused or contributed to, or knowingly or willfully permitted or acquiesced to, by the borrower, any affiliate or agent of the borrower, or any related party. In addition, in conjunction with the making, renewal, or modification of the loan, extension of credit, guaranty, or other obligation secured by the real property security, neither the borrower, nor any related party, affiliate, or agent can possess actual knowledge or notice of the release or threatened release. If such a person had knowledge or notice of the release or threatened release, the borrower must have made written disclosure thereof to the secured lender after the secured lender's written request for information concerning the environmental condition of the real property security, or the secured lender otherwise obtained actual knowledge thereof, prior to the making, renewal, or modification of the obligation.[41]

---

[41]For purposes of this law:

Affected parcel means any portion of a parcel of real property security that is (A) contiguous to the environmentally impaired parcel, even if separated by roads, streets, utility easements, or railroad rights-of-way, (B) part of an approved or proposed subdivision within the meaning of Section 66424 of the Government Code, of which the environmentally impaired parcel is also a part, or (C) within 2,000 feet of the environmentally impaired parcel.

Borrower means the trustor under a deed of trust, or a mortgagor under a mortgage, where the deed of trust or mortgage encumbers real property security and secures the performance of the trustor or mortgagor under a loan, extension of credit, guaranty, or other obligation. The term includes any successor-in-interest of the trustor or mortgagor to the real property security before the deed of trust or mortgage has been discharged, reconveyed, or foreclosed upon.

Environmentally impaired means that the estimated costs to clean up and remediate a past or present release or threatened release of any hazardous substance into, onto, beneath, or from the real property security, not disclosed in writing to, or otherwise actually known by, the secured lender prior to the making of the loan or extension of credit secured by the real property security, exceeds 25 percent of the higher of the aggregate fair market value of all security for the loan or extension of credit (A) at the time of the making of the loan or extension of credit, or (B) at the time of the discovery of the release or threatened release by the secured lender. For the purposes of this definition, the estimated cost to clean up and remediate the contamination caused by the release or threatened release shall include only those costs that would be incurred reasonably and in good faith, and fair market value shall be determined without giving consideration to the release or threatened release, and shall be exclusive of the amount of all liens and encumbrances against the security that are senior in priority to the lien of the secured lender. Notwithstanding the foregoing, the real property security for any loan or extension of credit secured by a single parcel of real property which is included in the National Priorities List pursuant to Section 9605 of Title 42 of the United States Code, or in any list published by the State Department of Health Services pursuant to subdivision (b) of Section 25356 of the Health and Safety Code, shall be deemed to be environmentally impaired.

Environmental Lien/Nuisance Statute

Under California law, the owner of any real property upon which an on-site waste-water disposal system is located, which system is subject to abatement as a public nuisance by the public agency, may request the public agency to replace or repair, as necessary, such system. If replacement or repair is feasible, the board may provide for the necessary replacement or repair work.[42]

The person(s) employed by the board to do the work must have a lien for work done and materials furnished, and the work done and materials furnished will be deemed to have been done and furnished at the request of the owner.[43] The zone, in the discretion of the board, may pay all, or any part, of the cost or price of the

Hazardous substance means (A) any "hazardous substance" as defined in subdivision (f) of Section 25281 of the Health and Safety Code as effective on January 1, 1991, or as subsequently amended, (B) any "waste" as defined in subdivision (d) of Section 13050 of the Water Code as effective on January 1, 1991, or as subsequently amended, or (C) petroleum, including crude oil or any fraction thereof, natural gas, natural gas liquids, liquefied natural gas, or synthetic gas usable for fuel, or any mixture thereof.

Real property security means any real property and improvements, other than a separate interest and any related interest in the common area of a residential common interest development, as the terms "separate interest," "common area," and "common interest development" are defined in Section 1351 of the Civil Code, or real property which contains only 1 to 15 dwelling units, which in either case (A) is solely used (i) for residential purposes, or (ii) if reasonably contemplated by the parties to the deed of trust or mortgage, for residential purposes as well as limited agricultural or commercial purposes incidental thereto, and (B) is the subject of an issued certificate of occupancy unless the dwelling is to be owned and occupied by the borrower.

Related party means any person who shares an ownership interest with the borrower in the real property security, or is a partner or joint venturer with the borrower in a partnership or joint venture, the business of which includes the acquisition, development, use, lease, or sale of the real property security.

Release means any spilling, leaking, pumping, pouring, emitting, emptying, discharging, injecting, escaping, leaching, dumping, or disposing into the environment, including continuing migration, of hazardous substances into, onto, or through soil, surface water, or groundwater. The term does not include actions directly relating to the incorporation in a lawful manner of building materials into a permanent improvement to the real property security.

Secured lender means the beneficiary under a deed of trust against the real property security, or the mortgagee under a mortgage against the real property security, and any successor-in-interest of the beneficiary or mortgagee to the deed of trust or mortgage.

Id. The California law cannot "be construed to invalidate or otherwise affect in any manner any rights or obligations arising under contract in connection with a loan or extension of credit, including, without limitation, provisions limiting recourse." *Id.*

The law continues:

(g) This section shall only apply to loans, extensions of credit, guaranties, or other obligations secured by real property security made, renewed, or modified on or after January 1, 1992, and before January 1, 2000.

Id.

[42]CAL. HEALTH & SAFETY CODE § 6979(a) (Deering 1994) (Owner's request of zone to replace or repair system subject to abatement as public nuisance; mechanic's lien).

[43]This lien will be subject to the provisions of subdivision (b) of Section 6978.

work done and materials furnished, and, to the extent that the zone pays the cost or price of the work done and materials furnished, the zone must succeed to and have all the rights, including, but not limited to, the lien, of such person or persons employed to do the work against the real property and the owner.[44]

The owner of any real property upon which is located a sanitary sewage, septic, or septic tank disposal system, which system is subject to abatement as a public nuisance by the district, may request the district to replace or repair, as necessary, such system. If replacement or repair is feasible, the board of directors, in its sole discretion, may provide for the necessary replacement or repair work.[45]

The person(s) employed by the board of directors to do the work must have a lien for work done and materials furnished, and the work done and materials furnished must be deemed to have been done and furnished at the request of the owner.[46] The district, in the discretion of the board of directors, may pay all, or any part, of the cost or price of the work done and materials furnished, and, to the extent that the district pays the cost or price of the work done and materials furnished, the district must succeed to and have all the rights, including, but not limited to, the lien, of such person or persons employed to do the work against the real property and the owner.[47] As an alternative power to the enforcement of the lien, the board of directors may, by ordinance adopted by two-thirds vote of the members:

> fix the costs of replacement or repair;
> fix the times at which such costs must become due;
> provide, prior to the replacement or repair, for the payment of the costs in installments over a period not to exceed fifteen years;
> establish a rate of interest, not to exceed 8 percent per annum, to be charged on the unpaid balance of the costs; and
> provide that the amount of the costs and the interest must constitute a lien against the respective lots or parcels upon which the work is done.[48]

With the written consent of the owner and the lienholder, if other than the district, the board of directors may issue an improvement bond to represent and be secured by the lien.[49] The bond may be delivered to the lienholder if other than the district or may be sold by the board of directors at public or private sale. The amount of the bond must be the amount of the lien, including incidental expenses

---

[44]§ 6979(b). *See generally* Cal. Jur. 3d (Rev) Drains and Sewers § 36.

[45]CAL. WATER CODE § 31148 (Deering 1994) (sanitary sewage, septic, or septic tank disposal system subject to abatement as public nuisance; request for replacement or repair; liens; payment of costs).

[46]Again, this lien will be subject to the provisions of subdivision (b) of Section 31147.

[47]§ 31148(b).

[48]CAL. HEALTH & SAFETY CODE § 6979(b) (Deering 1994). This lien will be subject to the provisions of subdivision (b) of Section 31147.

[49]The improvement bound, if it is to be issued, must be issued pursuant to the improvement bond provisions of the Improvement Act of 1911 (Part 5 [commencing with Section 6400] of Division 7 of the Streets and Highways Code).

allowable under the Improvement Act of 1911. The bond term and interest rate must be determined by the board of directors within the limits established by the Improvement Act of 1911 and other applicable provisions of law.[50]

Guam

Guam has no brownfields redevelopment program at this time.

Hawaii

In 1997, Hawaii amended the state equivalent of CERCLA. In so doing, the state legislature specifically authorized the state agency to use its response authority, including use of the state environmental response revolving fund, to solicit the cooperation of PRPs and encourage voluntary cleanup efforts.[51] The new law also expressly allows the state to enter into cleanup agreements,[52] presumably including, but not necessarily limited to, the voluntary response actions conducted under the state Voluntary Response Action Program.[53]

Voluntary Response Action Program[54]

The voluntary cleanup program is tied to the state equivalent of CERCLA. The program can be used on any site that has a release or threatened release for which the state has response action authority.[55] Prospective purchasers who complete a voluntary response action and receive a letter of completion from the department must be exempt from future liability to the department for those specific hazardous substances, pollutants, contaminants, media, and land area addressed in the voluntary response action. Certain sites are not eligible for participation. These include:

> NPL listed sites;
> those sites with respect to which an order or other enforcement actions has been issued or entered under CERCLA and is still in effect;
> a site where the United States Coast Guard has issued a federal Letter of Interest;
> a site that is the subject of RCRA corrective action;
> a site where the state has issued an order or is conducting a response pursuant to an enforceable agreement, unless the state agrees otherwise;
> a site which poses an imminent and substantial threat to human health,the environment, or natural resources as determined by the director; and
> a site where the director has determined that there is a significant public interest.[56]

---

[50]CAL. WATER CODE § 31148(d) (Deering 1994).
[51]HAW. REV. STAT. § 128D-4 (a)(3) (1997).
[52]Id.
[53]HAW. REV. STAT. § 128D-31 et seq. (1997).
[54]Id.
[55]§ 128D-33(a).
[56]§ 128D-33(b).

Parties applying for the program must provide the department with written consent from the property owner to conduct the voluntary response action including any restrictions of property rights.[57] In addition, the requesting party must pay a nonrefundable processing fee of $1,000 with each application to be eligible for a voluntary response action.[58]

The department must review each application in a timely manner and approve or deny the application. Each application must include the following information:

the requesting party's name, mailing address, telephone number, facsimile number, if applicable, or electronic mail address;

the property owners' names, mailing addresses, telephone numbers, facsimile numbers, if applicable, or electronic mail addresses;

the property location, mailing address, street or physical location, address, latitude and longitude, tax map key numbers, and telephone number for the requesting party;

a brief description of the site, its operational history, and any known or suspected contamination;

a listing of any permits obtained by any facility on the property;

a description of the intended scope of work;

a description of any civil, criminal, or administrative actions relative to the environmental matters of the subject property;

a written consent by the property owner supporting the proposed voluntary response action including any restrictions of property rights; and

the signature of the requesting party.

The requesting party must also provide the department with any and all reports and data pertaining to environmental investigations or response actions on the subject property.

Within sixty days after initial approval of the application, the requesting party and the department must negotiate an agreement for conducting the voluntary response action. This agreement must contain guarantees of completion, such as letters of credit, personal guarantees, insurance, or similar measures of guarantee. If, after sixty days, an agreement cannot be negotiated in good faith, the department may deny the application.[59]

In considering an application, the director may consider the following:

an administrative enforcement action has been initiated that concerns the remediation of the hazardous substance, pollutant, or contaminant;

site eligibility based on the criteria in section 128D-33;

completeness and accurateness of the application:[60]

---

[57]§ 128D-33(c).

[58]§ 128D-33(d).

[59]§ 128D-34.

[60]If, however, an application is denied because it is incomplete or inaccurate, the director, not later than forty-five days after receipt of the application, shall identify the omission or inaccuracy for the

inappropriate or inadequate scope of work;

pending litigation;

the capacity of the requesting party or the requesting party's agent to carry out the response action properly;

whether the department will receive a substantial benefit for cleanup or an indirect public benefit in combination with a reduced direct benefit to the department;

whether the continued operation of the site or new site development,with the exercise of due care, will aggravate or contribute to the existing contamination or interfere with the department's response action;

whether the continued operation or new development of the property will pose health risks to the community and those persons likely to be present at the site; or

the financial viability of the prospective purchaser.

If the director finally denies the application, the director must notify the requesting party that the application has been denied. In addition, the state must explain the reasons for denial of the application. The department's decision on an application is, however, considered final, with no right of appeal.[61]

Within thirty days of satisfactory completion of the voluntary response action, the director must issue a letter of completion for the response action completed by the requesting party. The letter of completion must identify the specific hazardous substances, pollutants, contaminants, media, and land area addressed in the response action. If contamination is left on the site, the letter of completion must identify land use restrictions and any required management plan.

The letter of completion must be noted on the property deed and must be sent to the county agency that issues building permits. The benefits and restrictions identified in the letter of completion must run with the land and apply to all future owners of the property. The exemption from liability does not apply to those persons who were liable under the state CERCLA equivalent prior to conducting the voluntary response action.

To qualify for an exemption from liability, a requesting party that is also a prospective purchaser must have obtained final approval to conduct a voluntary response action from the department prior to becoming the owner or operator of the property. The exemption from future liability to the department applies only to those specific hazardous substances, pollutants, and contaminants cleaned up to a risk-based standard of not more than one total lifetime cancer risk per one million and only to the specific media and land area addressed in the voluntary response

---

requesting party. A requesting party whose application has been denied because it is incomplete or inaccurate, may resubmit an application for the same response action without submitting an additional application fee. If a requesting party's application is denied a second time, the director may require an additional $1,000 processing fee for any subsequent submittal.

[61] § 128D-34.

action; provided that the exemption only applies to the contamination which occurred prior to conducting the voluntary response action.

A party who is exempt from future liability to the department is not be liable for claims for contribution or indemnity regarding matters addressed in the voluntary response action. The department reserves the right to take action consistent with the voluntary cleanup law against responsible parties.

The exemption from liability is not effective if a letter of completion is acquired by fraud, misrepresentation, or failure to disclose material information. Similarly, it is void where transactions were made for the purpose of avoiding liability.

There is no exemption from liability for other laws or requirements.

Voluntary Response Action Account and the Environmental Response Revolving Fund

The law established an account, to be called the voluntary response action account, within the environmental response revolving fund for the purpose of administration and oversight of the new law.

The $1,000 nonrefundable application fee must be deposited into the voluntary response action account. Upon initial approval of an application, the department may require a deposit of up to $5,000 to initiate a site-specific account. The department may require an additional deposit of up to $5,000, whenever the balance of the site-specific account falls below $1,000.

If a site-specific account balance is inadequate to support oversight, the department may discontinue oversight on the voluntary response action. The department may pursue enforcement action against the requesting party and any other liable person when an account balance is inadequate to support further oversight by the department.

At the completion of the voluntary response action, or at the termination of the agreement, the department must provide a final accounting of the site-specific account and return the balance to the requesting party. For purpose of assessing oversight costs, the legislature authorized the department to calculate costs at $100 for each hour of staff time plus actual expenses or one hundred twenty-five per cent of actual cost when contracting for oversight services. The department must provide each requesting party or parties a summary of the oversight costs for the party's specific site on an annual basis.

Specialized Stakeholder Laws

Hawaii is one of the states that amended its state equivalent of CERCLA to include some specialized lender liability protection.

Hawaii law excludes from the definition of *owner* or *operator* under its state CERCLA equivalent statute "a person or financial institution who holds or held a lien, encumbrance, security interest, or loan agreement that attaches or is attached to the facility, vessel, or real property; provided that the person or financial institution makes or made no decision or takes or took no action that causes or caused or contributes or contributed to a release or threatened release of a hazardous substance from or at a facility, vessel, or real property."

Nevada

Specialized Stakeholder Laws

Nevada is one of the states that amended its state equivalent of CERCLA to include some specialized lender liability protection. On May 5, 1993, Nevada's revised hazardous waste laws[62] became effective and applied "to all proceedings on or after that date whether an environmental provision was made, or a secured indebtedness was incurred, before or after that date."

---

[62]*Hazardous substance* was defined as:

1. an element, compound, mixture, solution, material or substance whose use, possession, transportation, storage, release, discharge, or disposal is regulated pursuant to chapter 444, 445, 459, 477, 590, or 618 of NRS or the Uniform Fire Code (1988 edition);
2. an element, compound, mixture, solution, material, or substance designated as a hazardous substance pursuant to 42 U.S.C. § 9602 and an element, compound, mixture, solution, material, or substance described in 42 U.S.C. § 9601(14);
3. an element, compound, mixture, solution, material, or substance listed as a hazardous waste in, or having the characteristics identified in, 42 U.S.C. § 6921 on January 1, 1993, except any waste for which regulation under the Resource Conservation and Recovery Act of 1976 (42 U.S.C. §§ 6901 *et seq.*) has been suspended by an act of Congress; and
4. petroleum, including crude oil or any fraction thereof, natural gas, natural gas liquids, liquefied natural gas, synthetic oil, synthetic gas useable for fuel, or any mixture thereof.

CHAPTER 13

Region X:
Alaska, Idaho, Oregon, and Washington

Oregon and Washington have the most developed brownfields legislation in Region X. This is likely because of the stage of development in these two states. As population density increases in California, more people have moved to Oregon and Washington. The brownfields redevelopment efforts in these states have, thus, become increasingly important. Although the state is not as old, and the manufacturing history not as long, the pressure to increase brownfields redevelopment in Washington is magnified in the state due to the severe water shortage. When Californians emigrated to Oregon and Washington from California, they brought increased pressure on water and other public resources in that state as well. Hence, the laws enacted in Washington have begun to be explored in Oregon as well.

The wide-open spaces of Alaska and Idaho have, however, until 1996 negated political pressure on these states to enact brownfields laws. These states refrained from early attempts and instead enacted some of the newer, perhaps more sophisticated brownfields style laws.

The following table describes the more recent programs developed in Region X to encourage voluntary cleanup:

States	Brownfields Financing Programs	VCP with Written Assurances	Incentives to Attract Private Development
Alaska	X	X	
Idaho		X	X
Oregon	X	X	
Washington	X	X	

Alaska

Lien Law

Under Alaska law, the state has a lien for expenditures by the state from the oil and hazardous substance release response fund or from any other state fund, for the costs of response, containment, removal, or remedial action resulting from an oil or hazardous substance spill or, with respect to response costs, the substantial threat of a release of oil or a hazardous substance against all property owned by a person who is determined by the commissioner to be liable for the expenditures under state[1] or federal law. The lien includes interest, at the maximum rate allowable under state law,[2] from the date of the expenditures. The state may file an action in a court of competent jurisdiction in order to foreclose on the lien.

A lien against real property will not be effective until (1) a certificate of lien is recorded in the district recorder's office for the district in which the property is located, describing the property and stating the amount of the lien, the name of the owner as grantor, and, if known, the name of the person causing the oil or hazardous substance release; and (2) the commissioner sends a copy of the certificate located, describing the property and stating the amount of the lien, the name of the owner as grantor, and, if known, the name of the person causing the oil or hazardous substance release; and (3) the commissioner sends a copy of the certificate of lien by certified mail return receipt requested, or actually delivers a copy of the certificate of lien, to the owner, the person causing the release, and to all other persons of record holding an interest in the property.

When any amount of the lien has been paid or reduced, the commissioner must, upon request of the property owner, issue a certificate discharging or partially releasing the lien. That certificate may be recorded in the office in which the certificate of lien was recorded.

The commissioner may, in the commissioner's discretion, reduce, discharge, or partially release a lien if a bond, or other security, in a form and an amount satisfactory to the commissioner is posted. The bond or other security must include an amount sufficient to cover the cost of execution, collection, or foreclosure, including attorney fees. A reduction, discharge, or partial release may not be granted if it would be contrary to the public interest. When a lien is reduced, discharged, or partially released under this subsection, the commissioner must, at the request of the property owner, issue a certificate to that effect.

A person with an ownership interest in property against which a lien is recorded may bring an action in a court of competent jurisdiction to require that the lien be released. The lien may be released to the extent of that person's ownership interest if the court finds that the person is not liable for the expenses incurred by the state in connection with the costs of response, containment, removal, or remedial action resulting from the oil or hazardous substance release or threat of release of oil or a hazardous substance.

Idaho

The state of Idaho has no brownfields redevelopment program at this time.

---

[1]*See, e.g.,* ALASKA STAT. §§ 46.03 and 46.04, 42 U.S.C. § 9607.
[2]*See Id.* § 45.45.010(a).

Oregon

Voluntary Cleanup Legislation

Oregon revised its Hazardous Waste and Material Removal or Remedial Action law to include provisions for the promotion of voluntary cleanup efforts.[3] One state official described the program in testimony before Congress as follows:

> ... Oregon has an active Voluntary Cleanup Program and has employed prospective purchaser agreements to facilitate redevelopment. We believe that a clean environment and a strong economy are not mutually exclusive, but actually complement one another. When we redevelop brownfields, we want to make sure that both the land and the water are put back into productive use. Often, lack of information is the biggest impediment to redevelopment. By making both grants and loans available for site characterization and assessment, we can "jump start" the redevelopment process.[4]

The new law allows the Department of Environmental Quality (DEQ) to issue a release from potential liability to the state to parties who assume the voluntary cleanup of their property.[5] The participants must meet the following criteria:

(1) the participant cannot be currently liable for an existing release of hazardous waste at the facility;
(2) the removal or remedial action must be necessary at the facility to protect human health or the environment;
(3) the proposed redevelopment or reuse of the facility will not contribute to or exacerbate existing contamination, increase health risks, or interfere with remedial measures necessary at the facility; and
(4) there will be a substantial public benefit as a result of the agreed cleanup (i.e., productive reuse of a vacant or abandoned industrial facility).[6]

The DEQ is to consult with local planning districts to determine the future land use plans of the area when considering a proposal for voluntary cleanup.[7] If the DEQ finds that such voluntary actions are feasible, an agreement between the

---

[3]Ch. 662, H.B. No. 3352, approved July 14, 1995, codified at OR. REV STAT. §§ 465.255, 465.325 (1995).

[4]Langdon Marsh, Director, Oregon Department of Environmental Quality, House Transportation Water Resources and Environment, Superfund Reauthorization, Federal Document Clearing House Congressional Testimony (March 5, 1997).

[5]H.B. 3352, § 4(1).

[6]*Id.*

[7]*Id.* § 4(2).

parties may be promulgated. There are two forms of agreement used, and they are depicted in the following table:

Document	Description
Letter of Agreement	Used on majority of sites; quite basic; intended for simple cleanup projects
Voluntary Cleanup Agreement	More detailed; used on complex sites & includes outline of work to be done

The form of agreement used depends in part on how the site is characterized: simple or complex.

Form	Description
Simple	Must have limited soil contamination; Have no groundwater contamination; Involve no sensitive ecological zones; and Contain contaminants for which numerical cleanup targets exist (residential and industrial)
Complex	Any sites that are not classified as "simple"

The voluntary cleanup agreement used in complex sites must include:

(1) a commitment by the participants to perform the remedial measures;
(2) DEQ oversight of the remedial cleanup operations;
(3) a waiver of the right to bring claims against the state that arise over contamination at the site that exist at the date of acquisition, ownership, or operation of the facility;
(4) an irrevocable right of entry given to the DEQ;
(5) reservation of rights to any party not a part of the agreement between the participants and the state; and
(6) a legal description of the party.[8]

In addition, a fee of $5,000 is due upon signing of the agreement to cover oversight costs. Costs in excess of $5,000 are billed monthly.[9]

At the successful completion of the remedial activities, the state absolves the participants from further liability for release of hazardous waste at the site as of the date of acquisition.[10] The participants have the burden of proving that the voluntary cleanup was completed successfully, and that any further releases

[8]Id.
[9]See Charles Bartsch & Elizabeth Collaton, COMING CLEAN FOR ECONOMIC DEVELOPMENT at 112.
[10]Id. § 4 (4).

resulted from activities prior to acquisition of the facility.[11] The agreement between the state and the participants is to be made a permanent part of the deed, and is to be recorded in the county records office.[12] This agreement will run with the land.[13]

Finally, the DEQ has the right to recover any and all costs and expenditures that resulted from such actions, including the costs of program development, rulemaking, and other administrative actions required. These costs may be recovered from any person who is undertaking remedial actions under the DEQ's oversight.[14]

Financing

In 1997, the state legislature added certain financing provisions to the voluntary legislature creating a revolving loan fund within the state treasury named the Brownfields Redevelopment Loan Fund.[15] The fund will be administered by the state economic development department. No more than 40 percent of funds used on a brownfields project can be financed from the fund.[16]

Specialized Stakeholder Laws—Lender and Trustee

Oregon is one of the states that amended its state equivalent of CERCLA to include some specialized lender and trustee liability protection. Under the Oregon mini-Superfund statute, a fiduciary[17] that acquires property through foreclosure will not be liable for remedial action costs incurred by the state or any other person that are attributable to or associated with a facility. In addition there is no liability for damages for injury to or destruction of any natural resources caused by a release. An exception to this general rule of no lender or trustee liability occurs if the fiduciary:

(1) obtained actual knowledge of the release and then failed to promptly notify the department and exercise due care with respect to the hazardous substance concerned, taking into consideration the characteristics of the hazardous substance in light of all relevant facts and circumstances; or

(2) failed to take reasonable precautions against the reasonably foreseeable acts or omissions of a third party and the reasonably foreseeable consequences of such acts or omissions.

---

[11]*Id.*

[12]*Id.* § 4 (5).

[13]*Id.*

[14]*Id.* § 8.

[15]1997 Ore. ALS. 738; 1997 ORE. LAWS 738; 1997 Ore. H.B. 3724.

[16]*Id.*

[17]To qualify, the fiduciary must be "exempted from liability in accordance with rules adopted by the commission under OR. REV. STAT. 465.440."

To establish that the person did or did not have reason to know, the person must have undertaken, at the time of acquisition, all appropriate inquiry into the previous ownership and uses of the property consistent with good commercial or customary practice in an effort to minimize liability.[18]

Thus, under the state law, if the fiduciary fails to take reasonable precautions or has actual knowledge of the release and fails to act according to the law, then the fiduciary will be liable for remedial action costs incurred by the state or any other person that are attributable to or associated with a facility. Similarly, the fiduciary will also be liable for damages for injury to or destruction of any natural resources caused by a release.

Washington

Voluntary Cleanup Program

The state of Washington, similar to the state of New York, has promulgated legislation to promote the encouragement of community-based strategies for the voluntary cleanup of property.[19] Washington's Department of Ecology (DOE) set three programs designed to coordinate cleanup efforts in the state:

Program	Description
Independent Remedial Action Program (IRAP)	Leaves most of the work to be done by private parties on their own; does not provide binding certification, but issues a NFA letter upon completion
Prepayment agreement policy	Interested parties pay the agency up front for oversight costs
Prospective purchaser policy	Allows parties that want to purchase a contaminated site to enter into a cleanup agreement

To enter into the IRAP program a volunteer must pay a flat fee of the greater of $1,000 or 1 percent of the total cleanup cost, with the fee total not allowed to exceed $15,000. To enter into the prepayment or prospective purchaser programs, the volunteer must enter into either an agreed order or consent decree with DOE. These agreements must contain a provision for DOE to recoup all its oversight costs and enforcement provisions for failure to comply with the terms of the agreement.

---

[18]The statute permits private indemnity agreements:

(5)(a) No indemnification, hold harmless, or similar agreement or conveyance shall be effective to transfer from any person who may be liable under this section, to any other person, the liability imposed under this section. Nothing in this section shall bar any agreement to insure, hold harmless or indemnify a party to such agreement for any liability under this section.

[19]1995 Wash. S.B. No. 6261, Washington 1996 Regular Session of the 54th Legislature.

The state has two different forms of certification on completion of site remediation:

Certification	Description
No Further Action (NFA) letter	IRAP sites: sites whose parties entered into an agreed order with DOE; prepayment agreement sites
Covenant Not to Sue	Parties that entered into a consent order with DOE (prospective purchasers)

Both documents contain standard reopeners for fraudulent conduct undertaken by the parties, discovery of contaminants not tested for in site remediation, remedy failures, and adopting of stricter cleanup rules by DOE in the future.

In testimony in Congress, one state official described the program as highly successful:

> The state program also has a voluntary cleanup element that many liable persons take advantage of. This has led to the successful completion of cleanups at a significant number of low priority sites that the state would not have been able to actively address for years due to budget constraints. In fact, over 90 percent of the sites that have been cleaned up under MTCA have been through the voluntary cleanup route. Once clean, these properties have moved into active use, thus avoiding the "brownfield" syndrome so common around the country.
>
> To further address the brownfields problem, the state commonly enters into "prospective purchaser agreements." These agreements allow purchasers of contaminated property to avoid liability under MTCA in exchange for substantial (but not necessarily complete) cleanup of the property in question. These sites are often abandoned industrial properties that would not otherwise be developed because the cost of cleanup exceeds the value of the property. The liability protection provided by these agreements provides incentive to purchasers to buy the property, clean up the worst of the contamination, and return the site to beneficial use.
>
> The state program has been very successful. At over 2,300 sites no further action is required. Most of these sites have been cleaned up voluntarily with little state involvement. We have accomplished this with almost no litigation. In sum, we have created a program that works. It gets sites cleaned up, keeps costs down, and protects human health and the environment.[20]

---

[20]Jay J. Manning, Senior Assistant Attorney General, Washington Attorney General, House Transportation Water Resources and Environment Superfund Reauthorization, Federal Document Clearing House Congressional Testimony (March 5, 1997).

The law provides for state funding of local brownfield revitalization efforts, appropriating $1,000,000 from the budget of the Department of Ecology to be given as grants to local governments for the encouragement of developing and implementing areawide strategies for the cleanup and reuse of industrial lands.[21]

Business Transfer Statute

The Washington Superfund[22] covers nonresidential real property on which a significant release of a hazardous substance has been found to have occurred by the Washington Department of Ecology (the Department).[23] The statute's disclosure requirements are triggered by the sale of the property found to have a significant release by the Department requiring presale notice.[24] The seller must inform the purchaser of any releases of hazardous substances on the property over the past twenty years prior to the sale of the property.[25] In addition, the owner of the property must file a notice in the property records that indicates that a release occurred on the property, the date of the release, and that further information can be obtained from the Department.[26] Unless the parties provide otherwise, the seller is liable for damages to the buyer for failure to inform the purchaser of a release.[27]

---

[21]*Id.* § 2.
[22]*Id.*
[23]*Id.*
[24]*Id.*
[25]*Id.*
[26]*Id.*
[27]*Id.*

CHAPTER 14

Current Federal Brownfields Initiatives

Overview of Federal Efforts

On February 5, 1997, President Bill Clinton announced in the State of the Union Address that, "We should restore contaminated urban land and buildings to productive use."[1] EPA defines *brownfields* as:

> Abandoned, idled, or under-used industrial and commercial facilities where expansion or redevelopment is *complicated by real or perceived* environmental contamination.

In May 1997, EPA announced its renewed Brownfields National Partnership Action Agenda. This program, EPA's Brownfields Economic Redevelopment Initiative, is "designed to empower States, communities, and other stakeholders in economic redevelopment to work together in a timely manner to prevent, assess, safely clean up, any sustainable reuse brownfields." A brownfield is a site, or portion thereof, that has actual or perceived contamination and an active potential for redevelopment or reuse. EPA's brownfields initiative strategies include:

> funding pilot programs and other research efforts,
> clarifying liability issues,
> entering into partnerships,
> conducting outreach activities,
> developing job training programs, and
> addressing environmental justice concerns.

The Pilot Projects

In August 1994, U.S. Environmental Protection Agency administrator Carol Browner announced the federal brownfields initiative. The EPA Brownfields Economic Redevelopment Project conducts studies on how to return old industrial sites to productive use and appropriates grants to qualified communities to help achieve this goal. The federal brownfields initiative serves numerous coextensive purposes:

---

[1] *See* http://www.epa.gov/swe.../html-doc/who51397.htm.

to revitalize urban economies in the Northeast and Midwest through manufacturing and other industries;

to prevent the unnecessary development of virgin land and resources; and

to relieve the increased demand for new manufacturing and industrial infrastructure by reusing and redeveloping brownfields.

EPA explained in a press release:

> The current Superfund law extends liability to both past and prospective own-
> ers of contaminated sites. As a result, the market value of older industrial sites,
> or "brownfields," has declined. To address the effects of these declining prop-
> erty values found in inner-city industrial areas, EPA has announced its Brown-
> fields Action Agenda, which commits funding for 50 pilot projects. Under this
> initiative, EPA will work with stakeholders to address liability barriers and
> encourage economic redevelopment of these industrial facilities.

EPA also announced the availability of limited federal funding (up to
$200,000 in funding to support two-year demonstration projects) for approxi-
mately fifty pilot projects in 1995 and 1996. The goal of the pilot program is to develop a national policy on brownfields development that will serve as guidance to
states and localities struggling with this issue. As such, the pilot programs will
help:

test redevelopment models by removing regulatory barriers without sacrific-
ing environmental protection;

encourage community groups, investors, lenders, developers, and other affected parties to work together to clean up contaminated sites and return
them to productive use;

create models for states and localities struggling with this issue; and,

establish guidance to cities cleaning up and returning contaminated, abandoned property to productive use.

To date, EPA has exceeded the initial number of fifty pilot projects promised.
EPA has awarded more than seventy-five brownfields pilots—each through cooperative agreements of $200,000 over two years. The first three were awarded
shortly after the program was announced in 1995 to Cuyahoga County in Ohio,
Richmond, Virginia, and Bridgeport, Connecticut. The following table depicts the
locations of the other regional and national brownfields pilots:

States by Region	Regional EPA Pilot Projects	National EPA Pilot Projects
Region I		
Connecticut	Naugatuck Valley; New Haven	Bridgeport (original); Burlington
Maine	Portland	

(continued)

Massachusetts	Boston; Sommerville	Chicopee; Lowell; Worcester
New Hampshire	Concord	
Rhode Island		
Vermont		
Region II		
New Jersey	Camden	Trenton; Newark
New York	Buffalo	Rochester; New York; Rome
Puerto Rico		
Region III		
Delaware		
Pennsylvania	Philadelphia; Pittsburgh	Phoenixville
Maryland		Baltimore
Virginia		Cape Charles—Northhampton County; Richmond (original)
West Virginia		
Region IV		
Alabama	Prichard	Birmingham
Florida	Clearwater; Miami	
Kentucky		Louisville
Mississippi		
South Carolina		
North Carolina		Charlotte
Tennessee		Knoxville
Region V		
Illinois	East St. Louis; State of Illinois	West Central Municipal Conference
Indiana	State of Indiana; Northwest Indiana cities	Indianapolis
Michigan	Downriver Community Conference; Kalamazoo	Detroit; Chippewa County/Kinross Township
Minnesota	State of Minnesota	
Ohio	Cincinnati	Cleveland (original); Lima
Wisconsin		
Region VI		
Arkansas		
Louisiana	Shreveport	New Orleans
New Mexico		
Oklahoma		
Texas	Dallas	Laredo; Houston
Region VII		
Iowa		
Kansas		Kansas City
Missouri	Bonne Terre	St. Louis; Kansas City
Nebraska		

(*continued*)

Region VIII
Colorado	Sand Creek Corridor	
Montana		
Utah	Murray City;	
	Provo;	
	Salt Lake City;	
	West Jordan	
Wyoming		

Region IX
Arizona		
California	Oakland;	Emeryville;
	San Francisco	Richmond;
		Sacaramento;
		Stockton
Guam		
Hawaii		
Nevada		

Region X
Alaska		
Idaho	Panhandle Health District	
Oregon		Oregon Mills Sites—Astoria;
		Coquille;
		Grants Pass;
		Merlin;
		Myrtle Creek;
		Philomath;
		Sweet Home
Washington	Bellingham;	Tacoma
	Duwamish;	
	Puyallup Tribe of Tacoma	

"[T]his is a laudable effort but it is a long way from addressing the over half million Brownfield sites in America."[2]

Taken as a group, the seventy-five pilots, along with several other initiatives, will gain for EPA the local-level experience that will enable the agency to help revitalize other neglected areas around the country. EPA has been charged to "learn a great deal about how we can help more urban communities."[3] The agency will use the results of these initiatives to develop a federal strategy for coping with urban brownfields from a federal perspective.

---

[2]Chairman Sherwood Boehlert, "Reforming Superfund: Opening Statement," Congressman, House Transportation Water Resources and Environment Superfund Reauthorization (March 12, 1997).
[3]60 Fed. Reg. 28610 (June 1, 1995).

The National Environmental Justice Advisory Council's (NEJAC) Waste and Facility Siting Subcommittee and the U.S. EPA Public Dialogues on Urban Revitalization and Brownfields: Envisioning Healthy and Sustainable Communities

The NEJAC Waste and Facility Siting Subcommittee and the U.S. EPA convened a series of public dialogues on environmental justice issues related to urban revitalization and brownfields. The NEJAC is a body of nonfederal experts established to advise EPA on environmental justice issues. The NEJAC recognizes the compelling need to address the environmental justice concerns that are pivotal to issues of urban revitalization and creating healthy and sustainable communities. The NEJAC has identified brownfields cleanup and redevelopment as potentially contributing to the urban revitalization that is occurring within the environmental justice communities.

The public dialogues, titled, "Urban Revitalization and Brownfields: Envisioning Healthy and Sustainable Communities" took place throughout the country during June and July 1995. These dialogues were designed to provide an opportunity for grassroots environmental justice activists and residents of impacted communities to articulate their aspirations, concerns, and recommendations for developing healthy and sustainable urban communities—including brownfields redevelopment. Upon completion of the dialogues, the NEJAC issued a formal response to the EPA Brownfields Initiative Action Agenda. "EPA has committed that the dialogues will have a demonstrable role in shaping its brownfields initiatives."[4]

Interagency Working Group

After receipt of the NEJAC report, EPA convened an Interagency Working Group on Brownfields, established in July 1996 as a forum for federal agencies to exchange information on brownfields-related activities and to develop a coordinated national agenda for addressing brownfields.

Item	Explanation
The need for a federal Interagency Working Group	Communities with brownfields sites often face other concerns such as unemployment, substandard housing, outdated or faulty public infrastructure, crime, and a poorly skilled local work force. Although federal and state programs may be in place to address these local issues, too often the programs operate in isolation; the brownfields initiative provides an opportunity for federal agencies to work together in a more integrated fashion toward sustainable redevelopment.

(continued)

---

[4]EPA press release (May 1995).

	The Interagency Working Group makes the brownfields initiative more effective through the coordination of agency efforts and resources.
	Representing a wide range of brownfields stakeholders, the collaboration of diverse backgrounds, expertise, and experience of each agency helps make the initiative a success.
Coordination promotes more efficient government	Through the brownfields initiative, communities can leverage a myriad of public and private sources of capital and technical support that can ensure successful redevelopment.
	Brownfields assessment and cleanup activities can be linked to health and work force development programs through the creation of temporary and permanent jobs.
	Brownfields projects can be coordinated with transportation planning, ensuring access to transportation for new workers in redeveloped areas.
	Reuse options can include not only new economic and industrial opportunity, but development of urban agriculture and green spaces.
	Close cooperation from the beginning of a brownfields pilot may also decrease the likelihood that agencies will duplicate efforts, work at cross purposes to each other, or confuse community leaders and civic groups.

The Interagency Working Group has developed a brownfields strategy that will link more effectively environmental protection with economic development and community revitalization programs and guide the brownfields initiative into the future. The Action Agenda includes more than 100 commitments from the members of the Interagency Working Group. Currently, the following federal agencies are participating in the Interagency Working Group:

Department of Agriculture (USDA)
Department of Commerce (DOC)
Department of Defense (DOD)
Department of Education (ED)
Department of Energy (DOE)
Department of Health and Human Services (HHS)
Department of Housing and Urban Development (HLTD)
Department of the Interior (DOI)
Department of Justice (DOJ)
Department of Labor (DOL)
Department of Transportation (DOT)
Department of the Treasury (Treasury)
Department of Veterans Affairs (VA)
Environmental Protection Agency (EPA)
Federal Deposit Insurance Corporation (FDIC)
General Services Administration (GSA)
Small Business Administration (SBA)

The National Action Agenda is grouped into four phases of the brownfields process for convenience of viewing the information. These phases are:

> community planning,
> assessment and cleanup,
> redevelopment support, and
> sustainable reuse.

The following table describes the contribution that certain federal agencies and national organizations have committed to making concerning the community planning phase of the federal brownfields process.

Agency/ Organization	Community Planning Activity
USDA	Will distribute information and provide guidance to field offices and other partners on EPA brownfields pilots.
	Will include brownfields in planning for forest land management, rural community economic revitalization, and state Strategic Development Plans.
DOC/Economic Development Administration (EDA)	Will coordinate with EPA in achieving an interagency agreement to establish a full-time brownfields coordinator.
	Will update policies to promote identification and redevelopment of brownfields by state and local economic development planning agencies and organizations eligible for EDA assistance.
DOD	Will coordinate economic adjustment initiatives in defense-impacted communities with brownfields redevelopment planning efforts.
DOD/Corps of Engineers (COE)	In cooperation with states—will prepare comprehensive plans for environmental restoration of water resources at brownfields pilots.
ED (Department of Education)	Will share information about brownfields redevelopment goals with state and local education programs, educational organizations, and institutions.
	Will encourage local partnerships at the brownfields pilot communities, which help to focus on education needs in areas such as literacy, vocational and technical training, school-to-career experiences, out-of-school youth, and civic participation.
EPA	Will conduct a brownfields inventory study in three states to quantify the number of brownfields properties.
	Will train 3,500 local emergency planning committees about the brownfields initiative.
	Will provide strategic and project-level support to the Governor's Commission on Sustainable South Florida for recovering brownfields for urban revitalization in southeast Florida.
GSA	Will identify federal brownfields properties in twelve cities and match related community efforts for redevelopment.
HHS agencies	Will issue a concept paper on public health issues as they relate to brownfields sites and redevelopment.

(continued)

HHS/ATSDR and NACCHO (National Association of City/ County Health officials)	Will convene a series of public dialogues with local health officials to educate and discuss brownfields issues. Will initiate and develop relationships with local public health contacts in every pilot community.
HHS/NIEHS	Will link its environmental justice grantees, EPA and NIEHS minority worker training grantees, and EPA, DOE, and NIEHS hazardous waste worker grantees with brownfields pilots to increase communication and collaboration.
HLJD	Will sponsor a policy roundtable to promote accessibility of public and private sources of capital for brownfields cleanup and redevelopment.
HM	Will review and revise Community Development Block Grants regulations to facilitate use for brownfields.
DOT	Will clarify that brownfields cleanups are an eligible activity for funding in transportation projects. Will ensure that brownfields are addressed in transportation planning by developing or modifying existing guidance and by offering technical assistance and training to metropolitan planning organizations, states, and local governments.
Treasury	Will provide information to the financial community on brownfields related to lender liability and Community Reinvestment Act issues.
ICMA (International City/County Managers Association)	Will distribute resource tool kit on brownfields to local governments.
AIO (Americans for Indian Opportunity)	Will disseminate information and provide technical assistance to federally recognized tribes on brownfields and federal pilot application process.
NGA (National Governors' Association)	Will continue research in brownfields redevelopment, provide information to governors' staffs, hold brownfields workshops, and publish an issue brief for their membership on brownfields revitalization.
ASTSWMO (Association of State and Territorial Solid Waste Management Officials)	Will continue to provide information to state waste managers and hold forums on voluntary cleanup and brownfields issues.
NACCHO	Will train local health agencies on brownfields-related health issues in pilot communities.

The following table describes the contribution that certain federal agencies and national organizations have committed to making concerning the assessment and cleanup phase of the federal brownfields process.

Agency/ Organization	Assessment and Cleanup Activities
USDA	Will provide technical advice on urban and community forestry and water quality to pilot communities. Will conduct studies of bioremediation technologies for use at brownfields pilots. Will revise existing policies and regulations to ensure brownfields cleanup and restoration can be part of development projects.

(*continued*)

DOC/EDA	Will provide technical assistance to EPA on the development of its Revolving Loan Funds at brownfields pilots.
DOC/NOAA	Will assist in the identification, evaluation and assessments of brownfields in coastal areas.
DOD	Will share its Base Realignment and Closure (BRAC) partnering, assessment, and cleanup tools with brownfields pilots.
	Will resolve problems associated with transfers of contaminated property between federal agencies.
DOD/COE	Will review and revise internal civil works policies and guidance so as to support brownfields cleanup and restoration.
	Will review inventory of Formerly Used Defense Sites (FUDS) and make inventory information available to brownfields pilots.
	Will provide planning assistance to pilot communities.
	DOD/COE laboratories will share their research and technology with brownfields pilots to expedite and improve local assessment and cleanup efforts.
	Will share its "area economic data" with brownfields pilots.
DOE	Will provide $240,000 in funding to assist in the transfer of DOE characterization, cleanup, and other relevant technologies to brownfields.
DOL	Will inform state and local Job Training Partnership Act stakeholders about the brownfields initiative and related job training activities to enhance local collaboration.
DOT	Will encourage transportation providers to partner with EPA brownfields pilots to leverage cleanup and development on transportation-related projects.
EPA	Will set aside $100 million to fund additional brownfields site assessment and cleanup at brownfields pilots.
	EPA's current budget (fiscal year 1997) will fund assessment pilots, revolving loan funds for cleanups in current pilots, and state infrastructure to support voluntary cleanup programs.
	Will encourage and support state Voluntary Cleanup Programs (VCP) by providing $25 million in financial assistance.
	Will hold focus group(s) with stakeholders to determine views on principles for what constitutes adequate state VCPs.
	Will provide information on presumptive cleanup remedies used by EPA to communities.
	Will clarify the applicability of RCRA underground storage tanks and facilities requirements to brownfields assessment and cleanup activities.
	Will examine its site assessment program and explore alternative approaches to make the process more relevant to brownfields.
	Will incorporate brownfields information into national guidance on source drinking water protection to states, communities, and other stakeholders.
	Will conduct a pilot demonstration on portable site characterization and remedies.
	Will delete at least 3,000 additional sites from Superfund inventory of potential hazardous sites.
	Will issue tool kit on "Innovative Technologies at Brownfields" for brownfields pilot project managers.
	Will publish final guidance on state voluntary cleanup programs.
	Will develop a technical guidance document to aid states, communities, and private parties to select technologies to characterize and cleanup brownfields.

(continued)

	Will develop a generic sampling and quality assurance plan for brownfields pilots.
	Will continue its efforts in negotiating agreements with prospective purchasers of contaminated property.
	Will explore and support the use of alternative dispute resolution (ADR) and public participation techniques to expedite cleanup and reuse of brownfields properties.
EPA's Dallas office	Will pilot a brownfields Information Data Base to geographically depict brownfields.
GSA	Will provide $1 million to fund environmental assessments on federal properties to expedite potential brownfields redevelopment.
HHS/ATSDR and NACCHO	Will train local health agencies on brownfields-related health issues in pilot communities.
HHS/ATSDR	Will provide public health evaluations and consultations to address public health concerns in all EPA brownfields pilot communities, through cooperative agreements with state health departments in some cases.
HHS/NIEHS	Will increase communication and collaboration among brownfields pilots, seven minority worker training program grantees, twenty EPA worker training grantees, and Environmental Justice Partnership to strengthen all four programs.
	Will link its basic research programs on hazardous substances, exposure assessments, and remediation technologies to all pilot projects.
HLTD	Will earmark $4 million from its lead-based paint hazard control grant program to respond to residential lead-based paint hazards on or near brownfields pilots.
Veterans Affairs (VA)	Will work to provide a significant number of job-ready homeless and disabled veterans from VA's Compensated Work Therapy and Vocational Rehabilitation programs for both cleanup and redevelopment opportunities at brownfields pilots.
ASTSWMO	Will conduct a comprehensive analysis of state programs designed to identify, assess, clean up, and redevelop brownfields, and define how to integrate state/federal efforts.
The Institute for Responsible Management (IRM)	Will develop a "progress" tracking matrix for brownfields pilots.
Hazardous Materials Training and Research Institute (HMTRI)	Will work with community colleges to develop environmental job training programs and expand this effort to at least 100 pilot communities.

The following table describes the contribution that certain federal agencies and national organizations have committed to making concerning the redevelopment phase of the federal brownfields process.

Agency/ Organization	Redevelopment Support
USDA	Will identify rural Empowerment Zone/Enterprise Communities (EZ/ECs) that contain brownfields pilots and coordinate federal, state, and local redevelopment efforts.

(continued)

	Will award preference points to EZ/EC applications that include brownfields pilot sites, and encourage EZ/ECs to attract businesses that do environmental cleanup.
DOC/NOAA	With EPA, will implement the agency's MOU to increase coordination to assess, clean up, and redevelop brownfields in coastal and waterfront areas.
	Will provide funding to coastal states for brownfields redevelopment as part of waterfront revitalization efforts.
	Will explore policy on liability release by natural resource trustee agencies on potential brownfields sites.
DOC/EDA	And EPA will update Memoranda of Understanding to coordinate brownfields project selection.
	Field staff will work with brownfields pilots to coordinate cleanup and redevelopment.
	Will identify brownfields pilots as EDA strategic priorities.
DOD	Will develop model guidelines to streamline the early transfer of federal facilities.
	Will develop policy options to encourage reuse of brownfields on DOD property rather than greenfields.
DOD/COE	Will share GIS and other data on brownfields pilots—within its project study areas—with pilots.
	Will provide market impact studies and cost benefit analysis to brownfields pilots.
DOI	Will fund two to three community based nonprofits to test long-term sustainability of brownfields pilots.
DOT	Will examine existing policies and guidance, currently requiring avoidance of contaminated properties, and revise as appropriate.
	With EPA, will explore issues related to liability and transportation projects.
	Will provide major transportation industry associations with information on the brownfields initiative.
	Will distribute information to field offices, metropolitan planning organizations, and state transportation agencies on the brownfields initiative and EPA brownfields pilots.
EPA	Will implement "comfort letter guidance" designed to encourage developers and lenders to invest in brownfields properties.
	Will conduct outreach to the insurance industry to promote access to brownfields insurance.
	Will implement new lender liability statute.
	EPA's Environmental Finance Advisory Board will continue research on financing of brownfields cleanup and redevelopment.
	Will facilitate the creation of Baltimore's Fairfield Ecological/Industrial Park through an XL Communities Pilot Project.
	Will explore mechanisms to specifically address underground storage tank liability issues (e.g., comfort letters and prospective purchaser agreements).
GSA and EPA	Will sign MOU fostering cooperation between brownfields and idle federal properties.
HHS/OCS	Will provide $500,000 to community development corporations and community action agencies for restoration of the physical environment, economic revitalization, and job training activities at brownfields pilots.

(continued)

HUD	Will revise Community Development Block Grant regulations to encourage use of funds for brownfields.
	Will fund a brownfields job training demonstration project in a HUD priority economic redevelopment area.
	Will request $25 million of Economic Development Initiative funds annually for brownfields redevelopment.
	Will provide $155 million in support of local brownfields cleanup and redevelopment activities through its Community Development Block Grant (CDBG), Economic Development Initiative (EDI), Housing and EZ/EC programs (supporting $165 million in section 108 loan guarantees).
Department of Treasury	Will work to ensure passage of the administration's proposed brownfields tax incentive.
	Will help educate the financial community on lender liability and the Community Reinvestment Act.
Treasury/Office of the Comptroller of the Currency	Will issue an advisory letter to review the liability limits for banks as nonmanaging financiers of the brownfields sites and the applicability of the Community Reinvestment Act to activities relating to brownfields sites.
	Will develop a fact sheet for national banks and examiners on brownfields redevelopment initiatives.
SBA	Will distribute brownfields information to Small Business Development Corporation networks.
U.S. Conference of Mayors	Will conduct roundtables on brownfields redevelopment issues.
NALGEP (National Association of Local Government Environmental Professionals)	Will convene work groups, composed of local officials, that will research and advance solutions to regulatory, environmental, and financial barriers to brownfields redevelopment.

The following table describes the contribution that certain federal agencies and national organizations have committed to making concerning the sustainable reuse phase of the federal brownfields process.

Agency/ Organization	Sustainable Reuse
USDA	Will help develop, or strengthen, Urban Resources Partnership efforts in three brownfields communities, forging stronger linkages among economic development, blighted property, community initiatives and natural resources.
	Will conduct studies on the economic impacts of revitalizing brownfields.
DOC/NOAA	Will provide technical assistance to state and local governments to give coastal decision makers resources to facilitate the redevelopment of brownfields.
	Will document successful brownfields waterfront revitalization approaches.
DOD/COE	Will provide appraisal, title, and deed restriction services to brownfields pilots.

(continued)

	Will carry out projects for the protection, restoration, and creation of aquatic and ecologically related habitats in connection with the disposal of dredged materials at brownfields pilots.
DOE	Will provide initial funding of $75,000 to local government organizations to sponsor a minimum of five peer exchange meetings on issues surrounding reuse of properties.
	Will publish one or more additional papers on property reuse, and conduct three or more workshops and video conferences to share lessons learned.
	Will investigate beneficial reuse potential at one or more Formerly Utilized Sites Remedial Action Program (FUSRAP) sites.
DOI/Park Service	With EPA, Memorandum of Understanding will link brownfields pilots activities with Park Service activities to create more attractive and sustainable communities.
	DOI/Park Service federal Lands-to-Parks program will coordinate its development efforts with brownfields pilot sites.
	DOI/Park Service National Heritage Corridors will coordinate its efforts with brownfields pilot sites to identify brownfields property suitable for cleanup and reuse.
DOT	Will include brownfields as one focus area in its $4.2 million in fiscal year 1998 ($25 million over five years) pilot research program on transportation, land use, and sustainability issues.
EPA	Will work with two states to conduct Permits Improvement Team Pilots at two brownfields pilots.
	Will maintain the Smart Growth Network to promote mechanisms and strategies for urban landfill and revitalization developed by multi-stakeholders.
	EPA/Office of International Activities will work with multilateral organizations (e.g., Organization for Economic Cooperation and Development [OECD] and ICMA) to share unique international approaches and experiences with brownfields pilots.
HHS agencies	Will assess and respond to child care needs.
HUD	With EPA, will issue Joint Study on Redevelopment of Brownfields.
	Will fund a job-training demonstration project in a HUD priority economic redevelopment area.
The Trust for Public Land (TPL) and EPA	Will, in accordance with their MOU, bring TPL's expertise in securing public green spaces to brownfields pilot communities. Will select three cities for initial collaboration.

Memoranda of Understanding (MOUs)

Agencies use Memoranda of Understanding (MOUs) to establish policies and procedures between agencies and support projects of mutual interest. EPA currently has five MOUs, one with each of the following agencies:

the Department of the Interior's Rivers, Trails, and Conservation Assistance Program (RTCA);
the Department of Housing and Urban Development;
the Department of Labor's Employment Training Administration (ETA);
the Department of Commerce's Economic Development Administration (EDA); and,
the National Oceanic and Atmospheric Administration (NOAA).

The following table lists other cooperative efforts between various federal agenices to support brownfields redevelopment.

Goal	Cooperative Agencies
Linking green spaces and brownfields	DOI and USDA support citizen groups and local and state governments to help ensure that redevelopment includes green spaces in areas that suffered from urban blight.
Tapping redevelopment resources	HUD, EDA, NOAA, and DOT are integrating brownfields into their planning processes, ensuring that brownfields cleanup and redevelopment are eligible expenses for their project funds. They are removing internal barriers and encouraging their field offices to work with communities on brownfields projects.
Creating jobs around brownfields through joint partnerships	DOL, ED, HHS, and VA are connecting their missions with brownfields activities to ensure sustainable communities. They help create and maintain job opportunities and provide the link between environmental cleanup and work force development.
Coordinating federal lands with brownfields	GSA, DOD, DOE, USDA, and DOI are responsible for managing vast tracts of property in this country. These agencies are taking actions to use brownfields principles to ease the reuse of federal properties benefiting the communities and the environment as well as furthering the mission of the responsible agencies.
Offering technical support to local cleanup efforts	EPA, GSA, and the U.S. Army Corps of Engineers offer their technical services to communities to assess and clean up property. Their specialized skills further support brownfields redevelopment.
Providing certainty to brownfields redevelopment	EPA and DOJ work closely with the states to clarify regulatory oversight roles and offer a variety of tools to assist brownfields redevelopment. This work helps to provide buyers, sellers, investors, and communities with the certainty they need to redevelop brownfields properties.
Financing brownfields	Financial agencies such as Treasury, FDIC, and SBA play a key role in the brownfields arena. Building on the president's tax incentive proposal, agencies that deal with financial policy are at the forefront of coming up with innovative ways to provide incentives and support for investing in the cleanup and reuse of brownfields.

Comprehensive Environmental Response, Compensation, and Liability Information System (CERCLIS) Site Deletion

In a demonstration of its commitment toward this initiative, EPA regional offices in partnership with the states have assessed and determined that about 25,000 to 38,000 potential hazardous waste sites do not require federal involvement and should be removed from the Superfund Inventory. Some sites may have received a No Further Action designation because state or local authorities had assumed enforcement responsibility. On February 15, 1995, EPA announced that it removed approximately 25,000 sites from the Site Tracking System known as CERCLIS. These sites were screened out of active investigation and designated as No Further Remedial Action Planned or NFRAP. This is the first time EPA has pruned the list of sites determined not to warrant Superfund cleanup.

Since all lenders check the CERCLIS as part of a Phase I environmental assessment, even these "No Further Action" sites were stigmatized, severely limiting financing opportunities and blocking most redevelopment activities. Pruning the CERCLIS may be only partially effective in reducing brownfields anxiety, because in many cases EPA has delayed final determinations of No Further Action while focusing its attention on remedial action sites. Although in the future EPA may be willing to clear more sites in response to specific inquiries, deletion of these sites from CERCLIS is not a guarantee that such sites are uncontaminated. Nor does it protect lenders or current property owners from third-party suits for cleanup of any contamination on site. For example, Cornell University successfully negotiated with EPA to be deleted from the CERCLIS but was later sued—and held liable for contribution costs—by Alcan, a private third-party plaintiff.[5]

Community Reinvestment Act (CRA)

The administration recently promulgated agency guidance for EPA dealing with lenders and prospective purchasers. On May 4, 1995, new regulations were amended to provide an incentive for bankers and developers to help rescue cities with polluted industrial properties. "For the first time, lenders subject to the federal Community Reinvestment Act—aimed at directing capital into poor neighborhoods—can claim CRA credit for loans made to help clean up and redevelop urban, industrial property." The new rule, orchestrated by EPA administrator Carol Browner, was designed to complement the EPA's new Brownfields Action Agenda.

Under this program, EPA is one of many federal agencies committing resources to economic revitalization of urban areas. EPA coordinates its efforts with the U.S. Department of Housing and Urban Development and the Department of Agriculture on designed "empowerment zone programs." The programs are designed to demonstrate that sustainable development is possible by linking redevelopment and job creation to environmental improvement and responsible stewardship.

EPA has not only worked to advise the two awarding departments about the environmental considerations in the applications but will also work with those communities that win Empowerment Zone or Enterprise Community (EZ/EC) status to help ensure the environmental sustainability of their activities. EPA explained:

> We are confident that the brownfields initiative and Superfund reforms will lower many of the barriers to reclamation and redevelopment of urban lands by eliminating disincentives to buying and selling well-remediated properties; by facilitating voluntary cleanups to put land back into use; and by introducing procedures that will lead more quickly to cleanups more appropriate for the site and its future use.

---

[5]*See* United States v. Alcan Aluminum Corp., 964 F.2d 252 (3d Cir. 1992).

EPA is charged with protecting human health and the environment. That does not mean we need to be the spoiler of the redevelopment of abandoned industrial property. After all, land redevelopment is one of the most important forms of recycling around. Brownfield redevelopment should not only lead to cleaned up inner city properties and preservation of our remaining green spaces, but should also stimulate local economic growth and thereby help reverse the cycle of decay that endangers our urban cores.[6]

Revised Prospective Purchaser Agreement Guidance

On May 25, 1995, EPA released new criteria that will govern consideration of candidates for a Prospective Purchaser Agreement.

Elements Dictated by the May 25, 1995, Guidance	Detailed Description
Enforcement action is anticipated by EPA	EPA will enter into a settlement agreement at sites where any form of federal involvement has occurred or is expected to occur and where there is a realistic probability of incurring Superfund expense. EPA intends to substantially expand the universe of sites qualifying for prospective purchaser agreements to include sites that have not yet been the subject of formal site investigation and remediation.
Direct cleanup benefit or indirect economic benefit will occur	EPA must now consider prospective purchaser agreements both where substantial direct benefit will occur to EPA in the form of a cleanup commitment or funds for cleanup, or alternatively, where a lesser indirect benefit may be conferred on the affected community in terms of cleanup, creation of jobs, and development of otherwise unusable property.
Risk posed to persons likely to be present at the site	EPA must now consider potential risks of exposure and adverse health effects for employees and the public.
Prospective purchaser is financially viable	EPA must consider whether it can recoup its costs in the event that the prospective purchaser breaches its obligations under the agreement.

The new guidance supersedes the agency's 1989 guidance, which stipulated that EPA was to enter into agreements only where enforcement action was anticipated and where the new owner provides the government with direct, substantial benefits. The new guidance is intended:

to provide EPA greater flexibility in offering Covenants Not to Sue to parties interested in buying contaminated land for cleanup or redevelopment;
to ease these earlier restrictions, alleviate financial pressures for the individual parties, and enhance incentives to clean up contaminated sites;

---

[6]Testimony of Timothy Fields, Jr., Deputy Assistant Administrator, U.S. Environmental Protection Agency, Before the U.S. House of Representatives, Committee on Commerce, Subcommittee on Commerce, Trade, and Hazardous Material (June 15, 1995).

to allow more frequent application of the prospective purchaser agreements; to allow EPA into the agreements where any agency activity is anticipated or under way.

to allow EPA to accept less direct benefit from the parties, provided that there is otherwise substantial benefit (which can include cleanup or response costs, or other activities considered to be justifiable benefits to the community, such as improving the quality of life or expanding employment opportunities).

EPA included a model Prospective Purchasers Agreement with the May 25 Guidance. This model agreement provides for a Covenant Not to Sue, release, and contribution protection with respect to subsequent third-party claims (such as those from prior landowners or adjoining property owners). The agreement would serve to release not only claims by EPA, but related claims by any other federal authority (including claims under other environmental laws such as those pertaining to clean water, clean air, and disposition of hazardous waste). The agreement would not, however, seek to provide a release from site conditions "unknown" at the time the agreement was signed. Nor would the protection of the agreement be assignable to other parties.

The following table depicts the prospective purchaser agreements negotiated to date.

States by Region	Federal Prospective Purchaser Agreements
Region I	
Connecticut	—
Maine	—
Massachusetts	Industri-Plex
New Hampshire	—
Rhode Island	Health-Tex
Vermont	—
Region II	—
New Jersey	—
New York	Warwick Landfill
Puerto Rico	—
Region III	
Delaware	—
Pennsylvania	Merit Products Site;
	Key West Connection;
	Commodore; Croydon TCE
Maryland	Mid-Atlantic Wood Preservers
Virginia	—
West Virginia	—
Region IV	
Alabama	—
Florida	Florida Steel Indiana Mill Town
Kentucky	—
Mississippi	—

(*continued*)

South Carolina	—
North Carolina	Sayles-Biltmore
Tennessee	Tennessee Chemical Company
Region V	
Illinois	D.C. Franche;
	Southeast Rockford
Indiana	Continental Steel
Michigan	Bofors Nobel, Inc.
Minnesota	—
Ohio	Vision Properties
Wisconsin	—
Region VI	
Arkansas	—
Louisiana	—
New Mexico	—
Oklahoma	—
Texas	Sol Lynn
Region VII	
Iowa	—
Kansas	Kansas City Structural Steel
Missouri	Jasper County
Nebraska	—
Region VIII	
Colorado	Denver Radium Site
Montana	Anaconda Smelter Site
Utah	—
Wyoming	—
Region IX	
Arizona	Indian Bend Wash
California	Sargent Fletcher, Inc;
	San Gabriel Valley
Guam	—
Hawaii	—
Nevada	—
Region X	
Alaska	—
Idaho	—
Oregon	—
Washington	Commencement Bay Site

Final Policy Toward Owners of Property
Containing Contaminated Aquifers

On May 25, 1995, EPA released guidance entitled Final Policy Towards Owners of Property Containing Contaminated Aquifers. This guidance was designed for owners of uncontaminated properties situated above groundwater systems that have been contaminated by sources outside the property.[7] The guidance was based,

---

[7]Neither the prospective purchasers or aquifer owner guidances had Office of Solid Waste and Emergency Response (OSWER) directive numbers printed on them.

in part, on the success of cooperative brownfields cleanup projects in Wichita, Kansas.

Where uncontaminated property may have been contaminated by subsurface groundwater migration in a nondrinking water source, the EPA guidance directs the enforcement staff not to take action against owners provided that the owner has not caused contamination or made the contamination worse. Such contribution could be through the handling and disposition of the same or similar substances at the property. This policy is based on EPA's enforcement discretion and is, of course, in no way binding on third parties that may want to sue for cleanup of subsurface contamination.

In order to be eligible for this program, EPA officials must be satisfied that the person who caused the release of the hazardous substance that contaminated the groundwater system is not an agent, employee, or party connected by contractual relationship or otherwise related to the owner whose liability is to be resolved. The guidance permits EPA to consider affording contribution protection to qualified owners by entering into a "de minimis settlement agreement." It is, however, discretionary on the part of the agency and beyond the control of the party seeking to take advantage of the guidance. According to EPA administrator Carol Browner, even with the administrative guidance in place, "we still need legislative reform of the superfund law to ensure that it makes economic and environmental sense for our cities, small businesses and communities."[8]

State Deferral Guidance

EPA released a state deferral guidance designed to speed site cleanups while increasing the involvement of states, Indian tribes, and local communities.[9] This guidance allows EPA to defer from adding sites to Superfund's NPL if: (1) states oversee cleanup activities or (2) cleanups will be supervised by the state and conducted and paid for by PRPs. The deferral program is designed to speed cleanups by encouraging PRPs to settle early by allowing them to avoid listing on the NPL.

To be eligible for state or tribal deferral, sites must be on the CERCLIS, but not yet listed on the NPL. The sites must have PRPs who agree to pay for and conduct cleanups. Cleanups must be as protective as other Superfund cleanups and comply with all applicable federal and state requirements. In addition, deferral of the site cleanup to the state must have the support of the affected community. Federal facilities are not eligible for deferral.

The state deferral policy differs from the "state empowerment" proposal also being considered by EPA under the administration's "reinventing government initiatives." The empowerment proposal would require legislative changes before

---

[8]*State Role, Liability, Cleanup Guidance among Major Reforms Announced by EPA,* BNA STATE ENVIRONMENT DAILY (May 30, 1995).

[9]OSWER Directive No. 9375.6-11 (May 3, 1995).

implementation is possible. It would allow states to run the Superfund program within their jurisdictions. The deferral policy needs no legislative change for implementation. It can be implemented under current law under EPA's enforcement discretion.

Land Use Directive

Another of the reform initiatives issued on May 25, 1995, is EPA's land use directive,[10] which sets forth EPA's policy regarding consideration of future site usage into account when developing a cleanup plan. Under the guidance document, discussions regarding future land use should take place as early as possible in the investigation and study phases of cleanup. Residents as well as local planning authorities and officials should be consulted, as disagreements over future land use often exist at sites.

The land use policy would allow for less stringent cleanups in cases where a site has been and would probably remain industrial. According to the directive, "EPA has been criticized for too often assuming future land use will be residential."[11] Some of the factors listed in the document to take into account when deciding future land use include:

current land use,
zoning decisions,
community master plans,
accessibility to existing infrastructure,
site location in relation to other areas,
development patterns,
cultural factors,
environmental justice issues, and
proximity to wetlands, critical habitats, and other sensitive environmental
    areas.

The guidance allows certain portions of a site to be cleaned up to meet certain standards to the future land use, but other parts may require deed restrictions and containment of contamination. In addition, because of cost or practicality issues, some anticipated land uses may not be possible, and a more restricted use must be implemented.

If the cleanup implemented is based on less than unrestricted land use (a residential or eatable dirt standard), the guidance calls for institutional controls to be included in the site cleanup and maintenance plan. Accordingly, in cases where waste is left on-site at levels that would require limited use and restricted exposure, EPA will conduct reviews at least every five years to monitor the site for

[10]OSWER Directive No. 9355.7-04 (May 25, 1995).
[11]*Id.*

changes. If landowners or others decide in the future to change the land use in a way that would require additional cleanup, EPA would retain the authority to take further response action to ensure protectiveness. EPA would not expect, however, to become actively involved in those cleanups.

Current Regulatory Relief for Lenders and Prospective Purchasers

The Clinton Administration urged Congress to revise CERCLA to include statutory protections for lenders and prospective purchasers. As one member of the Clinton Administration testified:

> EPA's brownfields initiative is dedicated to the economic renewal and redevelopment of communities with contaminated properties. Far too often, sites which once provided the heartbeat of economic vitality to thriving downtowns are being abandoned for fear of the contamination which might lurk in the soils or the groundwater. These "brownfields," or urban wastelands, have been shunned by new industry and prospective developers who are understandably afraid they might inherit expensive cleanup liabilities for contamination they did not create.
>
> EPA's Brownfields Economic Redevelopment Initiative is designed to turn that around, to encourage economic redevelopment through environmental cleanup. As Industrial enterprises have left urban centers, joblessness has increased along with the number of abandoned, contaminated properties, Both are dangerous to our communities. Our Brownfields effort is designed to help communities in their efforts to eliminate potential health risks, improve standards of living, and help restore economic vitality to areas where these sites exist.[12]

Federal regulators recognize that legislative changes are necessary to make brownfield redevelopment programs successful. Presently, these are severe statutory limits to EPA's authority to regulate in this area. For example, Lois Schiffer, assistant attorney general, Environment and Natural Resources Division, Department of Justice, recently testified:

> A statutory exemption for bona fide prospective purchasers, in addition to the statutory protection of lenders described above, would take care of this "brownfields" problem. However, in order to prevent speculators from receiving a windfall when the government cleans up a property at its expense and thereby makes the property more valuable, the prospective purchaser

---

[12]Prepared Statement of Carol M. Browner, Administrator, U.S. Environmental Protection Agency, before the House Committee on Transportation and Infrastructure Water, Resources and Environment Subcommittee (June 27, 1995).

exemption should be accompanied by a "windfall lien" provision that would use profits resulting from the cleanup to pay back the costs of cleanup.[13]

History of the Problem

In 1992, EPA issued its Final Rule on Lender Liability under CERCLA ("CERCLA Lender Liability Rule" or "Rule").[14] In addition to addressing lender liability, the Rule clarified the language of Section 101(20)(D) of CERCLA,[15] which provides an exemption from the definition of "owner or operator" for certain government entities that involuntarily acquire property, and Section 101(35)(A) of CERCLA,[16] which pertains to the "third-party" defense potentially available to government entities that involuntarily acquire property. The Rule was codified in the Code of Federal Regulations (CFR).[17]

In 1994, the U.S. Court of Appeals for the District of Columbia Circuit vacated the Rule.[18] Consequently, in 1995, EPA removed the Rule from the CFR.[19]

On September 30, 1996, the Asset Conservation, Lender Liability, and Deposit Insurance Protection Act (Asset Conservation Act or Act)[20] was enacted. This law amended CERCLA Section 107 by adding lender liability protection. The law also acted to reinstate regulations pertaining to liability under CERCLA for the involuntary acquisition of property by governmental entities.[21]

---

[13]Testimony of Lois J. Schiffer, Assistant Attorney General, Environment and Natural Resources Division, Department of Justice, House Transportation Water Resources and Environment Superfund Reauthorization (June 27, 1995).

[14]57 Fed. Reg. 18344 (April 29, 1992).

[15]42 U.S.C. 9601(20)(D).

[16]42 U.S.C. 9601(35)(A).

[17]The Rule was codified at 40 CFR 300.1100 and 300.1105.

[18]Kelley v. EPA, 15 F.3d 1100 (D.C. Cir. 1994), *reh'g denied,* 25 F.3d 1088 (D.C. Cir. 1994), *cert. denied,* American Bankers Ass'n v. Kelley, 115 S. Ct. 900 (1995).

[19]"Final Rule on Removal of Legally Obsolete Rules," 60 FR 33912, 33913 (June 29, 1995).

[20]110 Stat. 3009-462 (1996).

[21]*See also* Memorandum from Jerry Clifford to Gail C. Ginsberg Re: Municipal Immunity from CERCLA Liability for Property Acquired through Involuntary State Action (Oct. 20, 1995), reprinted in KEY ENVIRONMENTAL ISSUES IN U.S. REGION VI (Dallas, Texas, May 15-16, 1996). The regulations were originally codified at 40 CFR 300.1105 in 1992, but were subsequently vacated by the U.S. Court of Appeals for the District of Columbia.

Section 2504 of the Asset Conservation Act reinstated, effective September 30, 1996, the portion of the rule that addresses involuntary acquisitions by government entities. The rule states that "Governmental ownership or control of property by involuntary acquisitions or involuntary transfers within the meaning of CERCLA section 101(20)(D) or section 101(35)(A)(ii) includes, but is not limited to:

(1) acquisitions by or transfers to the government in its capacity as a sovereign, including transfers or acquisitions pursuant to abandonment proceedings, or as the result of tax delinquency, or escheat, or other circumstances in which the government involuntarily obtains ownership or control of property by virtue of its function as sovereign;

(2) acquisitions by or transfers to a government entity or its agent (including governmental lending and credit institutions, loan guarantors, loan insurers, and financial regulatory entities that ac-

The New Lender Liability Law

Specifically, the Asset Conservation, Lender Liability, and Deposit Insurance Protection Act of 1996[22] provides that the liability of a fiduciary under any provision of CERCLA for the release or threatened release of a hazardous substance at, from, or in connection with a vessel or facility held in a fiduciary capacity will not exceed the assets held in the fiduciary capacity. This principle is limited "to the extent that a person is liable under [CERCLA] independently of the person's ownership of a vessel or facility as a fiduciary or actions taken in a fiduciary capacity."[23] Nor does the law "limit the liability pertaining to a release or threatened release of a hazardous substance if negligence of a fiduciary causes or contributes to the release or threatened release."

The new law creates a safe harbor for lenders and other fiduciaries. It provides that a fiduciary will not be liable under CERCLA in his or her "personal capacity" for merely engaging in certain activities. The protected activities include:

> undertaking or directing another person to undertake a response action under CERCLA subsection (d)(1) or under the direction of an on-scene coordinator designated under the National Contingency Plan;
>
> undertaking or directing another person to undertake any other lawful means of addressing a hazardous substance in connection with the vessel or facility;
>
> terminating the fiduciary relationship;
>
> including in the terms of the fiduciary agreement a covenant, warranty, or other term or condition that relates to compliance with an environmental law or monitoring, modifying, or enforcing the term or condition;
>
> monitoring or undertaking one or more inspections of the vessel or facility;
>
> providing financial or other advice or counseling to other parties to the fiduciary relationship, including the settlor or beneficiary;
>
> restructuring, renegotiating, or otherwise altering the terms and conditions of the fiduciary relationship;

---

quire security interests or properties of failed private lending or depository institutions) acting as a conservator or receiver pursuant to a clear and direct statutory mandate or regulatory authority;

(3) acquisitions or transfers of assets through foreclosure and its equivalents (as defined in 40 CFR 300.1100[d][1]) or other means by a federal, state, or local government entity in the course of administering a governmental loan or loan guarantee or loan insurance program; and

(4) acquisitions by or transfers to a government entity pursuant to seizure or forfeiture authority.

The rule also states that, "Nothing in this section or in CERCLA Section 101(20)(D) or Section 101(35)(A)(ii) affects the applicability of 40 CFR 300.1100 to any security interest, property, or asset acquired pursuant to an involuntary acquisition or transfer, as described in this section." *See* 62 Fed. Reg. 34602 (June 26, 1997).

[22]H.R. 3610, 104 P.L. 208; 110 Stat. 3009.

[23]*Id.* Section 2501(2).

administering, as a fiduciary, a vessel or facility that was contaminated before
the fiduciary relationship began; or
declining to take any of the above listed actions.

For purposes of the safe harbor, the term *lender* means:

(1) an insured depository institution;
(2) an insured credit union;
(3) a bank or association chartered under the Farm Credit Act of 1971;
(4) a leasing or trust company that is an affiliate of an insured depository institution;
(5) any person (including a successor or assignee of any such person) that makes a bona fide extension of credit to or takes or acquires a security interest from a nonaffiliated person;
(6) the Federal National Mortgage Association, the Federal Home Loan Mortgage Corporation, the Federal Agricultural Mortgage Corporation, or any other entity that in a bona fide manner buys or sells loans or interests in loans;
(7) a person that insures or guarantees against a default in the repayment of an extension of credit, or acts as a surety with respect to an extension of credit, to a nonaffiliated person; and
(8) a person that provides title insurance and that acquires a vessel or facility as a result of assignment or conveyance in the course of underwriting claims and claims settlement.

For purpose of the safe harbor, the term *fiduciary* is defined as a person acting for the benefit of another party as a bona fide:

(1) trustee;
(2) executor;
(3) administrator;
(4) custodian;
(5) guardian of estates or guardian ad litem;
(6) receiver;
(7) conservator;
(8) committee of estates of incapacitated persons;
(9) personal representative;
(10) trustee (including a successor to a trustee) under an indenture agreement, trust agreement, lease, or similar financing agreement, for debt securities, certificates of interest, or certificates of participation in debt securities or other forms of indebtedness as to which the trustee is not, in the capacity of trustee, the lender; or
(11) representative in any other capacity that the administrator, after providing public notice, determines to be similar to the capacities described above.

The term fiduciary does not include a person who:

> is acting as a fiduciary with respect to a trust or other fiduciary estate that was organized for the primary purpose of, or is engaged in, actively carrying on a trade or business for profit, unless the trust or other fiduciary estate was created as part of, or to facilitate, one or more estate plans or because of the incapacity of a natural person; or
> acquires ownership or control of a vessel or facility with the objective purpose of avoiding liability of the person or of any other person.

The term *fiduciary capacity* means "the capacity of a person in holding title to a vessel or facility, or otherwise having control of or an interest in the vessel or facility, pursuant to the exercise of the responsibilities of the person as a fiduciary."

The law does not apply to a person if that person:

(1) Acts in a capacity other than that of a fiduciary or in a beneficiary capacity; and in that capacity, directly or indirectly benefits from a trust or fiduciary relationship;	or	(2) Is a beneficiary and a fiduciary with respect to the same fiduciary estate; and as a fiduciary, receives benefits that exceed customary or reasonable compensation, and incidental benefits, permitted under other applicable law.

Nor does the law preclude a claim under CERCLA against either (1) the assets of the estate or trust administered by the fiduciary or (2) a nonemployee agent or independent contractor retained by a fiduciary.

The law amends the definition of *owner or operator* in CERCLA Section 101(20) by adding specific language concerning the exclusion of lenders that do not participate in management. The new law defines the term *owner or operator* as not including "a person that is a lender that, without participating in the management of a vessel or facility, holds indicia of ownership primarily to protect the security interest of the person in the vessel or facility." For purposes of the law "indicia of ownership to protect security" expressly contemplates foreclosure:

> The term "owner or operator" does not include a person that is a lender that did not participate in management of a vessel or facility prior to foreclosure, notwithstanding that the person—
> "(I) forecloses on the vessel or facility; and
> "(II) after foreclosure, sells, re-leases (in the case of a lease finance transaction), or liquidates the vessel or facility, maintains business activities, winds up operations, undertakes a response action under section 107(d)(1) or under the direction of an on-scene coordinator appointed under the National Contingency Plan, with respect to the vessel or facility, or takes any other measure to preserve, protect, or prepare the vessel or facility prior to sale or disposition, if the person seeks to sell, re-lease (in the case of a lease finance transaction), or otherwise divest the person of the vessel or facility at the ear-

liest practicable, commercially reasonable time, on commercially reasonable terms, taking into account market conditions and legal and regulatory requirements."

The terms *foreclosure* and *foreclose* mean, respectively, acquiring, and to acquire, a vessel or facility through one of the following means:

Purchase at sale under a judgment or decree, power of sale, or nonjudicial foreclosure sale; A deed in lieu of foreclosure, or similar conveyance from a trustee; or Repossession, if the vessel or facility was security for an extension of credit previously contracted;	Conveyance pursuant to an extension of credit previously contracted, including the termination of a lease agreement; or	Any other formal or informal manner by which the person acquires, for subsequent disposition, title to or possession of a vessel or facility in order to protect the security interest of the person.

For purposes of interpreting lender liability *participation in management* means "actually participating in the management or operational affairs of a vessel or facility." The term "does not include merely having the capacity to influence, or the unexercised right to control, vessel or facility operations."

A person who is a lender and holds indicia of ownership primarily to protect a security interest in a vessel or facility will be considered to participate in management only if, while the borrower is still in possession of the vessel or facility encumbered by the security interest, the person does the following.

Control	Description
Decision-making control over environmental compliance	Exercises decision-making control over the environmental compliance related to the vessel or facility, such that the person has undertaken responsibility for the hazardous substance handling or disposal practices related to the vessel or facility.
Over all managerial control	Exercises control at a level comparable to that of a manager of the vessel or facility, such that the person has assumed or manifested responsibility: For the overall management of the vessel or facility encompassing day-to-day decision making with respect to environmental compliance; or Over all or substantially all of the operational functions (as distinguished from financial or administrative functions) of the vessel or facility other than the function of environmental compliance.

The term *participate in management* does not include performing an act or failing to act prior to the time at which a security interest is created in a vessel or facility. Similarly, the term does not include:

(1) holding a security interest or abandoning or releasing a security interest;
(2) including in the terms of an extension of credit, or in a contract or security agreement relating to the extension, a covenant, warranty, or other term or condition that relates to environmental compliance;
(3) monitoring or enforcing the terms and conditions of the extension of credit or security interest;
(4) monitoring or undertaking one or more inspections of the vessel or facility;
(5) requiring a response action or other lawful means of addressing the release or threatened release of a hazardous substance in connection with the vessel or facility prior to, during, or on the expiration of the term of the extension of credit;
(6) providing financial or other advice or counseling in an effort to mitigate, prevent, or cure default or diminution in the value of the vessel or facility;
(7) restructuring, renegotiating, or otherwise agreeing to alter the terms and conditions of the extension of credit or security interest, exercising forbearance;
(8) exercising other remedies that may be available under applicable law for the breach of a term or condition of the extension of credit or security agreement; or
(9) conducting a response action under Section 107(d) or under the direction of an on-scene coordinator appointed under the National Contingency Plan, if the actions do not rise to the level of participating in management.

The term *extension of credit* includes a lease finance transaction in which the lessor does not initially select the leased vessel or facility and does not during the lease term control the daily operations or maintenance of the vessel or facility. It also includes a transaction that conforms with regulations issued by the appropriate federal banking agency or the appropriate state bank supervisor[24] or with regulations issued by the National Credit Union Administration Board, as appropriate.

RCRA Secured Creditor Exemption

Further relief for lenders from RCRA liability for USTs was also afforded to lenders under first EPA's recently promulgated rule and, more recently, under the Asset Conservation, Lender Liability, and Deposit Insurance Protection Act of 1996.[25]

The rule, which parallels the CERCLA lender liability rule struck down by the District of Columbia Circuit, is designed to provide protection under RCRA

---

[24] As those terms are defined in Section 3 of the Federal Deposit Insurance Act, 12 U.S.C. 1813.
[25] H.R. 3610, 104 P.L. 208; 110 Stat. 3009.

with respect to security interests in underground storage tanks.[26] The RCRA rule is, however, expensive to implement, since it requires certain extraordinary measures—such as promptly removing the USTs in order to maintain the exemption. EPA reissued the lender liability guidance under CERCLA in December 1995. This guidance was issued, of course, under EPA's enforcement discretion and is not binding on third-party suits. Congress codified the rule promulgated by EPA concerning USTs in fall 1996. Thus, the relief lenders have been urging Congress to adopt for the past decade is now codified into law.

Financing Brownfields

On August 5, 1997, President Clinton signed the Taxpayer Relief Act.[27] This law created a "new tax incentive to spur the cleanup and redevelopment of brownfields in distressed urban and rural areas."[28]

Under the new Brownfields Tax Incentive, environmental cleanup costs for properties in targeted areas are fully deductible in the year in which they are incurred, rather than having to be capitalized. The $1.5 billion incentive is expected to leverage $6.0 billion in private investment and return an estimated 14,000 brownfields to productive use. Thus, making the Brownfields Tax Incentive "a valuable and potent tool that communities can now utilize in addressing brownfields."[29]

The tax incentive is applicable to properties that meet specified land use, geographic, and contamination requirements. To satisfy the land use requirement, the property must be held by the taxpayer incurring the eligible expenses for use in a trade or business or for the production of income, or the property must be properly included in the taxpayer's inventory, To satisfy the contamination requirement, hazardous substances must be present or potentially present on the property. To meet the geographic requirement, the property must be located in one of the following areas:

> EPA Brownfields Pilot areas designated prior to February 1997;
> Census tracts where 20 percent or more of the population is below the poverty level;
> Census tracts that have a population under 2,000, have 75 percent or more of their land zoned for industrial or commercial use, and are adjacent to one or more census tracts with a poverty rate of 20 percent or more, and
> Any Empowerment Zone or Enterprise Community (and any supplemental zone designated on December 21, 1994).

---

[26]59 Fed. Reg. 30,448.

[27]H.R. 2014/PL 105-34.

[28]BROWNFIELDS TAX INCENTIVE (EPA 500-G-97-155 August 1997), reprinted at, http://www.epa.gov/swe...f/html-doc/taxfs-2.htm.

[29]Id.

Both rural and urban sites may qualify for this tax incentive. The taxpayer must get a certification from the state environmental agency that his/her property is in a targeted area.

The Brownfields Tax Incentive sunsets after three years, thereby covering eligible costs incurred or paid from the date of enactment until January 1, 2001. Sites on EPA's National Priorities List are excluded.

The Brownfields Tax Incentive was designed to build "on the momentum of the Clinton Administration's Brownfields National Partnership Action Agenda," announced in May 1997.[30] According to the federal government, the Brownfields Tax Incentive "will help bring thousands of abandoned and under-used industrial sites back into productive use, providing the foundation for neighborhood revitalization, job creation, and the restoration of hope in our nation's cities and distressed rural areas."[31]

Federal tax law generally requires that those expenditures that increase the value or extend the useful life of a property, or that adapt the property to a different use, be capitalized—and, if the property is depreciable, that they are depreciated over the life of the property. This means that the full cost cannot be deducted from income in the year that the expenditure occurs. This capitalization treatment also applies to the cost of acquiring property. In contrast, repair and maintenance expenditures generally can be deducted from income in the year incurred. In the past, many environmental remediation expenditures had to be capitalized over time, and could not be fully deducted or expensed in the year incurred.

In 1994, the Internal Revenue Service (IRS) issued a ruling that stated that certain costs incurred to clean up land and groundwater could be deducted as business expenses in that same year. Unfortunately, the ruling only addressed cleanup costs incurred by the same taxpayer that contaminated the land. It did not address cleanup costs incurred by a party that had purchased contaminated property, or an owner who was interested in putting the land to new use. Further, the IRS ruling was unclear as to whether other remediation costs not specifically addressed in the ruling would be deductible in the year incurred or would have to be capitalized.

These unresolved issues created potential financial obstacles in the contaminated properties market. Specifically, owners of contaminated property could remediate their property and sell the clean property at its full market value, enabling them to fully recover the cost of remediation. Prospective purchasers of contaminated property had to purchase the property at its impaired value, attributable to the contamination, and capitalize the remediation costs. This arguably left prospective purchasers at a disadvantage in terms of environmental remediation expenditures. Additionally, property owners who wanted to remediate their property and put it to a different use were at a disadvantage because they were not able to fully deduct their remediation costs in the year incurred. The new law aims at reversing these disincentives.

---

[30]The National Partnership outlines a comprehensive approach to the assessment, cleanup, and sustainable reuse of brownfields, including specific commitments from fifteen federal agencies.
    [31]*Id.*

In addition to the Brownfields Tax Incentive included in the Taxpayer Relief Act, the federal government has used many existing programs to encourage the redevelopment of brownfield sites.[32] The following table describes some of these programs:

Federal Financing Programs	Applicable to Brownfields Redevelopment
Loans	SBA's microloans;
	SBA's § 504 development company debentures;
	EDA's Title IX 9—capital for local revolving loan funds;
	HUD's locally determined CDBG loans & floats;
	EPA's brownfield revolving loan funds (target to twenty-nine pilot communities);
Loan guarantees	SBA's § 7(a) and low-doc programs;
	HUD's § 108 loan guarantees;
Grants	HUD's community development block grants (locally determined projects);
	EDA's Title I (public works) and Title IX (economic adjustment);
	DOT (system construction and rehabilitation programs);
Equity capital	SBA's small business investment companies;
Tax incentives and tax exempt financing	Historic rehabilitation tax credits;
	Low-income housing tax credits;
	Industrial development bonds;
	Empowerment zone facility bonds;
	Targeted expensing of cleanup costs;
Tax advantaged zones	HUD/USDA empowerment zones;
	HUD/USDA enterprise communities.

---

[32]For a discussion of insurance products available for brownfields cleanup *see* EPA OSWER, POTENTIAL INSURANCE PRODUCTS FOR BROWNFIELDS CLEANUP AND REDEVELOPMENT: SURVEY RESULTS OF INSURANCE INDUSTRY PRODUCTS AVAILABLE FOR TRANSFERENCE OF RISK AT POTENTIALLY CONTAMINATED PROPERTY (EPA 500-R.96-001 June 1996).

CHAPTER 15

What Will It Take to Fix the Problem?
Policy Options

Administrative versus a Legislative Fix

The federal government has only recently begun to recognize and address brownfield liability issues of prospective purchasers:

> We recognize that fears of Superfund liability have had the effect of preventing the sale and redevelopment of Superfund properties, taking some of these properties out of productive economic use. Both prospective purchasers and lenders have been reluctant to invest in these properties, for fear of becoming liable as "owners" of a Superfund site. As EPA Administrator Carol Browner has testified . . . this Administration has made a concerted effort to address this problem and promote "brownfields" redevelopment, by undertaking a pilot program to provide assurances that the government will not take action at particular sites against "prospective purchasers."[1]

A major hurdle in EPA's ability to reduce "brownfields anxiety," however, is the EPA's lack of statutory authority to do so.

EPA was formed in 1970 by President Nixon through an executive reorganization. EPA has no one organic statute that clearly defines the agency mission. Rather, the agency derives its authority through a myriad of legislation, each with a separate and distinct purpose. EPA is the largest federal regulatory agency, employing over 18,000 individuals and with a budget of over $6.5 billion. EPA is wholly and partially responsible for administering the following major statutes:

Statute	Agencies with Shared Authority and Responsibility	Description
Clean Air Act (CAA)	Department of the Interior; Department of Energy; delegated state agencies	Requires EPA to set mobile source limits, ambient air quality standards, hazardous air pollutant

(continued)

---

[1]Testimony of Lois Schiffer, FDCH Congressional Testimony, April 27, 1995.

335

Statute	Agencies with Shared Authority and Responsibility	Description
		emission standards, standards for new sources, and require- ments to prevent significant deterioration and to focus on areas that do not attain NAAQS
Clean Water Act (CWA)	Army Corps of Engineers; Delegated state agencies	Establishes the sewage treatment construction grants program and a regulatory program for discharges into water of the United States
Ocean Dumping Act (ODA)		Regulates the intentional disposing of materials into ocean waters and establishes research on the effects of and alternatives to ocean dumping
Safe Drinking Water Act (SDWA)	Delegated state agencies	Establishes primary drinking water standards, regulates underground injection of hazardous wastes and establishes a limited groundwater control program
Resource Conservation Recovery Act (RCRA)	Delegated state agencies	Provides cradle-to-grave regulation for hazardous substances and waste
Comprehensive Environmental Re- sponse, Compensa- tion, and Liability Act (CERCLA)	Department of the Interior	Establishes a fee- maintained fund to clean up abandoned sites containing haz- ardous substances
Federal Insecticide, Fungicide, and Rodenticide Act (FIFRA)	Federal Trade Commission	Governs registration of pesticides
Toxic Substances Control Act (TSCA)	Federal Trade Commission	Requires the testing of chemicals and regu- lates their use
Environmental Research and Development Demonstration Act (ERDDA)		Provides authority for all EPA research programs

(continued)

National Environmental Policy Act (NEPA)	Council on Environmental Quality	Requires EPA to review environmental impact statements
Pollution Prevention Act of 1990	Department of Energy	

These statutes comprise the six major programs controlled by EPA, which are: pesticides, air protection, water protection, radiation, toxic substances, and hazardous wastes.[2]

In *Kelly v. EPA*,[3] the U.S. Court of Appeals for the District of Columbia addressed the question of the extent of EPA's authority to promulgate rules concerning liability issues under CERCLA. In that case, the precise question addressed was whether EPA has the statutory authority to issue its lender liability rules. The court's analysis is, however, equally applicable in the brownfields context.

The portions of CERCLA EPA relied on for specific authority were to interpret CERCLA's liability provisions:

> Section 105 of CERCLA, which provides that the agency has responsibility to promulgate the national contingency plan setting forth the actions and procedures to be taken in response to a contamination;
>
> Section 106(b)(2), under which a party that has cleaned up a contaminated site pursuant to an administrative order may petition the EPA for reimbursement of its reasonable costs; if EPA refuses, a federal court may order reimbursement if it determines that the party is not liable or, even if liable, that the party has demonstrated that the cleanup actions it was ordered to take were arbitrary and capricious or otherwise unlawful; and
>
> Those statutory provisions that grant it authority to seek enforcement.

The court rejected each of these provisions as giving EPA statutory authority to carve out a liability rule that limits the ability of *private parties* to bring actions for contribution. The court explained that the NCP was procedural in nature and pertained to cleanup standards, not liability issues. Although EPA can exercise its enforcement discretion not to prosecute, it cannot extend this mandate to third parties. Finally, the limit on judicial review of agency orders again pertained to cleanup issues and not liability questions.

The lack of statutory authority plagues EPA's brownfield initiatives. EPA can establish a procedure to issue a Covenant Not to Sue once a brownfield cleanup is complete under its enforcement discretion. Such a covenant could not, however, bind any third parties. Nor would it preclude future agency action if cleanup standards should change in the future. Similarly, if issued under enforcement discretion, such a covenant could not run with the land and enure to the benefit of future

---

[2]ROSEMARY O'LEARY, ENVIRONMENTAL CHANGE: FEDERAL COURTS AND THE EPA (Temple 1993).
[3]15 F.3d 1100 (D.C. Cir. 1994).

purchasers—it would be personal between the enforcement officer reviewing the case and the brownfield owner that cleaned up the site.

Administrative remedies such as EPA's pilot project are useful but of limited utility. EPA funds these projects out of its authority to study environmental problems. As explained, EPA has no statutory authority under CERCLA or any other federal environmental statute to promulgate legislation limiting the liability of lenders, developers, prospective purchasers or other brownfield stakeholders under CERCLA or other federal environmental law.[4] As such, EPA's pilot project could never evolve into any more than a voluntary program or a recommendation to Congress for statutory reform. EPA has recognized the limits to its authority:

> "The administrative reforms announced today go a long way toward our goal of a faster, fairer, more efficient Superfund. They encourage economic development, particularly in urban areas, and greater community involvement, among other things," said EPA Administrator Carol M. Browner. "Legislative action by Congress, however, is still required if we are to fully reform the program on behalf of the one-in-four Americans who live near a hazardous waste site. We will continue to work with Congress to achieve those reforms."[5]

If Congress is to wholly address the brownfields issues, it must do so by addressing all of the potential environmental liabilities an owner, lender, or developer could incur when redeveloping brownfields properties. *CERCLA is only one liability statute that threatens brownfields owners, lenders, prospective purchasers, and developers with uncertainty.* If Congress amends only CERCLA to address the brownfields questions, then only one small strain of the brownfields liability threat will be addressed. While CERCLA has probably received the greatest media attention as causing indeterminent liability risk in the brownfields context, there are many other statutory and common law causes of action that make liability (and litigation risk) almost as unpredictable. Amending the CERCLA statutory provision or any other single statute will reduce some legal uncertainty. It will not, however, completely eliminate the problem—which is unknown and unquantifiable environmental risk.

If Congress is to eliminate the problem it must come up with a new statutory scheme that addresses the collective brownfield issues. Modifying CERCLA to address brownfield issues, will only address a small portion of the problems that contribute to "brownfield paralysis."

---

[4]*See Kelly,* 15 F.3d 1100 (D.C. Cir. 1994).
[5]EPA press release (Feb. 22, 1995).

Global Issues

[1] Overview

The Superfund reform debate has largely been driven by the concerns of lenders, developers, and other industrial stakeholders. Although the industrial sector is a large part of the Superfund debate, it is essential to remember that at the core of Superfund are communities that must live, play, and work every day at or near these brownfield sites. The overarching mandate of Superfund (and other environmental statutes administered by EPA) is "to protect human health and the environment." While the legislative parameters to carry out this mandate are being debated, it should not be forgotten that there are people who could suffer from both real and perceived adverse physical, emotional, and financial effects that often result from living next door to a polluted site—large or small. Certain overriding concepts need to be considered in evaluating policy options.

[2] Difficulty in Quantifying Costs

First, the greatest difficulties in redeveloping brownfields properties arise due to the great uncertainty in valuing potential environmental liabilities associated with these sites. Simply put, if environmental liabilities could be quantified with certainty and reliability there would be no real impediment to redevelopment of brownfields. As one commentator explained:

> Rightly or wrongly, liability concerns are having an important impact on the economic development process. Both public and private officials emphasize that developers, investors, and companies "want to quantify their risk as quickly as possible when undertaking a project" and therefore seek a firm understanding of their potential liability at its outset.[6]

Uncertainty in quantifying the liability at brownfield sites arises in six critical ways:

> engineering estimates and costs to contain environmental contamination are unclear and not easily converted into realistic or reliable dollar projections;
> regulatory requirements (current and future) are in flux and are not easily translated into projected costs;
> the market's reaction to environmental impairment is unpredictable;
> third-party liability is impossible to quantify because it cannot be identified or terminated at any point despite best efforts;
> financing for cleanups is often difficult or impossible to obtain; and
> cleanup costs change over time as regulations and laws change.

---

[6]CHARLES BARTSCH ET AL., NEW LIFE FOR OLD BUILDINGS at 25 (1991).

The law currently does not specify a standard level of cleanup nationwide; instead, it establishes a complex cleanup framework biased toward permanent cleanups and the use of treatment technologies. Applicable, relevant, and appropriate state and federal standards often prescribe overly stringent cleanup levels. Consequently, cleanup costs are often high and cleanup goals, remedies, and costs can differ for comparable sites across the country.

The uncertainty surrounding engineering estimates and the cost to correct environmental contamination are often substantial. Typical costs include:

preliminary risk assessment costs;
site investigation costs;
remediation planning and budgeting;
the costs of designing the remediation plan;
the costs of monitoring the plan once implemented;
the costs of public communication and training; and
the costs of the emergency response plan.

In many cases, estimating a volume of contaminated soil or groundwater degrades into a process no more scientific than digging until no more contamination can be found. Even then, it cannot be certain that the property is "clean." Thus, even a property that has been remediated will be somewhat stigmatized by the potential of further remediation or the possibility of future standards holding that the cleaning conducted did not go far enough.

The uncertainty surrounding the standards applied by environmental regulators—state and federal—is a moving target. Regulatory decisions are contingent and can almost always be revised, based on changes in law, technology, or public perception. In addition, progressively more stringent environmental standards require revised, more intensive (and often more costly) remediation plans. The risk of a bad regulatory decision is almost always borne by those being regulated. If a decision is made contrary to science and then corrected, the regulated entity is often required to remediate to current scientific standards.

The uncertainty in the financial markets due to the past, present, and future existence of environmental impairment varies. Critical issues include: the financial strength of capital sources; the strength of the local market conditions; the nature and scope of the environmental impairment in question; the perceived quality of technical analysis of the impairment; the perceived status of the regulatory attitudes toward the impairment; and the visibility and quality of the ultimate remediation. These uncertainties lead to difficulties in obtaining financing, difficulties in selling property, and increased insurance costs.

Most of the private sector costs not directly associated with cleanup activities are considered "transaction costs." While transaction costs for the government have been relatively low, there is widespread agreement that Superfund generates high transaction costs for private parties. These especially arise in private party contribution litigation and in follow-up litigation between those parties and their insurance carriers. These costs can be very burdensome to small businesses.

The uncertainty surrounding the possibility of third-party law is also staggering. Factors affecting suits include: (1) the nature and extent of the problem, (2) its potential impact of neighboring property, (3) the technical capabilities to remedy the problem, and (4) the potential magnitude of claims. Whether such suits will develop is impossible to determine—as is quantification of the potential liability. The property must not only be cleaned up, but in many cases it must be monitored in the future as well. Contingent liabilities may include off-site injury as well as those found on-site. It may include personal injury claims brought under common law theories of damage.

[3] The Brownfields Problem Goes Beyond CERCLA

Although Superfund is a large part of the brownfields problem, it is not the only legislation casting shadows on inner city redevelopment opportunities. If Congress is truly to address the question, it needs to do so as separate legislation and not simply as an overlay to CERCLA reauthorization. CERCLA is not the only statute creating difficulties in quantifying brownfields liabilities. The common law and many other state and federal environmental statutes create tremendous regulatory uncertainty in the cleanup of brownfield sites.

[4] EPA Lacks the Statutory Authority to Cure the Brownfields Problem

Since EPA has no one organic statute and derives its authority from a myriad of statutes, including CERCLA, EPA must be given specific authority to address these issues. Otherwise, all EPA attempts to carve brownfields out of CERCLA and other EPA programs would be solely under the agencies prosecutorial discretion. While such informal policies could prevent suits by the federal government, it would do nothing to preclude suits by private parties.

[5] The Perceived Unfairness of the Environmental Programs—CERCLA and Wetlands in Particular—to Small Businesses and Landowners

CERCLA and many other environmental statutes are perceived, rightly or wrongly, as grossly unfair and overburdensome to the average small business— be it a real estate agent or broker, gas station or dry cleaner owner or operator, inner city developer, demolition company, or construction company. Whether or not this is in fact true, a perception has arisen among a sector of the American public that environmental laws have been misfocused and unfairly persecute the common man. For example, one commentator said:

> Almost without our realizing it, a whole new level of government has emerged in America. It is composed of a combination of layers and bureaucrats who

have come to dominate federal, state, and local government. Nobody likes it, except the practitioners. It is government by regulation. It is now the fourth branch of government. It functions as law maker; though unelected, its rules and regulations have force of law. It functions also as both judge and jury in cases involving its own rules. The accumulation of legislative, executive, and judicial power in the same hands is a real definition of tyranny. . . . [I]t has become the essence of environmental overkill.[7]

This issue—the question of perceived unfairness—is as important a problem as the actual deterrents of brownfields redevelopment. Even EPA administrator Carol Browner acknowledged the significance of the problem of perceived unfairness:

The current liability regime is also criticized as being burdensome or unfair by many parties. Small businesses and municipalities complain that the liability system can impose significant financial burdens on them. Larger businesses resent having to pay the "orphan share" at sites where other responsible parties cannot pay their fair share. Lenders and trustees fear that they will incur liability if they become involved in Superfund sites.[8]

[6] The Perceived Unfairness of the Environmental Programs Is Exacerbated by the Fact that PRPs Include Those Who Did Not Cause the Pollution and Whose Activities, If They Did Cause the Pollution, Were Legal, Permitted, and Often Directed by the State When Undertaken

No distinction is made in law between industrial site pollution and contamination due to illegal activities. The present law also neither requires nor encourages industrial site owners to clean up their own properties without the state (or some third party) taking an enforcement action. Property owners are similarly neither required nor encouraged to examine their own properties for unknown, potential environmental or health risks.

---

[7]DIXY LEE RAY AND LOU GUZZO, ENVIRONMENTAL OVERKILL: WHATEVER HAPPENED TO COMMON SENSE at 161 (Harper Perennial 1993). *See also* Nancie G. Marzulla, *The Property Rights Movement: How It Began and Where It Is Headed,* in BRUCE YANDLE, LAND RIGHTS: THE 1990'S PROPERTY RIGHTS REBELLION (Rowan & Littlefield 1995); Hertha L. Lund, *The Property Rights Movement and State Legislation,* in BRUCE YANKLE, LAND RIGHTS: THE 1990'S PROPERTY RIGHTS REBELLION (Rowan & Littlefield 1995). *But see* Natural Resources Defense Council, BREACH OF FAITH: HOW THE CONTRACT'S FINE PRINT UNDERMINES AMERICA'S ENVIRONMENTAL SUCCESS 3-4 (Feb. 1995) ("Essentially, we believe that Americans want more environmental protection and less bureaucracy—something we have tried to bring about with our reform proposals for better and more efficient Safe Drinking Water, Superfund, and clean water programs. . . . The untold environmental protection story in the United States is success").

[8]STATEMENT OF CAROL M. BROWNER, ADMINISTRATOR, U.S. ENVIRONMENTAL PROTECTION AGENCY, BEFORE THE U.S. SENATE, COMMITTEE ON ENVIRONMENT AND PUBLIC WORKS, SUBCOMMITTEE ON SUPERFUND, WASTE CONTROL, AND RISK ASSESSMENT (March 10, 1995).

Legislation encouraging the reuse of polluted urban industrial sites is not only vital to the economy, it is good for the environment. It promotes cleanup of underutilized sites and strengthens existing tax bases and infrastructure without causing increased contamination of new sites. Reuse of brownfield sites is the "ultimate recycling program."

[7] Overlapping Federal/State Relationship

Currently, the federal government has primary responsibility for implementing the Superfund program and it has exclusive access to the money in the CERCLA Trust Fund. States, however, play a significant role in the program's implementation. States perform much of the site assessment functions and have taken the lead for managing the design or cleanup at over seventy-five NPL sites. State standards apply to all cleanups. As such states often have significant input in selecting cleanup remedies. In addition, states must pay a share of any Fund-financed remedial cleanup and states must also pay for all operation and maintenance at Fund-financed cleanups. This overlapping authority and responsibility often results in both federal and state agencies overseeing cleanup activity at the same site. This inefficiency and redundancy contributes to the cost and duration of some cleanups and can result in confusion among stakeholders.

In addition, since many states have developed their own state "Superfund" programs and are capable of site remediation without intervention by EPA, many brownfield sites have only state oversight and no way to limit federal liability. Some qualified states have the infrastructure to support a delegation of specified activities. Other states do not have adequate infrastructure to assume complete authorization of the state's Superfund program, nor do some states want to assume wholesale responsibility for the cleanup of NPL caliber sites. Regardless of the level of development of the specific state program, the federal-state relationship must clearly delineate remedial responsibilities to avoid duplication and delay in cleaning up brownfield sites.

Under the current statute, EPA and the state meet to decide who will take the lead responsibility for each NPL listed site. For a federal-led response (including federal facilities), EPA is the lead agency; it assigns responsibility for managing and conducting the work. In this instance, the state or Indian tribe generally functions as a support agency. As a support agency, the state or tribe may hold some key responsibilities and perform specific parts of the cleanup, but it does not take on the major tasks for the response.

For a state- or tribal-led response, EPA functions as a support agency. A state or tribal lead in cleanup of NPL listed sites can mean one of three things. The state is:

overseeing a PRP cleanup;

carrying out most aspects of the cleanup but the response is trust fund-financed; or

given lead responsibility and is financing the response.

Even if the state takes the lead, EPA must still approve all remedy selection decisions that require trust fund financing.

States currently have specific responsibilities under CERCLA. These include: (1) identifying applicable, relevant and appropriate requirements (ARARs)—that is, the state environmental requirements that are applicable or relevant and appropriate to the remedial action, and (2) providing certain assurances.

Before a state can engage in a Trust Fund-financed remedial action, whether EPA or state led, CERCLA mandates that a state must assure it will:

Pay Part of the Cleanup Costs. A state is required to pay 10 percent of the costs of a remedial action if the site was privately operated at the time of the hazardous substance release. A state is required to pay 50 percent of all response costs if the state or locality operated the site.

Ensure Adequate Off-Site Storage. A state is required to ensure that hazardous substances removed from a site will receive adequate off-site storage, disposal, or treatment.

Assume Responsibility for Operation and Maintenance Activities. The state is required to assume responsibility for all O&M activities for a remedial action. The state assumes ultimate responsibility for performing O&M of the selected remedy, even though a political subdivision may manage the actual site.

Accept Interest in Real Estate. The state is required to document its commitment to accept interest in real estate that may need to be acquired for a Superfund response.

Maintain Capacity for Disposal and Treatment of Wastes. The state is required to maintain the capacity for the disposal or treatment of all hazardous wastes expected to be generated within the state for twenty years. This assurance consists of the state's capacity assurance plan, which must be approved by EPA before Trust Fund-financed remedial actions take place in the state.

These assurances can be provided through site-specific remedial cooperative agreements or Superfund state contracts. In addition, there may be other site specific agreements between EPA and the states. The principal vehicle for supporting states and Indian tribes financially is through Core Program Cooperative Agreements. With Core Program funding, EPA helps the states and tribes build and maintain basic program capabilities that cannot be accounted for on a site-specific basis, such as management, accounting and record keeping, legal counsel, and training. Since the inception of the Core Program in 1987, as part of SARA, EPA has awarded the states over $103 million in Core grant funding. Currently, forty-six states and Indian tribal governments or consortia are receiving this assistance.

EPA has documented state and tribal Superfund capabilities since 1989 through a study generally referred to as the "50-State Study." Findings show that:

forty-five states have some dedicated cleanup funds and some enforcement capabilities, although the size of these funds and the degree of enforcement capability varies widely among those states;

thirty-six states use site inventories or priority lists;

thirty-five states provide public notice of identified sites or completed remedial action plans;

twenty-three states solicit public comments during the site management process;

thirty-five states have provisions for public meetings or hearings;

six states currently make use of citizen advisory groups in planning cleanups and revising cleanup policies;

at least thirty states have retroactive liability provisions;

thirty-nine states impose strict liability;

thirty-four states impose joint and several liability;

twenty-four states have all three components (strict, retroactive, and joint and several liability) of a CERCLA-like liability system.

Most states seek a larger role in implementing the Superfund program than the current statute allows. Critics of the current process complain that CERCLA's provisions relating to active federal and state involvement at sites results in substantial duplication of effort and often delay of cleanup. While much progress has been made in improving coordination and relations, the appropriate state role for the future of the Superfund program needs to be better defined. Examination of state authorities and capabilities indicates that some states have large, well-developed Superfund programs that could support a delegated program. Others with emerging or currently inactive programs need federal technical and financial assistance to become fully capable.

EPA has historically supported the delegation of certain site assessment, allocation, and remedy selection authority to qualified states. For example, EPA has supported proposals to reform Superfund by providing that states with adequate legal authority, capability, resources, and experience to carry out the federal program would be able to receive delegation of the full range of response activities at Superfund sites. Under that proposed program, states that enter into cooperative agreements with EPA would receive funding from the Trust Fund to carry out response activities to meet all applicable state and federal cleanup requirements. States would be responsible for paying the incremental differences for attaining any cleanups that exceed federal standards and state applicable requirements.

This approach minimizes the overlap between federal and state governments by delegating to capable states more responsibility, and therefore provides them with more autonomy. Furthermore, by reducing uncertainty regarding the appropriate role either state or federal government is to play, the most important net effect will be faster cleanup of contaminated sites—large and small.

[8] Inadequate Community Involvement

Many communities near Superfund sites, including low income, minority, and Native American communities, do not feel they are given an adequate opportunity to participate in the Superfund decision-making process. These and other communities believe the program does not address the concerns of those living closest to the hazardous waste site when evaluating risk or determining the method and level of cleanup. Consequently, communities often conclude that the resulting cleanup is either overly conservative or insufficiently protective. This perception is often, although not always, based on lack of information and involvement in the site cleanup process.

Specific Concerns

[1] Cleanup of Hazardous Waste Sites Is a Local Land Use Issue

Cleanup of hazardous waste is a local land use issue and as such it should be delegated to the states through a state and federal partnership modeled after the Clean Air Act, Clean Water Act, and RCRA. As one commentator recently testified:

> The recycling of inner city properties is, at heart, a local process. EPA is trying to find ways to keep federal environmental laws, regulations and policies from interfering in that process where possible, and we would certainly welcome advice from this Committee on other efforts we might consider undertaking.
>
> We would do the public a disservice by promising economic revitalization only at the price of less protection. For our part, the Administration's initiatives advance economic reuse of Brownfields without compromising our ability to protect public health or the environment. We hope the states will be similarly successful developing their own Brownfields efforts.[9]

A state-federal partnership achieved through a state implementation process approved by EPA headquarters in Washington would ensure that a minimum standard would be achieved nationwide but would allow states the flexibility to provide added environmental protections to their residents. The state implementation process would allow more community involvement. States could better control development around hazardous waste sites by viewing the cleanup decision in the context of total planning and zoning.

States would set applicable cleanup standards for permitted uses—e.g., residential and nonresidential cleanup standards. Local inhabitants would have a

---

[9]TESTIMONY OF TIMOTHY FIELDS, JR., DEPUTY ASSISTANT ADMINISTRATOR, U.S. ENVIRONMENTAL PROTECTION AGENCY, BEFORE THE U.S. HOUSE OF REPRESENTATIVES, COMMITTEE ON COMMERCE, SUBCOMMITTEE ON COMMERCE, TRADE, AND HAZARDOUS MATERIAL (June 15, 1995).

streamlined development process because cleanup and land use would be considered together at the local level.

> Contaminated sites are essentially local problems and should be handled by a level of government that is closer to the problem. States have the responsibility and the authority to act to protect the health and safety of their citizens. ... The federal EPA is less efficient in managing cleanups. ... Most states have the capability, expertise, authority and funding to deal with contaminated sites. One need only look at some of the figures in the EPA's "Fifty State Study" to recognize the maturity to which many state programs have risen.[10]

Since the enactment of CERCLA in 1980, roughly 1,300 sites have been placed on the National Priorities List of Superfund sites, and the list could grow to as many as 2,800. In addition to these high-risk, high-priority sites, there are tens of thousands of lower-priority sites for which states have primary cleanup responsibility. With limited resources, states have had to focus on those sites posing the most risks, leaving some lower-risk sites yet to be cleaned up. Many responsible parties are willing and able to clean up the site, but they refuse to proceed with cleanup in the absence of direction from the state that the cleanup is being performed to the state's satisfaction and standards. Investors and developers, also uncertain of the costs and liabilities associated with these properties, generally avoid getting involved at all and look instead to develop more pristine properties. This migration from "brownfields" to "greenfields" is particularly problematic to certain urban and industrial areas, where the loss of investment capital and jobs further exacerbates existing economic and social conditions.

In response to these limitations and problems, roughly twenty states have enacted their own voluntary cleanup laws and programs. These programs allow responsible parties, as well as prospective purchasers to enter into agreements with the state to voluntarily clean up the site, pay for state oversight, and generally receive some sort of acknowledgment or sign-off from the state. State programs have made substantial progress in cleaning up brownfield sites. A program that, delegated to the states, would provide technical and other assistance to states to establish and expand voluntary response programs under specific conditions. Elements necessary to such state delegated programs would include providing:

opportunities for technical assistance;
oversight and enforcement of the voluntary cleanup program;
adequate opportunities for public participation, and
that the state would finish cleanups if the volunteering party fails to do so.

---

[10]STATEMENT OF THE HONORABLE MICHAEL G. OXLEY, CHAIRMAN, U.S. HOUSE OF REPRESENTATIVES, COMMITTEE ON COMMERCE, SUBCOMMITTEE ON COMMERCE, TRADE, AND HAZARDOUS MATERIALS (June 15, 1995).

Federalism of the nation's growing number of voluntary cleanup programs, through a state by state delegation process, would increase the attractiveness of these programs. A codified state-federal partnership like that enjoyed under the CWA, CAA, and RCRA, would benefit a wide spectrum of the community—bankers, developers, environmentalists, and all other elements of a viable and thriving community. To correct a great deal of the brownfield problem, at a minimum Superfund should be amended to provide flexibility and to encourage states to institute voluntary cleanup programs.

State cleanup programs are developing and maturing. These efforts are evidenced by the buildup of staffs and infrastructure, as well as by the enactment of both statutory and regulatory frameworks. The sophistication of these programs can also be demonstrated by the simple day-to-day experience gained in directing cleanups. In Superfund, as in most other endeavors, the states are developing skills and experience and they are applying that knowledge through both traditional and innovative approaches. States should have that flexibility to continue and expand these efforts with the added benefit of additional protection for parties entering the program against federal CERCLA actions.

Delegation offers the PRPs, states, environmental justice community, and other members of the public the assurance that the reforms crafted into a comprehensive reauthorization program will give states the tools to operate the program with reasonable, but modest federal oversight, funding, and assistance. The program would be similar to that currently in existence in the other major media statutes—air, water, and solid and hazardous wastes.

[2] Create Low Interest Loans and Other Economic Incentives for Cleanup

Tax deductions, abatements, and credits should be provided to encourage environmental cleanup. This can be accomplished in part by allowing the states to take portions of the federal Superfund throughout the state implementation process and allow the states to use the fund not just to clean up sites, but also to provide the basis for a loan or tax abatement program. States should be encouraged to set up a voluntary site remediation trust fund and a voluntary remediation assistance program. Fees charged for remediation assistance should be applied to the fund and used to finance loan programs as well as the remediation assistance program. Fees charged must be consistent and measurable (perhaps a percentage of the approved cleanup plan).

[3] Restrict Property Uses Based on Cleanup Standards

Restrict property uses based on cleanup standards and provide adequate notice of contamination. This ensures that cleanup need not be more expensive than the land use requires and yet protects against health hazards as land uses change over time. Once cleanup is complete to the appropriate standard, a Covenant Not to Sue or

No Further Action letter should be issued that would be binding against the state, the federal government, and all third-party suits. This certificate should be recorded, transferable to future property owners and otherwise run with the land.

Considering future land use in remedy selection should not be viewed as a remedial alternative that assumes the residual risk level at a site is consistent with all current and future on-site and off-site land uses. Cleanup of hazardous waste sites should not be an unfair paradox between environmental and economic development. Land use management is a catalyst in putting contaminated industrial lands back into use as productive economic assets. Despite advocating land use management, it should still be noted that there are instances where variable land use considerations are not appropriate. Congress must clearly lay out the parameters when land use determinations are and are not appropriate.

Remedial decisions that consider land use criteria will require greater and earlier involvement in the remedial decision-making process by all stakeholders, particularly local communities and local government. Since land use is essentially a local issue, it is imperative that stakeholder involvement includes citizens who live near hazardous waste sites and are directly affected by the cleanup decision and local and state officials.

Land use considerations should not circumvent CERCLA's mandate: protection of human health and the environment. Whenever a remedy relies on land use restrictions to be protective, there must be appropriate safeguards to ensure that land designations remain consistent throughout the use of the property. Appropriate institutional controls that restrict land use include zone redesignation and covenants on the deed restricting site usage. Moreover, it is imperative that there be an identified agency responsible for enforcing the land use determination.

[4] Create a Limited Environmental Audit Privilege and Define the Term *All Appropriate Inquiry* for Purposes of the CERCLA Innocent Landowner Defense

Privilege must be applicable to state and federal government as well as private discovery. Guidelines must be established as to how much time will be considered reasonable to correct problems discovered in an audit and still preserve the privilege.

[5] Privatize the Cleanup of Contaminated Properties

Create licensing standards for environmental consultants. Consultants should be subject to state testing. There should be a bureau that can collect complaints and discipline environmental consultants. Fees should be assessed to defray all direct and indirect costs of the voluntary cleanup program.

Since there are no federally accepted standards for certifying environmental consultants, currently it is incumbent upon each state to draft a legislative response. Due to the various approaches adopted by individual states, it is difficult to predict with certainty the consultants' appropriate standard of conduct. This lack of coordination makes it difficult to pin liability on a consultant because the courts have no means to measure their negligence.[11] In order to provide stability to this area, and create an incentive for consultants to perform their assessments in a commercially reasonable manner, Congress needs to implement federal guidelines for certification.

There have been several congressional attempts to pass a bill concerning the licensing of environmental professionals. One approach that could be used to license the consultant could come from the proposed Weldon Bill, H.R. 2787. This bill tried to set out what constituted "all appropriate inquiry" to meet the innocent landowner defense of CERCLA. The same standards that were debated in that bill could be used to license an environmental professional.[12] Although introduced in differing forms over many years, this bill has failed to pass Congress.

State bills have achieved more success. In Illinois, a legislative equivalent of the Weldon bill was passed to amend the state CERCLA equivalent. In Ohio,[13] the Real Estate Reuse Act, Ohio Senate Bill 221, includes a requirement that a certified environmental professional be used in order to participate in the voluntary cleanup program. If all of the necessary criteria are met, the state will issue a Covenant Not to Sue, in exchange for a voluntary cleanup of the site. In order for a consultant to qualify for this program, the consultant must meet the following requirements:

> demonstrate at least five years of experience in the investigation and remediation of hazardous substances and be either a "diplomate" of the American Academy of Environmental Engineers and also a professional engineer;
> be a geologist certified by the American Institute of Professional Geologists; or
> be a person holding a master's degree or above in toxicology or a related field.[14]

Federal and state authorities are not the only entities proposing certification and licensing requirements for environmental consultants. These private efforts only add to the confusion. Several private organizations are working on new multidisciplinary approaches to certify environmental/hazardous waste professionals. Some of the more recent guidelines have been proposed by the Air and Waste Man-

---

[11]Barnes, *Efforts Aimed at Requiring Certification of Consultants,* Report on Defense Plant Wastes, October 23, 1992.

[12]H.R. 2787, 101st Congress, 1st Session (1989).

[13]There are other states that are proposing definitions for consultants. Michigan has a proposed definition for consultants who remove underground storage tanks. *See* MSA § 299.801. Minnesota also sets out standards for a consultant in the Petroleum Tank Release Cleanup Act, chapter 115C.02. For a complete listing of state certifications of hazardous waste professionals, *see Report on Defense Plant Wastes,* October 23, 1992.

[14]Proposed Senate Bill 221, Section 3746.07.

agement Association and the Board of Certified Safety Professionals. The National Association of Environmental Professionals (NAEP) certifies environmental professionals in environmental planning, documentation, assessment, operation, research, and education. NAEP is currently working on developing a certification program for environmental auditors.

One of the more popular standards for establishing environmental due diligence under CERCLA was adopted in 1993 by the American Society for Testing and Materials (ASTM). Their procedure sets out what needs to be done for an acceptable environmental site assessment for purposes of satisfying the legal standard of making "all appropriate inquiry." The ASTM practice includes the following tasks: (1) obtaining and reviewing records that will help identify areas of potential environmental concern associated with the site; (2) visiting the site to identify obvious physical conditions that would reveal areas of environmental concern; (3) conducting interviews with knowledgeable parties to identify the nature of past on-site operations, management practices, or aspects of the site history that may be relevant to the site's environmental status; and (4) preparation of a written report of findings based on the information collected and evaluated during the site-assessment process.

The lack of uniform standards creates more uncertainty in an already uncertain industry. Enacting a federal standard could go a long way to help quantify risk—the ultimate issue in brownfields paralysis.

[6] Create a No-action Process

A federal no-action process must be created, whereby the regulated community can write to state and federal agencies and get a determination that the proposed cleanup or control technology upgrades (or both) meet legal requirements and that the respective agencies will take no-action enforcement action against the regulated entity.

Many states have used this process effectively to expedite the cleanup process. On a federal level, the Internal Revenue Service and the Securities and Exchange Commission, through its Division of Corporate Finance, have for years used a no-action letter process. The regulated entities write into the agency, describe the facts and the proposed remedy and then the agency responds with a nonbinding letter detailing whether it would likely take an enforcement action or not on such a position.

In addition to expediting the environmental cleanup and compliance process, these no-action procedures go a long way toward increasing the appearance of fairness to small businesses and landowners. These small entities, with limited resources, can approach the agency looking for help—rather than resenting and fearing the agency since the only present contact is in the enforcement setting.

Note that creating a no-action process is consistent with EPA's current policy of making its enforcement and compliance division coequal. It would not require either hiring or firing within the agency but retraining enforcers to work in a cooperative atmosphere with the states and the regulated community. The enforce-ment division would then be left with the cases of malfeasance, not accidental non-compliance with an intricate, ambiguous, and often overwhelming environmental law and regulatory regime.[15]

[7] Repair the Constitutional Deficiencies in the State and Federal Environmental Lien Laws

The constitutional problems identified by the courts can easily be cured by pro-viding an opportunity for notice and comment prior to imposing a lien unless ex-igent circumstances exist that would require the notice and comment period to im-mediately follow the filing of the lien. Notice procedures must be standardized across the states so that they are searchable and can provide actual notice. Notice should be to environmental protection agencies, prospective purchasers, or lessors of property and should be placed in the deeds.

[8] Shield Developers, Lenders, and Prospective Purchasers of Contaminated Property from Potential Liability from Previous Contamination by Creating a New Category of Owner in CERCLA Called a *Bona Fide Prospective Purchaser*[16]

Congress should modify the liability provision of CERCLA to allow EPA to fur-ther define the secured creditor exemption. Congress should also adopt standard-ized audit practices for purposes of satisfying the CERCLA innocent purchaser de-fense. A *bona fide prospective purchaser* is defined as one who knowingly pur-chases contaminated property but is not otherwise a responsible party or affiliated with any responsible party.

Prospective purchasers (including landlord/tenants) and lenders must be af-forded limited protection in order to encourage cleanup and redevelopment of con-taminated sites. Covenants Not to Sue must be issued so that they bind the state and federal governments. Guidelines must be set for lenders so that environmen-tal compliance and cleanup is encouraged and lenders are not scared away from projects that require environmental cleanup.

[9] Summary Chart of Specific Remedies

The following chart summarizes the policy options that Congress should consider in evaluating and remedying the brownfields problem.

---

[15]For a detailed discussion of this concept *see* Elizabeth Glass Geltman and Andrew E. Skrowback, *Reinventing the EPA to Conform with the New American Environmentality,* 23 COLUM. J. OF ENVTL. L. 1 (1998). *See also* National Environmental Policy Institute, GETTING BACK ON THE COMPLIANCE TRACK: DRAFT DISCUSSION REPORT OF THE COMPLIANCE AND ENFORCEMENT SECTOR (Fall 1996).

[16]Superfund Reform Act of 1994, § 605(i), S. 1834 (February 3, 1994).

If Congress wishes to . . .	Then it could . . .
Improve the effectiveness of established state brownfield programs	Amend CERCLA to allow delegation to the states, retaining federal oversight
Distinguish between brownfield sites and NPL sites	Amend CERCLA to allow for voluntary registration with EPA or the state delegated enforcement authority to certify compliance with a brownfield cleanup program
	Amend CERCLA to allow for private parties to petition for delisting of brownfields sites from the CERCLIS based on meeting certain cleanup sites or entering into an MOU to do same
Provide prospective purchasers (including landlord/tenants) with limited protection in order to encourage cleanup and redevelopment of contaminated sites	Provide statutory authority for expanded prospective purchaser agreements, including the authority to enter into a Covenant Not to Sue (which is subject to public notice and comment) that would bind the state and federal governments (and private party suits)
	Amend CERCLA to define "all appropriate inquiry" for purposes of the innocent purchaser defense
Encourage brownfield redevelopment in urban centers	Provide low-interest loans and other economic incentives for cleanup (such as tax deductions, abatements, and credits) to encourage environmental cleanup
Avoid the issue of unfunded mandates	Allow the states to take portions of the federal Superfund throughout the state delegation process and allow the states to use the funds not just to clean up sites but also to provide the basis for a loan or tax abatement program
	Allow for imposition of permit fees (which would be retained by the regulatory authority) upon entry into voluntary cleanup programs and receipt of a Covenant Not to Sue or Certificate of Completion, which would make the program at least partially self-funding
Decrease the cost of cleanups	Restrict property uses based on cleanup standards and provide adequate notice of contamination
Ease prospective purchaser and lender anxiety about environmental liens or restrictions on property that could cause devaluation	Standardize recording of environmental liens and other deed restrictions so that they are easily searchable and can provide actual notice
	Mandate duplicate copies of deed restrictions and environmental liens be sent to U.S. EPA and recorded in a public-accessible data base (other than CERCLIS)
Ease lender and prospective purchaser concern about maintenance of	Create a limited environmental audit privilege that is conditioned on promptly remedying

(continued)

If Congress wishes to . . .	Then it could . . .
property to appropriate cleanup standards during normal use	any discovered environmental problems that are applicable to state and federal government as well as private discovery
	Provide statutory authority to promulgate guidelines to (1) define an environmental audit and (2) establish how much time will be considered reasonable to correct problems discovered in an audit and still preserve the privilege
Avoid overburdening the state or federal government with supervisory responsibilities in brownfield properties	Create licensing standards for environmental consultants engaged in brownfield remediation
	Subject such consultants to standardized testing
	Establish a bureau that can collect complaints and discipline environmental consultants
Eliminate the fear of prosecution and current frustrations with government	Create a no-action process whereby the regulated community can write to state and federal agencies and get a determination that the proposed cleanup strategies or control technology upgrades meet legal requirements and that the respective agencies will take no-action enforcement action against the regulated entity
Eliminate the unfairness of EPA taking over a site, precluding PRP input into site cleanup, and placing an environmental lien on the property without opportunity for judicial review	Amend CERCLA to repair the constitutional deficiencies in the state and federal environmental lien laws—specifically providing an opportunity for notice and comment prior to imposing a lien
Bring properties more quickly into commerce	Amend CERCLA to allow EPA to promulgate regulations that would establish presumptive cleanup standards based on future use
	Despite presumptive cleanup standards, allow use of no-action process (describing cleanup standards, future use, and deed restrictions, if any) for site-specific remedies
Reduce the number of brownfield properties	Revise CERCLA to allow EPA to promulgate regulations that exempt landowners who did not cause or contribute to contamination but whose property contains groundwater contamination due to others where groundwater is not used as a drinking source
	These regulations should reflect scientific risk
Provide greater certainty in valuing brownfield sites	Amend CERCLA to allow EPA or the state delegated authority to issue some form of certificate of completion once the site is cleaned up
	Such certificates should be recordable or otherwise transferable to future purchasers and "run with the land" provided the land use does not change

(continued)

Encourage private development of brownfield sites without governmental oversight or drastically amending the program	Amend CERCLA to use a portion of Superfund money to set up a public/private insurance company (similar to OPIC) that will insure redevelopment of brownfield sites
Increase financial incentives to redevelop brownfields	Encourage use of industrial development bonds for brownfields properties
	Establish a federal tax credit for brownfields redevelopment
	Allow for low-interest loans for redevelopment of brownfields properties
Eliminate generator and arranger liability for municipalities	Create a municipal waste exemption that parallels the exemption under RCRA
Conduct a pilot program to evaluate the effectiveness of the program	Include a sunset provision in the statute
	Require periodic review to Congress of the progress of the programs
Avoid fraud in the voluntary cleanup program	Make it a criminal act to submit false or fraudulent information in connection with a voluntary cleanup
Evaluate the effect of CERCLA on environmental cleanup practices	Ask EPA and the states to report compliance data rather than enforcement data to Congress; EPA's appropriations should be based on how many sites were cleaned up, not how many actions were brought to recoup expenses
Allow flexibility in cleanup standards	Create a no-action procedure whereby the regulated community and those volunteering can have their cleanup efforts approved in advance in exchange for limited immunity based on compliance with the plan
	Allow private parties to suggest cleanup and evaluate them based on results not based on strict adherence to regulations (i.e., parties should be free to propose new technologies that are cheaper and more effective than those listed in regulations, encouraging the regulated community to take advantage of evolving technologies)
Eliminate the perceived unfairness of CERCLA to small business and present owners (i.e., those who are not polluters)	Completely exempt the smallest contributors at Superfund sites and prospective purchasers of contaminated property from liability (i.e., cut the nonpolluters out of this "polluter pays" liability scheme)
Reduce the transaction costs of de minimus contributors	Provide guaranteed opportunities for early settlement for de minimus parties and parties with ability-to-pay problems
Enhance the ability of municipalities to subcontract MSW activities	Cap the liability of generators and transporters of municipal solid waste
Increase the availability of insurance	Create a system where liability is quantifiable
	Reduce the cost of Superfund to insurance and reinsurance carriers
Eliminate the adverse impact of joint and several liability for parties who cooperate	Create a standardized procedure for expedited allocation of responsibility

(continued)

If Congress wishes to . . .	Then it could . . .
and clean up sites by assuring that they would pay no more than their allocated fair share	
Enhance state voluntary cleanup laws and programs	Allow states with voluntary cleanup programs to encourage PRPs, as well as prospective purchasers, to enter into agreements with the state (serving as the federal implementing agency) to voluntarily clean up the site, pay for state oversight, and generally receive some sort of acknowledgment or sign-off from the state (again acting under federal authority or at least in concert with the federal government)

Conclusion

Economic redevelopment of abandoned industrial sites commonly referred to as "brownfields" will be the cornerstone in the revitalization of some economically distressed urban and rural areas. Productive reuse of existing abandoned industrial sites is pollution prevention at its best. Industrial and commercial ventures which consider the purchase and redevelopment of previously used sites will foster economic redevelopment and productive reuse of those existing industrial sites. It will also stimulate economic growth of the surrounding areas and will preclude siting of industrial developments on limited "greenfields" or pristine property. Although Superfund has been effective in fostering pollution prevention and waste minimization, the law has inadvertently produced a chilling affect which has stymied prospective purchasers and lenders from investing in the renewal of abandoned contaminated waste sites. In some instances, this situation has led to financial institutions redlining communities that are situated near or adjacent to contaminated waste sites. These sites are shunned by prospective developers who are afraid they might inherit exorbitant cleanup liabilities for contamination they did not create. Sites which once provided the lifeline of economic vitality and jobs to thriving communities have been abandoned for fear of the contamination which might be present.

This migration from "brownfields" to "greenfields" is particularly problematic to certain urban and industrial areas, where the loss of investment capital and jobs further exacerbates existing economic and social conditions.[17]

The existence of brownfields is not a new problem in many metropolitan areas, especially older, central cities and suburbs, where much of the economy was

---

[17]TESTIMONY OF PATRICIA WILLIAMS, NATIONAL WILDLIFE FEDERATION, HOUSE TRANSPORTATION WATER RESOURCES AND ENVIRONMENT SUPERFUND REAUTHORIZATION (June 21, 1995).

historically based on industrial and commercial activity. Where industry has closed or moved, land and buildings are left behind, idled, or underutilized, leading to job loss and reduction in local tax revenues. As many of these areas try to reverse this decline and promote economic development, significant attention has focused on brownfields and the barriers associated with their cleanup and reuse.

While more attention has been focused on brownfields in the past few years, information on the extent of the problem and the level of contamination at these sites is limited. Estimates range from tens of thousands to 450,000 brownfields sites in the country. In addition, their condition may vary from zero, low, or moderate contamination to extremely hazardous conditions, while many sites have not been evaluated.

Although the exact nature and extent of the problem is difficult to assess, most sites considered brownfields are not associated with extreme levels of contamination and will never be considered for addition to CERCLA's NPL or similar state priority lists. Since these sites do not pose a serious threat or warrant immediate federal attention, they become the responsibility of the states or municipalities where they are located. For this reason, states have taken an active role in identifying and confronting the barriers to promoting brownfield cleanup and reuse.

Congress should amend CERCLA or draft an entirely separate environmental statute to codify this trend and officially delegate brownfield cleanup to states with the capability and inclination to undertake such programs. A federal statute coupled with state delegation will allow the best of state and federal input. Like the CAA, CWA, and RCRA, the federal government would set the floor on brownfield cleanup requirements. States would, however, be free to have programs allowing more stringent requirements. States would also be free to consider the local nature of hazardous waste siting and cleanup, taking into account total land use planning. The federal oversight would preclude states from ignoring the rights of neighboring states and their citizenry in the brownfield cleanup process.

A number of major issues related to brownfields cleanup and redevelopment must be addressed in this federal legislation. These include:

> technical issues related to site assessment and cleanup;
> liability concerns associated with environmental contamination;
> financial barriers to cleanup and reuse;
> community concerns; and
> protection for stakeholders as a means of encouraging redevelopment.

Each of these issues require some attention and resolution in order to promote greater interest by developers and businesses in brownfield sites.

To date, the states have been the laboratories experimenting with cures for brownfield paralysis. As the states continue to debate issues relating to brownfields, a number of significant concepts emerge. These concepts, taken together, represent a significant paradigm shift in the approach to environmental law in the United States.

First, voluntary cleanup programs and environmental audit privilege statutes are each particularly popular because they encourage environmental compliance. In so doing, these voluntary programs reward good environmental citizens with a less confrontational and potentially more efficient alternative to achieve cleanup and reuse of sites than did the traditional command and control enforcement model. It is important to note that these voluntary programs in no way supplant or undermine the enforcement abilities of the states. Bad actors may still be punished both civilly and criminally. What these programs add to the enforcement model is a mechanism by which good environmental citizens can attain agency help in undertaking their efforts.

Second, in an attempt to reduce uncertainties related to liability for cleanup at brownfields and other environmental compliance and cleanup issues, some voluntary cleanup programs and environmental audit privilege statutes provide for limited enforcement immunity or other liability assurances to particular parties or sites. This limited enforcement immunity is conditioned upon proper completion of the cleanup (in the case of voluntary cleanup programs) or immediately coming into environmental compliance (in the case of environmental audit privilege legislation). Currently, this type of legislation can only offer liability protection as it is specifically defined by the individual state law and only applies to activities recognized by that state. Assurances by state programs do not shield stakeholders from liability under federal law or third-party lawsuits brought under state, federal, or common law. As a result, even though in many cases the likelihood of federal or state enforcement under Superfund provisions is low for these sites, there is still some question about whether these assurances can provide the needed security to increase the level of brownfield cleanup and redevelopment. While there is currently significant interest in removing or limiting liability for many stakeholders at the federal and state level, there is still only minimal experience with this practice through brownfields.

The last issue that may require attention is the nature of the relationship between the federal government and the states with respect to brownfields cleanup. As states continue to develop and improve voluntary programs they exercise their own discretionary authority for such matters as the adequacy of cleanup standards, expectations for future use of a site, and liability protection for particular parties. These programs would likely gain more credibility with the private sector if they received some form of agreement or approval from the U.S. EPA. Although this could come in the form of a Superfund Memorandum of Agreement (SMOA), such as those held between EPA Region V and the states of Illinois and Minnesota, such policies would not eliminate the issue of liability from third-party suits. History has shown that it is litigation from third parties (and not the federal government) that is most likely to be the problem in brownfield sites. The verification could also come through a certification process as outlined by legislation introduced in the 103rd Congress. Considerable variation among state voluntary programs could make development of SMOAs or certification difficult.

Clearly federal legislation is needed to address the brownfields question.

Without such legislation, the ability of EPA to address the problem of brownfields redevelopment would be limited to voluntary programs developed under EPA's enforcement discretion. Such policies could not bind third parties. Thus, EPA is currently without power to deal with the critical issues of brownfields redevelopment without authorization from Congress to do so.

Index

362   Index

Barnett, Harold C., 235n. 1, 281n. 1
Barrett, 142n. 29
Bartsch, Charles, 2n. 5, 4nn. 8, 9, 5n. 11,
7n. 16, 107n. 86, 128n. 152, 134n.
175, 135n. 1, 138n. 10, 147n. 58,
153n. 1, 175n. 1, 199n. 2, 235n. 4,
249n. 1, 269n. 1, 272n. 18, 281n. 3,
285n. 22, 286n. 26, 300n. 9, 339n. 6
Beckett, 172n. 93
Berger, Robert S., 147n. 54
best demonstrated available technology
(BDAT), 31
Board of Certified Safety Professionals,
351
Boehlert, Sherwood, 308n. 2
bona fide prospective purchaser, 352
Boyer, Kevin R., 190n. 72
Bozarth, Robert S., 149n. 73, 168n. 70
Brieger, Heidi, 79n. 108
Brown, Johnine, 7n. 15
Brown, Juliet, 96n. 47
Browner, Carol M., 305, 325n. 12, 342n. 7
Brownfields Action Agenda (BAA), 57
Brownfields financing programs. *See*
financing programs
Brownfields National Partnership Action
Agenda, 333
"brownfields paralysis," 3, 67
Brownfields Pilot Project, 305, 338
brownfields tax incentive, 333
empowerment zone, 333
enterprise community, 333
"brownfields trap," 5
business transfer programs, 69
and CERCLA, 69
Colorado, 272–73
Connecticut, 88
Connecticut Hazardous Waste Establish-
ment Transfer Act, 88
Connecticut Transfer Act, 75, 76
Connecticut Transfer of Hazardous
Waste Establishment Act, 75
Illinois Responsibility Property Transfer
Act, 75, 76, 203–4
Indiana Environmental Hazardous Dis-
closure and Responsible Party
Transfer Law (RPLA), 209, 212,
213–14

Indiana Responsible Property Transfer
Law of 1988, 75, 209–14
Iowa, 76, 255
Louisiana, 236
Michigan, 221
Minnesota, 225
Missouri, 76, 266
New Jersey Environmental Cleanup
Responsibility Act (ECRA), 75,
139–46
New York, 76, 148–49
1995 Amendments to Connecticut's
Transfer Act, 82
North Carolina, 194–95
Pennsylvania, 167–68
and RCRA, 69
Rhode Island, 130
South Carolina, 195
Washington, 304

California, 78, 281, 282–92
Certificate of Completion, 284, 286
deed restrictions, 78
Department of Toxic Substances Control
(DTSC), 285
DTSC's Annual Work Plan sites, 285
environmental lien, 290–92
Expedited Action Remedial Program
(ERAP), 82, 284
Expedited Site Remediation Trust Fund,
283
federal facilities, 285
lender protection statute, 288–89
lessee, 287
mini-CERCLA, 286
NPL, 285
nuisance statute, 290–92
petroleum products, 285
Private Site Management Program,
286–88
Prospective Purchaser Policy, 82
Remedial Action Certification, 285
Remedial Action Plan (RAP), 285,
286
Remedial Action Work Plan (RAW),
285, 286
remedial plan, 283
renter, 287